Modern Flow Analysis

Modern Flow Analysis

Editor
Paweł Kościelniak

MDPI • Basel • Beijing • Wuhan • Barcelona • Belgrade • Manchester • Tokyo • Cluj • Tianjin

Editor
Paweł Kościelniak
Jagiellonian University,
Faculty of Chemistry
Poland

Editorial Office
MDPI
St. Alban-Anlage 66
4052 Basel, Switzerland

This is a reprint of articles from the Special Issue published online in the open access journal *Molecules* (ISSN 1420-3049) (available at: https://www.mdpi.com/journal/molecules/special_issues/Flow_Analysis).

For citation purposes, cite each article independently as indicated on the article page online and as indicated below:

LastName, A.A.; LastName, B.B.; LastName, C.C. Article Title. *Journal Name* **Year**, *Article Number*, Page Range.

ISBN 978-3-03936-738-2 (Hbk)
ISBN 978-3-03936-739-9 (PDF)

© 2020 by the authors. Articles in this book are Open Access and distributed under the Creative Commons Attribution (CC BY) license, which allows users to download, copy and build upon published articles, as long as the author and publisher are properly credited, which ensures maximum dissemination and a wider impact of our publications.

The book as a whole is distributed by MDPI under the terms and conditions of the Creative Commons license CC BY-NC-ND.

Contents

About the Editor vii

Paweł Kościelniak
Modern Flow Analysis
Reprinted from: *Molecules* **2020**, *25*, 2897, doi:10.3390/molecules25122897 1

Marek Trojanowicz
Flow Chemistry in Contemporary Chemical Sciences: A Real Variety of Its Applications
Reprinted from: *Molecules* **2020**, *25*, 1434, doi:10.3390/molecules25061434 7

Apichai Intanin, Prawpan Inpota, Threeraphat Chutimasakul, Jonggol Tantirungrotechai, Prapin Wilairat and Rattikan Chantiwas
Development of a Simple Reversible-Flow Method for Preparation of Micron-Size Chitosan-Cu(II) Catalyst Particles and Their Testing of Activity
Reprinted from: *Molecules* **2020**, *25*, 1798, doi:10.3390/molecules25081798 59

Bruno J. R. Gregório, Ana Margarida Pereira, Sara R. Fernandes, Elisabete Matos, Francisco Castanheira, Agostinho A. Almeida, António J. M. Fonseca, Ana Rita J. Cabrita and Marcela A. Segundo
Flow-Based Dynamic Approach to Assess Bioaccessible Zinc in Dry Dog Food Samples
Reprinted from: *Molecules* **2020**, *25*, 1333, doi:10.3390/molecules25061333 73

Jixin Qiao
Dynamic Flow Approaches for Automated Radiochemical Analysis in Environmental, Nuclear and Medical Applications
Reprinted from: *Molecules* **2020**, *25*, 1462, doi:10.3390/molecules25061462 85

Víctor Vicente Vilas, Sylvain Millet, Miguel Sandow, Luis Iglesias Pérez, Daniel Serrano-Purroy, Stefaan Van Winckel and Laura Aldave de las Heras
An Automated SeaFAST ICP-DRC-MS Method for the Determination of ^{90}Sr in Spent Nuclear Fuel Leachates
Reprinted from: *Molecules* **2020**, *25*, 1429, doi:10.3390/molecules25061429 113

Burkhard Horstkotte and Petr Solich
The Automation Technique Lab-In-Syringe: A Practical Guide
Reprinted from: *Molecules* **2020**, *25*, 1612, doi:10.3390/molecules25071612 127

Antonios Alevridis, Apostolia Tsiasioti, Constantinos K. Zacharis and Paraskevas D. Tzanavaras
Fluorimetric Method for the Determination of Histidine in Random Human Urine Based on Zone Fluidics
Reprinted from: *Molecules* **2020**, *25*, 1665, doi:10.3390/molecules25071665 149

Jucineide S. Barbosa, Marieta L.C. Passos, M. das Graças A. Korn and M. Lúcia M. F. S. Saraiva
Enzymatic Reactions in a Lab-on-Valve System: Cholesterol Evaluations
Reprinted from: *Molecules* **2019**, *24*, 2890, doi:10.3390/molecules24162890 161

Kanokwan Kiwfo, Wasin Wongwilai, Tadao Sakai, Norio Teshima and Kate Grudpan
Determination of Albumin, Glucose, and Creatinine Employing a Single Sequential Injection Lab-at-Valve with Mono-Segmented Flow System Enabling In-Line Dilution, In-Line Single-Standard Calibration, and In-Line Standard Addition
Reprinted from: *Molecules* **2020**, *25*, 1666, doi:10.3390/molecules25071666 173

Joanna Kozak, Justyna Paluch, Marek Kozak, Marta Duracz, Marcin Wieczorek and Paweł Kościelniak
Novel Approach to Automated Flow Titration for the Determination of Fe(III)
Reprinted from: *Molecules* **2020**, *25*, 1533, doi:10.3390/molecules25071533 185

Justyna Paluch, Joanna Kozak, Marcin Wieczorek, Michał Woźniakiewicz, Małgorzata Gołąb, Ewelina Półtorak, Sławomir Kalinowski and Paweł Kościelniak
Novel Approach to Sample Preconcentration by Solvent Evaporation in Flow Analysis
Reprinted from: *Molecules* **2020**, *25*, 1886, doi:10.3390/molecules25081886 197

J. Jiménez-López, E.J. Llorent-Martínez, S. Martínez-Soliño and A. Ruiz-Medina
Automated Photochemically Induced Method for the Quantitation of the Neonicotinoid Thiacloprid in Lettuce
Reprinted from: *Molecules* **2019**, *24*, 4089, doi:10.3390/molecules24224089 211

Thitirat Mantim, Korbua Chaisiwamongkhol, Kanchana Uraisin, Peter C. Hauser, Prapin Wilairat and Duangjai Nacapricha
Dual-Purpose Photometric-Conductivity Detector for Simultaneous and Sequential Measurements in Flow Analysis
Reprinted from: *Molecules* **2020**, *25*, 2284, doi:10.3390/molecules25102284 221

About the Editor

Paweł Kościelniak was born in Krakow, Poland, in 1952. He received his Ph.D. in 1981 from the Faculty of Chemistry, Jagiellonian University in Krakow. He is a full professor (since 2000) and head of the Department of Analytical Chemistry at Jagiellonian University (since 1997). In 1993–2020, he managed the research group Team of Analytical Flow Techniques. In 2013, he received the scientific award of the Japanese Association for Flow Injection Analysis. He is a member of the Committee of Analytical Chemistry of the Polish Academy of Sciences. Prof. Kościelniak is an author, co-author and editor of more than 300 scientific articles, books and book chapters. His main areas of scientific interest are flow analysis, forensic chemistry and environmental analysis, with special attention paid to such issues as analytical calibration, the examination and elimination of interference effects, and the optimization and validation of analytical procedures.

Editorial

Modern Flow Analysis

Paweł Kościelniak

Faculty of Chemistry, Jagiellonian University, Gronostajowa 2, 30-387 Krakow, Poland; pawel.koscielniak@uj.edu.pl; Tel.: +48-1268-62411

Received: 22 June 2020; Accepted: 23 June 2020; Published: 24 June 2020

Abstract: A brief overview of articles published in this Special Issue of Molecules titled "Modern Flow Analysis" is provided. In addition to cross-sectional and methodological works, there are some reports on new technical and instrumental achievements. It has been shown that all these papers create a good picture of contemporary flow analysis, revealing the most current trends and problems in this branch of flow chemistry.

Keywords: flow analysis; flow chemistry

The idea to process an analytical sample through its flow to a measuring apparatus was one of the milestones in the development of the chemical analysis. The creator of this idea is considered Leonard T. Skeggs, Jr., who was the first to construct an analytical flow system and in 1957 presented its application in clinical analysis [1]. The motive for his action, as he wrote himself, was quite simple: "The staffs of laboratories of clinical chemistry are confronted with an ever-increasing number and variety of determinations", and it is also current, maybe even more so nowadays. Therefore, it is no wonder then that flow analysis has grown over the years and has found more and more followers.

Since "Skeggs times", many different flow techniques, systems and modules have been proposed, providing the possibility of analyzing many samples in a short time with increased safety of analytical work and with results of very good sensitivity and precision. These fully mechanized, automated and often miniaturized systems are increasingly using new methodologies and materials to improve the quality and speed of sample processing. The flow mode has been shown to favor various ways of manipulating sample and reagents, facilitating analytical operations such as calibration, titration or multi-component analysis. It also turned out that the hydrodynamic conditions of the flowing solutions create new possibilities for conducting and controlling chemical reactions.

The latter feature of flow systems has been noticed and exploited in other chemical fields, particularly in organic chemistry. Constructed slightly later than analytical systems, the first flow systems dedicated to the synthesis of organic compounds proved to be so useful and effective that, over the years, flow synthesis has acquired the established general name "flow chemistry", bypassing flow analysis. This situation makes it difficult to objectively assess the current state of flow analysis and its proper role and place in the development of chemistry as applied science.

Unfortunately, the aforementioned problem usually escapes the attention of flow analysts. More important and valuable is the review article written by Trojanowicz entitled "Flow Chemistry in Contemporary Chemical Sciences: A Real Variety of Its Applications" [2]. In this article, the author presents the in-depth comparative characteristics of flow analysis and flow synthesis, paying special attention to the chronology of inventions of physico-chemical operations and appropriate instrumental devices, which are widely employed in both areas of flow chemistry. It reveals many of their mutual, sometimes surprising similarities in terms of both the construction and operation of flow systems and their stages of development, primarily in the spheres of mechanization, automation and miniaturization. He proves his thesis with many literature examples (357 references!), including selected items from recent years, giving an excellent picture of contemporary, modern flow analysis and flow synthesis.

He finishes his work with a very important statement: "...it seems to be fully justified to use the term flow chemistry to represent all other chemical processes carried out under flow conditions of the reacting mixture, a sample to be analyzed, and other media chemically transformed under flow conditions". In my opinion, the article should be a must-read for analysts interested in flow chemistry, and above all for those who are starting their scientific and didactic work in this field.

A very good example of how to take advantage and practice the common features of flow chemistry is presented in [3]. The flow-based system working on the principle of segmented injection analysis (SIA), very popular and currently widely used in flow analysis, was used for synthesis of uniform micron-size CS-Cu(II) catalyst particles. It was also exploited to fast and effective monitoring of the catalytic activity of the synthesized particles for the reduction of p-nitrophenol with excess borohydride. The flow-based method provided advantages over the manual method in terms of throughput for preparation of the particles (100 drops min^{-1}), size distribution of the particles (150–210 µm) and their uniformity.

Another example justifying the concept of flow chemistry are the flow-based studies of bioaccessibility of bioactive compounds from food. The bioaccessibility test provides valuable information to choose the right dose and source of food matrices and thus to ensure the nutritional efficiency of food products. A valuable advantage of flow systems is the ability to quickly and precisely test the rate of absorption of compounds. Such a system was used for the implementation of dynamic extractions, aiming at the evaluation of bioaccessible zinc and the characterization of leaching kinetics in dry dog food samples [4]. This dynamic procedure was proved to be more flexible than the static traditional batch methods, allowing, in addition, the natural non-equilibrium processes to be much better imitated.

Dynamic flow procedures are also the main subject of the review paper written by Qiao [5]. In this case, the author focused on the implementation of various flow techniques for the determination of radionuclides. Typical analytical challenges involved in this area include, for instance, very low radioactivity of a sample, matrix effects and need to separate exactly the radioisotopes from the sample matrix. Based on the literature study author proved that the flow analysis meets all these requirements. He showed that the versatile flow approaches can be utilized in different steps for radiochemical analysis, including sample pretreatment, chemical separation and purification, as well as source preparation and detection. The flow mode makes the analysis fast, low laborious and-what is perhaps most important in this case-safe for operators because of less exposure to radioactivity. In conclusion, the author stated that "...continuous development of more advanced flow approaches is necessary to cope with the growing demands for radiochemical analysis in different fields...".

As if in response to this expectation, in [6], a commercially available fully automated flow-based device (SeaFAST) is presented that is dedicated to the determination of ^{90}Sr at trace levels in nuclear spent fuel leachate samples. Isotope ^{90}Sr is a fast released and hard to measure fission product and is of great interest due to its toxicity and high energy emission. The system, composed of an autosampler, series of syringes and a valve module, is coupled to ICP-MS combines the use of a Sr-specific resin and the reaction with oxygen as reaction gas in a dynamic reaction cell (DRC). As experimentally proved, strontium was possible to be determined in a single operational sequence of separation, pre-concentration and elution avoiding sophisticated and time-consuming procedures. In addition, the developed method was revealed to be safe, rapid, selective, and sensitive, showing a good agreement in terms of measurement uncertainties when compared with the classical radiochemical method.

Syringe-based flow systems are one of the examples of very useful and efficient flow devices that provides the possibility to establish different sample pathways and to assure a very stable and reproducible flow rate. Multisyringe Flow Injection Analysis (MSFIA) has been introduced in 1999 [7] and was intensively developed in the following years. In 2012 it was shown that the syringe can be used not only as a kind of pump, but also as a chamber in which the sample is fully processed before measurements [8]. From now on, this technique, called Lab-in-Syringe (LIS), is gaining more and more interest, constantly undergoing new modifications, modes of operation and technical improvements.

In this context, a very needed and helpful article is work [9], aimed to bring the LIS technique closer to newcomers and users of other flow techniques. The article reviews the different options for instrumental configurations and operations possible to be performed using LIS, including syringe orientation, in-syringe stirring modes, in-syringe detection, additional inlets, and addable features. Great attention is paid to LIS applications in the field of automation of the sample pretreatment procedures, especially of extraction processes in the liquid phase. The possible contributions of 3D printing techniques to LIS are also mentioned. In addition to the unmistakable advantages of this technique, some of its limitations are discussed that arise mainly from the large dead volume of a syringe. The article as a whole is not only a very good guide to LIS, but also gives a more general picture of new conceptual, technical and instrumental tools contributing to the development of flow analysis.

A common feature of different flow techniques developed in the last years is the minimization of sample volume, reagent consumption and waste production. These features are consistent with the requirements of the recently fashionable and very needed policy named "green analytical chemistry" (GAC). Its purpose is obviously to reduce the risk from analytical laboratories to the environment and human health. Flow analysis naturally favors these aspirations, although different efforts are still underway to improve flow techniques in this direction.

An example of a GAC-oriented flow technique is the Sequential Injection Analysis (SIA) [10]. It consists of gradually introducing small segments of the sample and reagents into a separate conduit, mixing them and delivering to the detector after bringing the reaction to a certain stage. Such a technique, named zone fluidics by authors, was used for the spectrofluorimetric determination of histidine in the urine samples [11]. Before reaching the detector, the reacting sample was stopped for a time selected as a compromise between sensitivity and sampling throughput. The method allowed histidine to be determined directly with minimum sample preparation and with very good precision and accuracy. The method ensured also minimal consumption of reagents and generation of waste compared to continuous flow techniques.

The "green" idea can be also implemented by completely processing a sample with a very small volume using the selection valve (Lab-on-Valve), which is an integral part of the SIA system. Such a technique (μSIA-LOV) was applied to the determination of cholesterol in serum samples [12], which is a widely relevant in clinical diagnosis, since higher values of cholesterol in human blood are an important risk factor for cardiovascular problems. The analytical method was based on the implementation of enzymatic reactions performed in the μSIA-LOV system by cholesterol esterase, cholesterol oxidase and peroxidase. The results obtained were shown to be reliable and accurate, confirming the usefulness of this methodology for the routine determinations of cholesterol and for other clinical determinations. The automation and the miniaturization of the analytical procedure leads to the reproducibility improvement and the reduction of reagents. As the authors emphasize, the revealed advantages are relevant when the methodology developed is compared with other automatic methodologies used in the flow analysis.

The modification of the segmented injection analysis (SIA) towards GAC may also consist of merging the sample and reagents in the form of liquid segments limited by air segments of very small volume (on the order of several or several dozen microliters). After some time, a single segment of reaction products is formed that is of limited dispersion and undiluted by the carrier stream. An example of the successful application of this mono-segmented technique is the simultaneous determination of albumin, glucose, and creatinine (the key biomarkers for diabetes mellitus) in the urine samples [13]. Due to the flow methodology, the fast reaction (for albumin) and slow reactions (for glucose and creatinine) were appropriately synchronized. From the analytical reliability point of view, what was important was the possibility of calibrating carefully the determinations by both external and standard addition methods without the change of the system configuration.

In work [14], the mono-segments were created for the titration purpose with use of three syringe pumps equipped with nine-position selection valves. During the titration process, two syringe pumps dispensed well-known decreasing and increasing volumes of the sample and titrant, respectively, which were then introduced simultaneously into the system, joined at the confluence point, and merged in the mixing coil to complete the reaction. Before and after introducing the defined volumes of sample and titrant, a segment of air was inserted using the third pump to form a monosegment. This simple and fully automated procedure, imitating traditional titration way, was used for the determination of iron(III) in the presence of iron(II), allowing one analysis to be performed for 6 min with very good precision and accuracy, and consuming as little as 2.4 mL of both sample and titrant solution.

Yet another approach to GAC requirements is to develop methods of sample processing that do not require a large number or amount of reagents. One of the ways used very often for the sample preparation is the sample preconcentration coupled with the analyte isolation from the sample matrix. It has been shown that a sample can be effectively preconcentrated in the mechanized sequential injection system on the basis of the membraneless evaporation [15]. The main element of the system was the preconcentration module working under high temperature and diminished pressure. Using different evaporation conditions, various values of the signal enhancement factor (from several to 20) could be achieved. The applicability of the method was positively verified on the example of the determination of Zn in certified reference materials of drinking water and wastewater using the capillary electrophoresis method.

The purpose of sample preconcentration is to enhancement the analytical signal, which consequently gives the opportunity to determine the analyte with increased precision and diminished limit of quantification. The signal enhancement can be performed by various ways. A very interesting way has been displayed with the use of the simple flow manifold equipped with multicommutated solenoid valves and a spectrofluorimetric detector [16]. The analyte was online irradiated with UV light to produce a highly fluorescent photoproduct that was then retained on a solid support placed in the detector flow cell. By doing so, the pre-concentration effect of the photoproduct was achieved in the detection area, increasing the sensitivity. The method was demonstrated on the example of the determination of insecticide thiacloprid, one of the main neonicotinoids, in lettuce samples. The analytical results were characterized with very good precision and accuracy, and a low detection limit of 0.24 mg kg^{-1}.

One more modification of the flow manifold that aimed to improve the detection capability is shown in [17]. This work presents a dual detector that consisted of a paired emitter–detector diode (PEDD) and a capacitively coupled contactless conductivity detector (C4D) for flow-based photometric and conductivity measurements, respectively. They have been incorporated in a single flow cell of an original design. In different flow configurations, the system was able to be adapted to either sequential or simultaneous determination of two analytes in a sample. In particular, the urine samples were analyzed in regard to the conductivity and creatinine concentration for monitoring the health problems in the human body.

It is seen that the articles included in this Special Issue of Molecules titled "Modern Flow Analysis" very well reflect the current state of flow analysis, which to a large extent is also a picture of modern flow chemistry. At the same time, these works give a picture of extraordinary ingenuity and creativity in solving problems and creating new, more and more perfect analytical approaches. Once again, it turns out that flow analysis is the area of analytical chemistry, in which imagination and scientific courage play a great role. One should hope that this is a guarantee of its further development in the future.

Funding: This research received no external funding.

Conflicts of Interest: The author declare no conflict of interest.

References

1. Skeggs, L.T., Jr. An automatic method for colorimetric analysis. *Am. J. Clin. Path.* **1957**, *28*, 311–322. [CrossRef] [PubMed]
2. Trojanowicz, M. Flow chemistry in contemporary chemical sciences: A real variety of its applications. *Molecules* **2020**, *25*, 1434. [CrossRef] [PubMed]

3. Intanin, A.; Inpota, P.; Chutimasakul, T.; Tantirungrotechai, J.; Wilairat, P.; Chantiwas, R. Development of a simple reversible-flow method for preparation of micron-size chitosan-Cu(II) catalyst particles and their testing of activity. *Molecules* **2020**, *25*, 1798. [CrossRef] [PubMed]
4. Gregório, B.J.R.; Pereira, A.M.; Fernandes, S.R.; Matos, E.; Castanheira, F.; Almeida, A.A.; Fonseca, A.J.M.; Cabrita, A.R.J.; Segundo, M.A. Flow-based dynamic approach to assess bioaccessible zinc in dry dog food samples. *Molecules* **2020**, *25*, 1333. [CrossRef] [PubMed]
5. Qiao, J. Dynamic flow approaches for automated radiochemical analysis in environmental, nuclear and medical applications. *Molecules* **2020**, *25*, 1462. [CrossRef] [PubMed]
6. Vilas, V.V.; Millet, S.; Sandow, M.; Pérez, L.I.; Serrano-Purroy, D.; Van Winckel, S.; Aldave de las Heras, L. An automated SeaFAST ICP-DRC-MS method for the determination of ^{90}Sr in spent nuclear fuel leachates. *Molecules* **2020**, *25*, 1429. [CrossRef] [PubMed]
7. Cerdà, V.; Estela, J.M.; Forteza, R.; Cladera, A.; Becerra, E.; Altimira, P.; Sitjar, P. Flow techniques in water analysis. *Talanta* **1999**, *50*, 695–705. [CrossRef]
8. Maya, F.; Estela, J.M.; Cerdà, V. Completely automated in-syringe dispersive liquid-liquid microextraction using solvents lighter than water. *Anal. Bioanal. Chem.* **2012**, *402*, 1383–1388. [CrossRef] [PubMed]
9. Horstkotte, B.; Solich, P. The automation technique Lab-in-Syringe: A practical guide. *Molecules* **2020**, *25*, 1612. [CrossRef] [PubMed]
10. Růžička, J.; Marshall, G. Sequential injection: A new concept for chemical sensors, process analysis and laboratory assays. *Anal. Chim. Acta* **1990**, *237*, 329–343. [CrossRef]
11. Alevridis, A.; Tsiasioti, A.; Zacharis, C.K.; Tzanavaras, P.D. Fluorimetric method for the determination of histidine in random human urine based on zone fluidics. *Molecules* **2020**, *25*, 1665. [CrossRef] [PubMed]
12. Barbosa, J.S.; Passos, M.L.C.; Korn, M.A.; Saraiva, M. Enzymatic reactions in a Lab-on-Valve system: Cholesterol evaluations. *Molecules* **2020**, *25*, 2890. [CrossRef] [PubMed]
13. Kiwfo, K.; Wongwilai, W.; Sakai, T.; Teshima, N.; Grudpan, K. Determination of albumin, glucose, and creatinine employing a single sequential injection Lab-at-Valve with mono-segmented flow system enabling in-line dilution, in-line single-standard calibration, and in-line standard addition. *Molecules* **2020**, *25*, 1666. [CrossRef] [PubMed]
14. Kozak, J.; Paluch, J.; Kozak, M.; Duracz, M.; Wieczorek, M.; Kościelniak, P. Novel approach to automated flow titration for the determination of Fe(III). *Molecules* **2020**, *25*, 1533. [CrossRef] [PubMed]
15. Paluch, J.; Kozak, J.; Wieczorek, M.; Woźniakiewicz, M.; Gołąb, M.; Półtorak, E.; Kalinowski, S.; Kościelniak, P. Novel approach to sample preconcentration by solvent evaporation in flow analysis. *Molecules* **2020**, *25*, 1886. [CrossRef] [PubMed]
16. Jiménez-López, J.; Llorent-Martínez, E.J.; Martínez-Soliño, S.; Ruiz-Medina, A. Automated photochemically induced method for the quantitation of the neonicotinoid thiacloprid in lettuce. *Molecules* **2020**, *25*, 4089. [CrossRef] [PubMed]
17. Mantim, T.; Chaisiwamongkhol, K.; Uraisin, K.; Hauser, P.C.; Wilairat, P.; Nacapricha, D. Dual-purpose photometric-conductivity detector for simultaneous and sequential measurements in flow analysis. *Molecules* **2020**, *25*, 2284. [CrossRef] [PubMed]

© 2020 by the author. Licensee MDPI, Basel, Switzerland. This article is an open access article distributed under the terms and conditions of the Creative Commons Attribution (CC BY) license (http://creativecommons.org/licenses/by/4.0/).

Review

Flow Chemistry in Contemporary Chemical Sciences: A Real Variety of Its Applications

Marek Trojanowicz [1,2]

[1] Laboratory of Nuclear Analytical Methods, Institute of Nuclear Chemistry and Technology, Dorodna 16, 03–195 Warsaw, Poland; trojan@chem.uw.edu.pl
[2] Department of Chemistry, University of Warsaw, Pasteura 1, 02–093 Warsaw, Poland

Academic Editor: Pawel Kościelniak
Received: 17 February 2020; Accepted: 16 March 2020; Published: 21 March 2020

Abstract: Flow chemistry is an area of contemporary chemistry exploiting the hydrodynamic conditions of flowing liquids to provide particular environments for chemical reactions. These particular conditions of enhanced and strictly regulated transport of reagents, improved interface contacts, intensification of heat transfer, and safe operation with hazardous chemicals can be utilized in chemical synthesis, both for mechanization and automation of analytical procedures, and for the investigation of the kinetics of ultrafast reactions. Such methods are developed for more than half a century. In the field of chemical synthesis, they are used mostly in pharmaceutical chemistry for efficient syntheses of small amounts of active substances. In analytical chemistry, flow measuring systems are designed for environmental applications and industrial monitoring, as well as medical and pharmaceutical analysis, providing essential enhancement of the yield of analyses and precision of analytical determinations. The main concept of this review is to show the overlapping of development trends in the design of instrumentation and various ways of the utilization of specificity of chemical operations under flow conditions, especially for synthetic and analytical purposes, with a simultaneous presentation of the still rather limited correspondence between these two main areas of flow chemistry.

Keywords: flow analysis; flow synthesis; flow reactors; flow-injection analysis

1. Introduction

Monitoring and controlling the progress in the course of a given chemical reaction is a fundamental issue in various applications of chemical science. This concerns fundamental investigations into both the composition and the properties of various materials, as well as studies on various phenomena and processes occurring on micro- and macro-scales in the natural environment and in living organisms. It also concerns the optimization of various technological processes involving specific chemical reactions. As indicated by the progress in different areas of chemical science achieved over the past century (or more), one of the contributing factors in the yield of chemical reactions is the movement of reagents depending on various mechanisms. In both laboratory and industrial practices, the most commonly employed processes are those under conditions of the forced flow of reagents. Changes in the transport rate of reagents under flow conditions can be utilized in the physico-chemical examination of the kinetics of the reaction, as well as for the improvement of the efficiency of different steps in analytical procedures or for carrying out chemical synthesis with favorable yields.

In spite of the large number of various applications of carrying chemical reactions under flow conditions, the term "flow chemistry" can only recently be found in the chemical literature, and it is used almost exclusively for the description of chemical syntheses carried out under flow conditions. The search of literature databases indicated its first uses in the 1970s in various fields such as modeling of chemical laser operations [1], fabrication of materials for the nuclear industry [2], transport of pollutants associated with irrigation [3], or the aerothermodynamic analysis of a stardust sample return capsule in the National Aeronautics and Space Administration (NASA) mission [4]. First examples of the use of the term "flow chemistry" in chemical synthesis or analytical fields were found at the turn of 1990s and 2000s in work on the pulsed generation of concentration profiles in flow analysis [5] and in the description of continuous-flow microreactors with fluid propulsion achieved by magnetohydrodynamic actuation, which was employed for the amplification of DNA through the polymerase chain reaction [6]. This can be considered to be an analytical device; however, it simultaneously works via the synthesis of the desired product.

In the case of papers dealing with syntheses under flow conditions, so far generally considered in the present chemical literature as the only field of flow chemistry, the term "flow chemistry" started to be used from the middle of the 2000s onward [7]. The term "flow synthesis", in turn, is commonly used in papers on organic synthesis since the 1970s [8].

The volume of published papers on widely recognized flow chemistry, including not only chemical synthesis but also flow analysis, fundamental physico-chemical investigations under flow conditions, or flow reactors in industrial applications, can be estimated to be 30 to 40 thousand papers in both scientific and technical journals. It should be taken into account that a very arbitrary selection of published works had to be made for the present review. Therefore, for instance, certain areas of measurements or studies under flow conditions are not included, for example, analytical process monitoring by dedicated industrial instrumentation or industrial processes carried out on a technological scale in flow reactors. This should be mentioned in order to not restrict the term "flow chemistry" in the context of modern chemical science and, simultaneously, to point out the importance of this field.

The intention of the author of this review is to present, for the first time in the literature, various fields of modern chemistry which should be considered within *flow chemistry*. A special emphasis is focused on the presentation of the development and chronology of inventions of numerous physico-chemical operations and appropriate instrumental devices, which are widely employed in both flow synthesis and flow analysis.

2. Milestones in the Development of Various Areas of Flow Chemistry

Although it is hard to underestimate the benefits of using numerous literature databases, tracing the evolution of various methodologies of conducting measurements or carrying out chemical syntheses under flow conditions is a very challenging task. The results of different applications of flow chemistry are distributed very broadly in hundreds of chemical journals and they are not always reported in the available databases. The general intention of preparing this section of the review is to present (at least roughly) the chronology of the development of the selected areas of flow chemistry.

2.1. Flow Analysis

It seems that the first chemical phenomenon to be observed under flow conditions and employed for analytical purposes was the separation of a mixture of chemical compounds on a flow-through column packed with a solid sorbent, which initiated the development of chromatographic methods. This is commonly attributed to Cwett's works conducted at the University of Warsaw at the beginning of the 20th century [9,10], although a similar observation was published also earlier [11]. It is necessary, however, to admit that, regardless of the similarity of both operations, the commonly used term "flow analysis" is rather associated with the much later invention of continuous-flow analysis with the segmentation of flowing stream by Skeggs in the middle of the 1950s [12,13], where the segmentation of a flowing stream with air bubbles was essential for limiting the analyte dispersion along the tubing. The developed pioneering system consisted of several instrumental flow-through modules, which allowed performing different physico-chemical operations for the clinical determination of urea in blood with photometric detection in a mechanized manner. It means then that conventional flow analysis follows instrumental set-ups (schematically presented as manifolds), where determinations with different detection techniques can be carried out with various operations of on-line sample processing, with or without the segmentation of the flowing liquid. This understanding of flow analysis is confirmed by a large number of published books and review papers in various journals [14]. This also means that, despite involving flowing conditions, chromatographic methods, electrophoretic methods, mass spectrometry, or atomic spectrometry methods, where the flow of analytes from the sample introduction to the detector takes place, do not fall into the flow analysis category. The concept of constructing a flow analyzer with a segmented stream was used to develop many set-ups for numerous applications in various areas of analytical chemistry [15], as well as for commercial instruments by specialized manufacturers.

In further modification of that concept, the segmentation of a flowing liquid was eliminated from the measurement system [16], a very small sample volume (20–200 µL) was used [17], and a transient signal was used (as an analytical one) instead of steady signal values recorded in the segmented flow analyzers. An extraordinary impetus for the further development of that version of flow analysis called flow-injection analysis (FIA) was given by a series of papers published in the middle of the 1970s by Ruzicka and Hansen [18–20], as well as some parallel ones by Stewart et al. [21,22]. Fast development of that methodology resulted in the availability of numerous commercial instruments [23], as well as further development of the modified versions of FIA such as sequential injection analysis (SIA) [24] or lab-on-valve (LOV) systems, which integrated the injection modules with detection and some operations of sample treatment on a renewable bed of solid sorbents [25].

Already at the end of the 1970s pioneering microfluidic systems were designed and produced on silicon wafers, firstly employed as a capillary column in gas chromatography [26]. Then, since the beginning of the 1990s, their first applications in flow analysis were developed, e.g., in hyphenation with the surface plasmon resonance analyzer for the determination of immunoglobulins [27], or in a miniaturized system with detections by solid-state electrochemical sensors or small-volume optical detectors [28]. The currently observed development of measuring systems involves further miniaturization and instrumental integration of hydraulic, detecting, and sample processing operations. This is also associated with tapping into current achievements of nanotechnology and new technologies for the transmission and processing of measured signals.

2.2. Flow Synthesis

Organic chemistry is an exceptionally broad area of modern chemistry embracing fundamental investigations of mechanisms of reactions, identification of natural compounds, and optimization of laboratory syntheses, as well as their scaling up to a technological level. Utilizing flow conditions for carrying out chemical syntheses is a very important part of the development of flow chemistry. Although, as already mentioned, the term "flow chemistry" was used only recently, i.e., in the last two decades, the real beginnings of this methodology were much earlier. According to the Web of Science database, the first contribution to flow synthesis was a short report on the use of a flow reactor with a phosphoric acid catalyst on silica gel for dehydration of diethylcarbinol, published in 1932 [29]. An evident increased interest in the use of flow reactors, especially those with an immobilized catalyst, was noted no earlier than the 1960s [30,31]. Numerical methods were employed to evaluate the participation of diffusion, the convection of a reactant, and the kinetics of homogeneous and heterogeneous reactions in catalytic tubular reactors [30], while experimentally diffusional and chemical effects were examined for fast reactions of cyclopropane in flow reactors [32]. Flow synthesis in an open flow-system was reported for producing layers of gallium phosphide for some electronic applications [33]. A continuous-flow salt gradient was utilized in the preparation of polynucleotide–polypeptide complexes [34], while excellent yields in cryptate synthesis could be obtained via a cyclization reaction carried out with efficient mixing in a suitable flow cell [35]. The 1980s brought further development in designing automated systems where the progress in flow synthesis was monitored on-line and was increasingly controlled by the computer, which was reported, e.g., for solid-phase synthesis of peptides [36,37] and oligodeoxyribonucleotides [38]. In the latter case, more than 600 oligomers were produced in approximately 15 min. Alpha-gliadin peptides were synthesized in a continuous-flow set-up under ultrasonic field conditions, which accelerated the coupling reaction [39].

A substantial increase in the number of published papers on various innovations in flow synthesis was observed since the beginning of the 2000s. It seems that this can be essentially attributed to an increasing interest in this technology and its great potential for speeding up drug development by leading manufacturers of pharmaceutical industries [40], as well as by very active research groups from top academic institutions such as, e.g., the University of Cambridge in the United Kingdom (UK) [41], the California Institute of Technology [42], or the Massachusetts Institute of Technology (MIT) [43]. Numerous flow syntheses can be successfully enhanced with microwave activation [41,44], while conventional pumping in the flow systems with piston pumps can be replaced with a magnetohydrodynamic actuator [6] or by the use of electroosmotic flow [45]. Flow syntheses can be also carried out under supercritical conditions, which was first employed when producing narrow-size distribution quantum dots [46].

As of the middle of the 2000s, flow syntheses with the use of microfluidics were performed more and more commonly, allowing for the reduction of the amounts of reagents and solvents—a pioneering example of gas–liquid–solid hydrogenation [47] or the production of nano amounts of a ^{19}F-labeled imaging probe for positron emission tomography (PET) [42]. Yet another direction in this trend was the use of droplet-based microfluidics, which was demonstrated, for instance, for the synthesis of anisotropic gold nanocrystal dispersions [48].

As it was shown, e.g., in the synthesis of the natural alkaloid called oxomartidine, complete synthesis of natural products can be carried out in a multistep procedure in the system with several packed columns containing immobilized reagents, catalysts, scavengers, or catch-and-release steps [7]. This work initiated intensive development of some other, similarly complex systems, also performed in microreactor networks, e.g., in the multistep synthesis of carbamate [49] or in the continuous-flow synthesis of carboxylic acids with the use of so-called tube-in-tube gas-permeable membrane reactors [50].

The most recent instrumental innovations in flow syntheses are, among others, the use of three-dimensional (3D) printed advanced reactors for Ag nanoparticles [51] or the creation of a robotic system for flow synthesis [52].

2.3. Fundamental Physico-Chemical Measurements under Flow Conditions

One of the areas where measurements under flow conditions were applied for decades is the investigation into the kinetics of chemical reactions, especially in the case of fast reactions taking place in solutions. Reports on such measurements can be found in the literature since the 1950s [53], while the stopped-flow methodology for this purpose was reported even earlier [54]. In the latter method, instead of the measurements of concentration changes in the monitored reaction along the length of the reaction tube, the detection system is located in the proximity of the mixing chamber into which the solutions of reagents are forced from the syringes. Kinetic measurements were also carried out in a typical flow analyzer with a segmented stream in studies on peptide bond hydrolysis [55]. Kinetic measurements in a stirred flow reactor for the alkaline bromination of acetone allowed evaluating both reaction orders, as well as determining the reaction-rate constants in a system of consecutive reactions [56].

Contemporary kinetic measurements are carried out with different detection techniques and in different configurations of flow systems. For instance, a successful application of this kind was reported for continuous-flow microfluidic devices with laser-based mid-infrared chemical imaging in studies on fast organometallic chemistry [57], while a microfluidic system—with in situ X-ray and fluorescence detection—was used in studies on hydrogelation kinetics [58]. In a microreactor set-up, kinetic data on exothermic synthesis by Michael addition was examined with fast in-line monitoring by Raman spectroscopy [59]. Then, two recently reported examples described kinetic studies devoted to free-radical chemistry under flow conditions with chemiluminescence detection. Infrared chemiluminescence detection was employed to monitor the kinetics of the reaction of •OH radicals with formamide and chemically activated carbamic acid [60]. The chemiluminescence of a methyl- cypridina luciferin analogue was employed to examine the kinetics of the reaction of superoxide radical ions with dissolved organic matter in a typical flow-injection analysis system [61].

In stopped-flow measurements, where the dead-time of a fraction of a millisecond can be reached in modern instrumentation, various spectral and electrochemical methods are used to investigate the kinetics of very fast reactions in solutions. Absorption or fluorescence spectroscopy are the two most commonly used detection techniques in dedicated commercial instruments. In a very early example of such works, the kinetics of unstable compounds in a biochemical reaction (e.g., the formation of catalase–hydrogen peroxide complex) was reported with spectrophotometric detection in a time span from milliseconds to several minutes [62]. Then, a very recent example was the examination of the kinetics of aroxyl-radical scavenging rates by important lipid-soluble antioxidant α-tocopherol and catechins [63]. An example of some recent stopped-flow kinetic investigations with the use of fluorimetric detection was a study on the binding of biotin and biocytin to avidin and streptavidin, which is broadly used in biotechnology and bioanalysis [64]. Creating a simple manifold that can be attached to a commercial stopped-flow apparatus, allowing kinetic studies to be performed from −12 °C to 45 °C, constitutes an instrumental improvement in such measurements reported in recent years [65]. The sensitivity of kinetic absorption measurements in ultraviolet (UV)–visible wavelengths can be enhanced by the use of optical cavity-based techniques. The reported low-cost experimental set-up providing broadband cavity enhancement of sensitivity by coupling to commercial stopped-flow instruments allows obtaining even 78-fold enhancement for the measurements at 434 nm, which was employed in the investigation of the kinetics of a potassium ferricyanide reaction with sodium ascorbate [66].

Among other spectroscopic detection techniques used under flow conditions, numerous applications in stopped-flow kinetic measurements were reported for circular dichroism (CD). This technique enables simultaneous monitoring of chiro-optical and absorbing transients, which was shown, for instance, in the binding of sulfonamide to bovine carbonic anhydrase [67]. Recently, CD detection is commonly employed to study protein folding kinetics [68], while ultraviolet CD was used to study the folding mechanism of the outer surface protein A. Kinetic measurements in a stopped-flow mode are also used with Fourier-transform infrared (FTIR) detection [69], whereas NMR spectroscopy was reported, for instance, in the examination of metallocene-catalyzed poly- merization of 1-hexene [70].

Electrochemical detections are used much less commonly in detection of stopped-flow measurements than in spectroscopy techniques. At the very early stage of the development of stop- ped-flow methodology, conductivity detection and potentiometric pH measurements were employed to study the following reaction: $OH^- + CO_2 \rightarrow HCO_3^-$ [71]. One recent example showed, e.g., the use of voltammetry in stopped-flow studies on the kinetics of reduction of cop- per-containing enzyme peptidylglycine monooxygenase [72]. The above-mentioned examples clearly show a crucial role of flow measurements in the evaluation of kinetic parameters for many different chemical processes.

2.4. Chemical Processing under Flow Conditions for Non-Analytical and Non-Synthetic Purposes

This section of the present review only very briefly highlights the importance of yet another field of investigations and processes which should also be considered as a substantial part of *flow chemistry*. It is based on research studies and processes oriented most commonly toward technological applications in flow reactors, but they are also commonly initiated in laboratory investigations under flow conditions. Moreover, they are usually investigated for final technological applications in environmental protection or the processing of various materials in different branches of industry, including food processing. The examples presented below were randomly and arbitrarily selected to roughly indicate the variety of areas and problems associated with this particular field of flow chemistry.

As already mentioned, these approaches do not follow the rules of the field of chemical engineering on a technological scale. A vast amount of literature on this subject can be found in technological journals, patents, and specialized books [73,74]; thus, in this section, only some examples from the current literature are provided.

Applications of a flow-through reactor with immobilized α-galactosidase, which was used in the reduction of raffinose concentration in beet sugar molasses [75], or in the application of a flow-through reactor for the photocatalytic decomposition of ethylene applied to control tomato ripening [76], are two examples of employing biotechnology in the food processing field. A dynamic flow approach was reported to examine the leaching of antioxidants from solid foods [77], while, for ore and mineral processing, a reflux flotation flow cell was designed to achieve fast flotation [78].

The processing of biomass of various origin, which can be carried out in flow-through reactors, is an important sector of modern biotechnology. For instance, the flash pyrolysis of microalgal or lignocellulose biomass in a flow reactor may lead to the production of bio-oils of a very different composition, depending on the type of an initial substrate and the conditions of pyrolysis [79]. The obtained products can be further employed as phenolic-rich bio-oils, liquid fuels, or fine chemicals. The hydrothermal conversion of a wet waste feedstock in a continuous-flow reactor may result in obtaining liquid or gas final products with acceptable residual organic contamination [80]. The production of bio-crude oil using hydrothermal liquefaction in a continuous-flow reactor was also reported from wastewater microalgae [81].

A particularly large number of research papers on both the investigation and the application of flow-through reactors were devoted to the development of different processes for environmental protection, involving various methods of chemical processing. The biological anaerobic removal of nitrate from wastes was developed with the use of a bioreactor packed with sludge carbonaceous material [82]. In yet another similar method for the biodegradation of 2-chlorophenol, adsorption and biodegradation processes were carried out in a flow-through reactor with suspended biomass on activated carbon [83]. The mechanism of removal included the sorption of a pollutant, its transport by diffusion in a biofilm, and biodegradation by suspended biomass. Just recently, the efficiency rates of bacterial aerobic granular sludge and algal–bacterial granular sludge packed in a 20-L continuous-flow reactor were compared for saline wastewater treatment [84]. Since high salinity enhances algae growth, the second material exhibited slightly higher total nitrogen and phosphorus removal efficiency.

Typical chemical processing methods were also widely investigated for water and wastewater treatment in flow-through reactors. These approaches include, first of all, various photochemical methods, involving (mainly) UV irradiation under different conditions with or without catalysts [85]. For instance, the photocatalytic degradation of bio-resistant dyes was demonstrated in a flow reactor with a TiO_2 catalyst embedded into a cement matrix and deposited at the bottom [86]. Regarding other methods, the catalytic wet air oxidation of industrial wastewater was carried out in a flow reactor packed with copper nanoparticle-doped and graphitic carbon nanofiber-covered porous carbon beads [87]. A complete chemical oxygen demand (COD) reduction in industrial wastewater was observed under optimized conditions. A significant reduction (50% to 80%) of COD and the removal of numerous inorganic contaminants from drinking water could be obtained with the use of a flow reactor packed with a mixture of anionic and cationic ion-exchange resins [88].

The processes such as sonolysis, electrochemical oxidation, or irradiation with a beam of accelerated electrons can be employed in continuous-flow reactors for the efficient decomposition of organic pollutants in waters and wastewaters. For instance, separate or simultaneous sonolysis and ozonation were examined under flow conditions for the degradation of antibiotics in wastewater [89]. Electroactive pollutants such as residues of different pharmaceuticals can be efficiently removed by electrochemical oxidation, which was demonstrated, e.g., for the residues of the widely used anti-inflammatory drug called naproxen using different porous materials of anode [90]. A flow reactor with a boron-doped diamond anode enabled efficient degradation of azo dyes occurring in synthetic textile effluent [91], whereas an ultra-nanocrystalline, boron-doped diamond anode coated on a niobium substrate, also employed in a flow reactor, was successfully used for the minera- lization of a persistent pollutant perfluorooctanoic acid [92]. An aerosol flow reactor developed for the efficient decomposition of organic pollutants by electron beam irradiation can be mentioned as a final example of different flow reactors investigated for environmental purposes [93].

The above-mentioned examples are only a few randomly selected examples that can be found in thousands of papers published in scientific journals and patents on the use of flow reactors for this field of research activity. It is impossible to provide a more complete review of these applications or to discuss the evolution of their construction, the optimization of their operation, or the broad fields of applications. It is doubtless, however, whether these types of chemical processes should also be considered a part of flow chemistry. It is also impossible not to admit that the processes investigated and optimized in such flow reactors are often utilized in flow systems developed for both analytical and synthetic purposes, contributing to the final results of their functioning.

3. Similarities in Development Trends in Flow Analysis and Flow Chemistry

Each of the four above-mentioned areas of flow chemistry underwent several decades of development, and, behind each, there is vast subject literature and numerous applications. It seems, however, that, in terms of the scale of reported processes, the choice of the employed physico-chemical unit operations, and the adoption of similar instrumental constructions, the closest similarity can be found between flow analysis and laboratory flow synthesis. Although this was already pointed out in an earlier review [94], as well as in the Website review [95], there is still almost a complete lack of interrelation between scientists contributing to each field, i.e., organic chemists working on flow synthesis and analysts working on flow analysis, which inclined us to present the development trends in these two areas of flow chemistry once again.

The development/evolution of flow analysis methods takes place in three basic ways. The first one is the instrumental development of detection systems, pumping devices, and on-line sample treatment modules. The second one deals with the development of methodologies of flow measurements and signal processing, while the third one concerns both the search for and the selection of the most suitable chemistry for the most successful determination of particular analytes. As mentioned before, according to the well-established custom, the flow analysis field does not include chromatographic and electrophoretic methods in the further discussion below.

The development of flow analysis in these three aspects aims at establishing the most efficient instrumentation and procedures, providing the most accurate, precise, and selective determination, while requiring the simplest instrumentation and smallest possible human effort during its routine use. The evolution of detecting devices tends to obtain the best possible accuracy and selectivity, appropriate detectability for particular applications, and the shortest response time. A significant contribution to obtaining satisfactory accuracy and selectivity was brought by the employed on-line pretreatment steps. The evolution of measuring methodologies and their selection for particular determinations is connected to the mode of signal acquisition, its processing into analytical information, and the choice of a calibration method. Moreover, the evolution of "chemistry of analytical determination" involves the interaction of an analyte with a detector or an employed analyte/sample pretreatment method, which provides the best sensitivity of detection with satisfactory selectivity.

To a great extent, similar factors, although with certain shift of accents, seem to be taken into consideration in flow systems for chemical syntheses. Obviously, in synthetic flow systems, the heart of the whole instrumentation is not the detector itself but the flow-through reactor(s). Instead of the sensitivity of detection, the most important factor is the yield of the conducted reaction. Synthesis time, necessary supervision by a human operator, and optimization of both the design and the functioning of different modules for carrying out some intermediate steps (mixing, cooling, heating, etc.) are also of critical importance. It seems that one of the most crucial differences between analytical and synthetic flow systems is the fact that, in the case of the former, a transient signal received from the detectors can be used as a useful analytical piece of information (assuming their satisfactory precision), while, in the case of flow synthesis, a crucial parameter is the completeness of the desired product and its purity, which play a significant role in the scavenging steps of many procedures.

3.1. Detectors, Reactors, Manifolds

More than six decades of the development of flow analysis can be found in over 20 thousand publications in various scientific journals. A list of books published on this subject is shown in Table 1 (see Conclusions and Perspectives section), while a large selection of numerous review papers can be found in an earlier review [14]. Such a vast literature provides a selection of examples for an in-depth discussion on the comparison of progress in flow synthesis. The development of very efficient flow-through detectors is a fundamental task for analytical systems. Generally speaking, they should be able to generate a signal of the largest possible sensitivity with very fast response time at the smallest possible dead-volume.

From the very beginning of the development of flow analysis, the greatest attention was given to spectrophotometric absorption detection in the UV–visible region, molecular and electrochemical luminescence, and the application of atomic spectrometers. The evolution of the simple cost-effective construction of detectors is associated with parallel development, provided mostly by the manufacturers of commercial instrumentation and detectors for liquid chromatography. In the case of spectrophotometric detectors, important progress was made in the replacement of simple flow cuvettes with long pathway detectors, utilizing multiple internal reflection [96], in conducting absorption measurements in cuvettes packed with solid sorbents on the optical pathway [97], or in the use of optoelectronic elements as a light source [98], as well as a detector. The simultaneous use of several different light-emitting diodes (LEDs) as a light source allows carrying out multi-analyte detection [99], while a suitable design of the detector cell enables conducting absorptive, fluorometric [100], nephelometric, and turbidimetric detections [101]. The development of common types of electrochemical detectors (thin-layer, wall-jet, wire, cascade, or multi-array detectors) was accompanied by the search of new electrode materials and surface modification of the working electrode surface, especially with biochemical receptors and nanomaterials [102]. The possibility of conducting simultaneous multi-analyte detection (without chromatographic separation) was achieved by appropriate polarization of the working electrode in voltammetric measurements [103], reporting a multi-pulse amperometric detection or the application of various sensors such as a potentiometric sensor with limited selectivity but with appropriate signal processing [104]. An example of such an advanced attempt was a rapid solid-phase fluoroimmunoassay [105]. There are specific instruments for flow analysis; however, as presented below, some flow detectors can be directly incorporated into flow synthesis systems for real-time monitoring of the progress of the conducted reaction.

A different strategy was used in the development of flow reactors for flow synthesis systems. In conventional laboratory systems (mini- or meso-fluidic), made of glass, quartz, or metallic coiled tubes of an internal diameter of a few mm, the volume of a reactor is not very important. In most cases, good heat transfer, simultaneous irradiation (in the case of photochemical reactions) [106], and microwave heating [107] are of bigger importance. A separate group of tools includes microreactors made of stainless steel or perfluorinated polymer tubes, or micro-structured devices [108]. A particular advantage of carrying out flow syntheses in microreactors is using very tiny amounts of substrates and products, which is crucial when using hazardous chemicals [109]. Microfluidic reactors can be manufactured using glass, silica glass, ceramics, or stainless steel by micro- machining. One reactor made of silicon was employed in the above-mentioned set-up with fluid propulsion by alternating current (AC) magnetohydrodynamic actuation [6]. Teflon microreactors were also developed into a "click-system" version, comprising two separate plates of Teflon fabricated by computerized numerical control milling [110]. A particular reported design was a microcapillary reactor made in the form of a fluoropolymer disc comprising 10 parallel capillary channels [111]. Using several of these discs in parallel enables scaling laboratory syntheses up to a small production scale. Although it is difficult to find the use of microreactors alone in flow analytical systems, microfluidics holds a solid position in flow analysis, which is shown later on in this paper.

Yet another trend in this field is the use of monolithic reactors made of different types of macroporous materials such as polymers, polymer/glass, silica, or zeolites in flow synthetic systems [112]. As compared to batch or packed-bed reactors, they provide higher yields and improved selectivity, and, in many cases, they exhibit improved mechanical and thermal stability. Some other materials used for this purpose may also induce some additional catalytic properties.

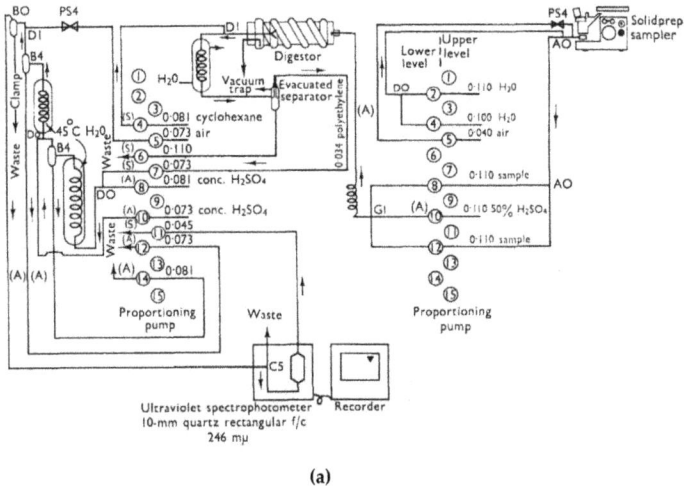

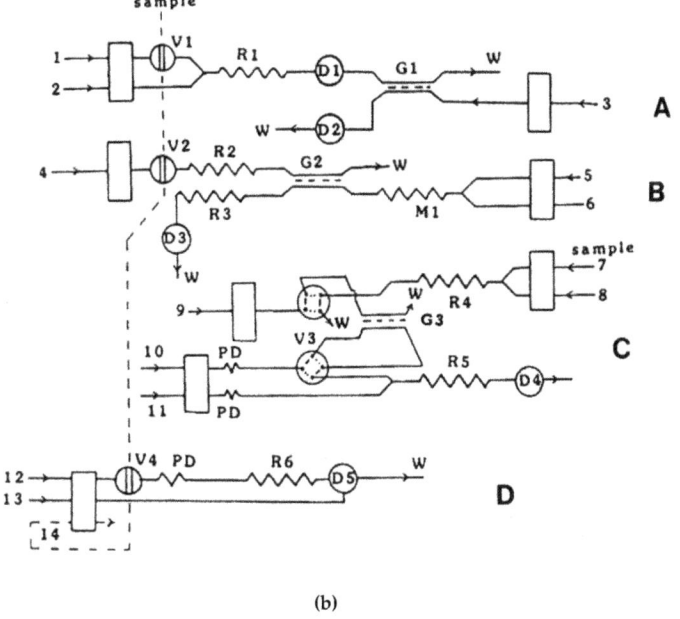

Figure 1. Schematic diagrams of multistep flow analytical systems: (a) air-segmented continu- ous-flow system for the determination of biphenyl pesticide in citrus fruit rind involving in-line distillation and ultraviolet (UV) detection [113]; (b) flow-injection system for simultaneous photometric determination of sulfide, polysulfide, sulfite, thiosulfate, and sulfate [114]. A—manifold for determination of sulfide and polysulfide; B—manifold for determination of sulfite; C—manifold for determination of thiosulfate; D—manifold for determination of sulfite.

It is worth mentioning that, since the late 1990s, monolithic columns were employed to carry out high-performance chromatographic separations in both low-pressure FIA and SIA systems [115].

The common element which connects the instrumentation for flow analysis and flow synthesis is the design and construction of multistep manifolds, which are complete flow systems for the determination of the target analyte or for the production of target products from different substrates using different synthetic steps. Here, it is easy to find numerous similarities, although the symbols for their graphical presentation in the schemes published in the subject literature can slightly differ.

Expanded manifolds for analytical determinations are constructed when multistep on-line operations of a sample treatment are necessary, as well as when analyte derivatization is necessary for a particular detection. The determination of the residues of pesticide known as biphenyl in the samples of citrus fruits, including on-line distillation among other operations, is a good example of a continuous-flow analytical procedure, reported in 1966 [113] (Figure 1a). The application of in-line distillation in continuous-flow chemical synthesis can be illustrated by the system with immobilized organic base catalysts developed for semi-continuous nitro alkene formation and Michael addition [116]. In this case, in-line distillation was used to remove the excess of nitromethane from the reaction mixture.

Yet another type of analysis requiring widely expanded manifolds with sample treatment steps is speciation analysis, where different compounds of the same element are simultaneously determined in the same sample. Some examples of such advanced flow analytical systems include those reported for the determination of four sulfur anions [114] (Figure 1b), chlorine-containing anions [117], or speciation of nitrogen compounds [118,119]. Some other expanded systems also comprise, for instance, an SIA system with atomic fluorescence detection employing cold vapor generation, constructed for the determination of labile and non-labile fractions of mercury in environmental solid samples [120], or flow systems with a sophisticated calibration procedure, allowing a reduction or elimination of the effects of interferences in the determination of calcium in dairy products with solenoid-type mini-pumps and flame atomic absorption spectrometry detection [121].

Multi-analyte flow systems are different kinds of determinations carried out under flow conditions in multiline systems. They were constructed in continuous-flow mode with stream segmentation (see, e.g., the system for determination of pharmaceuticals in tablets [122]), and they were further developed in the case of flow-injection methods [123,124]. In most of the above-mentioned FIA systems, the propulsion of fluids was carried out mainly with peristaltic pumps; however, syringe pumps are also quite widely employed for this purpose (see, e.g., multi-syringe systems [118,120,125]). They are especially indispensable in SIA systems including those with LOV and renewable sorbent beds where different directions of flow are commonly used. One more alternative, which was frequently used in recent years, involves solenoid minipumps and valves, especially advantageous for the design of fully computer-controlled flow analytical systems [121].

Multistep syntheses are common practice in organic laboratories. One type of widely used procedure in flow synthesis is the so-called telescoped reaction sequence, which involves consecutive reactions of products formed in the previous step via an addition of new reagents or catalysts to the reactor [126]. Piston pumps, commonly used in HPLC instrumentation, are usually used as a pumping device in such systems, and an example of such a configuration, developed for the bromine–lithium exchange reaction with *o*-bromotoluene [127], is shown in Figure 2a. All the reactions taking place here are very fast and exothermic; hence, the heat transfer is a critical issue. This is the case where, due to the high surface-area-to-volume ratio, conducting the reactions in microreactors is especially advantageous. Numerous multistep flow synthesis systems involve various physico-chemical processes carried out in different flow-through modules assembled in the whole set-up. Many of them can be found in comprehensive reviews [128–131]. A pioneering contribution to the development of such systems is usually attributed to Ley's research group from the University of Cambridge (UK), including their work on the synthesis of peptides [132] and the multistep flow system developed for the first continuous synthesis of an alkaloid natural product called oxomaritine [7]. The latter study consisted of seven steps involving the use of packed columns with immobilized reagents, catalysts, and scavengers. A scheme of all reactions carried out in that particular flow synthesis is shown in Figure 2b. The possibility of creating such systems was well illustrated in a recent paper on reconfigurable systems [133]. The analogy to the above-mentioned works on analytical flow systems with various steps of sample processing or for multicomponent determinations seems to be evident.

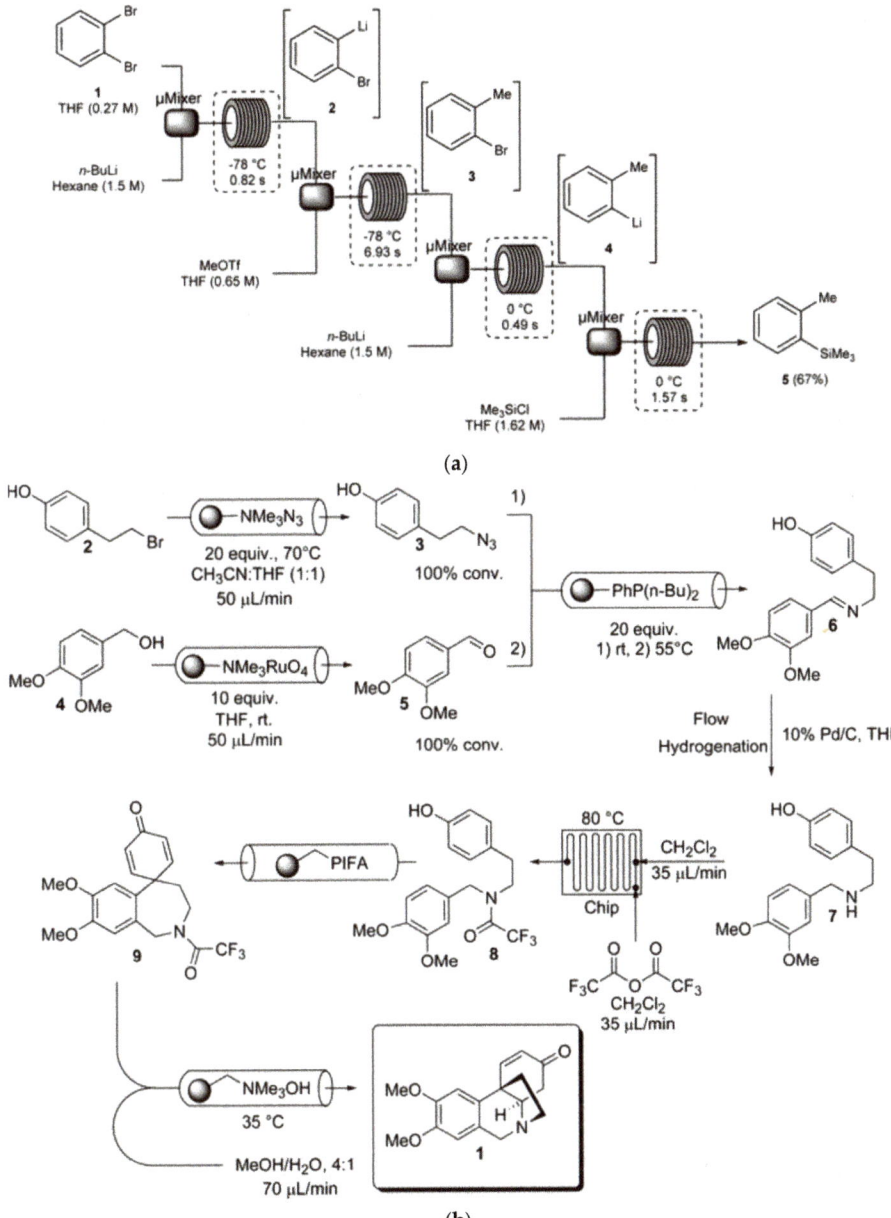

Figure 2. Cont.

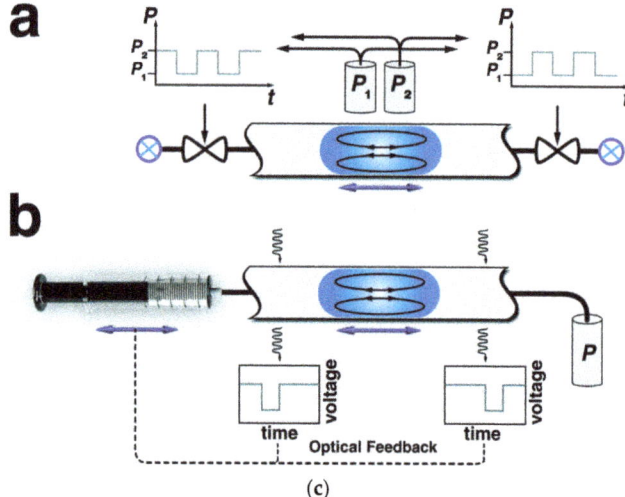

Figure 2. Schematic diagrams of two example multistep systems (**a,b**) and an oscillatory system (**c**) developed for flow synthesis: (**a**) a flow system with a so-called telescoping reaction sequence which involves consecutive reactions of products formed in the previous step via addition of new reagents or catalysts to the reactor [127] (adapted from [128]); in the system composed of four micro-mixers and four micro-reactors, sequential reactions of o-dibromobenzene with two electrophiles were carried out; (**b**) a flow system with various packed columns containing immobilized reagents, catalysts, and scavengers developed for the synthesis of the alkaloid natural product (±)-oxo- maritidine [7]; (**c**) basic types of oscillatory multiphase flow systems driven by valves (a) and a syringe pump (b) [134].

The segmentation of the carrier stream in pioneering Skeggs's continuous-flow analysis set-ups [12,13], together with those constructed and reported over the next two decades, served as a basic tool for the reduction of dispersion of a sample zone in the flowing stream. The process required the effective removal of air bubbles prior to the photometric detection cell; however, later on, debubbling was replaced by electronic controlling of photometric measurements without removing air bubbles. Such flow systems were usually constructed as multiline manifolds (Figure 1a); however, in a Technicon Chem-1 advanced commercial analyzer, segmentation was also used in a single-line system without debubbling [135]. This was performed in such a way that air bubbles were introduced only between each sample and the flowing flush solution [136], or into a so-called mono-segmented system where the sample was introduced between two air bubbles [137]. This also includes the set-ups in segmental flow-injection analysis [138], where the flow sample zone is surrounded from both sides by air bubbles to obtain the same effect of controlling dispersion along the tubing. Among some more recent examples of utilizing flow segmentation in analytical systems, one can find a microfluidic system designed for detecting bacteria and determining their susceptibility to antibiotics [139]. Plugs with aqueous solutions of different drugs are separated by fluorinated carrier fluid and mixed with a stream of examined bacterial solution and staining reagent stream. A degree of interaction was observed after appropriate incubation in a stopped-flow mode.

A similar idea to maintain a discrete entity within plugs to carry out partial reactions is employed in flow synthesis systems. If the aqueous segments of reagents with the catalyst in fluorinated fluid in a Teflon microreactor are separated by hydrogen bubbles, this provides a successful set-up for carrying out a regioselective hydrogenation reaction [140]. Such a procedure, described as a plug-flow approach, enables carrying out chemical reactions on a μL–mL scale. Moreover, in the segments of different organic solvents transported in a perfluorinated carrier solvent, one can carry out the optimization of reaction conditions that can be then transformed into a large-scale batch reactor [141]. The efficiency of catalytic oxidation of methanol was examined in a segmented flow system with zones of different catalysts in a perfluorinated carrier, where, under high pressure, the mixture of substrate and oxygen was introduced through the Teflon membrane [142]. It can be concluded that the commercial success of clinical analyzers with segmented flow between 1960 and 1980 found a kind of continuation/replication in the process of designing segmented flow synthesizers to accelerate the generation of compounds in pharmaceutical discovery [143].

A particular mode of exploiting the stream segmentation in flow systems is also the so-called oscillatory flow strategy, where, as a result of the periodic reversal of a flow direction, a droplet of the reaction mixture in an immiscible inert carrier oscillates within the tube with, for instance, in situ UV monitoring [134], in either valve-based or syringe pump-based flow set-ups (Figure 2c). Such oscillatory multiphase systems can be used both for screening and for conducting chemical processes with a long processing time, and they are also suitable for carrying out kinetic investi- gations.

The basic part of each analytical or synthetic flow system is a fluid propulsion device. Since the creation of the first analytical systems in the 1950s, the use of peristaltic and syringe pumps predominated over the use of gas-pressurized reservoirs, gravity flow, or piezoelectric micropumps. A different alternative initiated in the early 1900s was the use of electroosmotic flow (EOF) [144]. However, it has its limitations due to the fact that the direction of EOF depends on the composition of the solution. On the other hand, the line with EOF can be hydrostatically coupled to propel other solutions. The applications of this concept in micro-flow-injection analysis systems were reviewed [145]. Then, some years later, this concept was also adopted for synthetic purposes within microreactors, and these approaches were also reviewed [146]. It was demonstrated that, for instance, EOF can be employed in systems with a continuous-flow reactor for solid-supported synthesis [45]. The solid-supported catalyst bed is placed in the microreactor where EOF is generated, and one of the reagents is immobilized on functionalized silica gel. With EOF is used in the capillaries, the volumes of samples/reagents are limited to nL–µL volumes; however, in the case of tubes packed, e.g., with solid catalysts, much larger diameters of tubes can also be employed without back-pressure, which allows building such flow systems up to a semi-preparative scale [147].

Yet another kind of analytical and synthetic flow system which functions as a result of the sequence of several operations carried out under flow conditions is the configuration including one operation conducted in a non-flow discrete mode. These systems are described as flow-batch systems and their development was initiated by describing an automated micro-batch analyzer with the sample loaded into an injection loop and blowing into the reaction chamber using compressed gas [148]. This was developed for various determinations using different detection techniques, followed some years later by the creation of flow-batch titrators for photometric titrations [149], as well as a spectrophotometric flow-batch system for the determination of aluminum in plant tissues [150]. In such systems, for instance, a series of standard solutions can be prepared in a discrete mixing vessel and introduced into the flow measuring system, which was reported in the set-up designed for trace determination of Mn with electrothermal atomic absorption spectroscopy (AAS) detection [151]. The flow system can be stopped with a non-flow detector, operating also as a mixing vessel, a phase separator, and a detector, which was reported for the set-up designed for the photometric determination of Fe(III) in oils [152]. One of the most recent reports dealt with the creation of a flow-batch system for the speciation of Cr(III) and Cr(VI) with chemiluminescence detection, based on the role of Cr(III) as a catalyst in the reaction of luminol with hydrogen peroxide, and Cr(VI) as an oxidant in the luminol reaction using chemically modified carbon nanotubes in the separation process [153].

Such a method of creating flow-batch systems is even more frequently employed in designing flow systems for organic syntheses, which was very well illustrated by a recent comprehensive review on the development of flow synthetic methods for the pharmaceutical industry [154]. In numerous cases, a 2–3-step telescoping process is ended in a batch reactor with a quenching step. This mostly concerns the works published in recent years, including, for instance, the preparation of boronic acid 6 [155], the generation of dichloromethyllithium [156], or the running of tube-in-tube reactions with diazo-methane in a batch reactor [157], among many others. As a particular achievement in this area, the development of an automated platform integrating batch and flow reactions should be pointed out [158]. This was based on the modularity of the flow process, where, in a more favorable situation, the flow module can be replaced by a batch reactor. Based on the complex synthesis of the precursor of an anticancer drug candidate, the creation of three phases of the whole synthesis procedure, incorporating batch and flow steps configured in a telescoped manner, took place. The whole system was computer-monitored/controlled and cloud-based with the server and the server–equipment interaction occurring via the internet. The selected citations from the literature indicate the development of very similar solutions in the design of flow systems for both analytical and synthetic purposes.

A common feature of measurements and operations under flow conditions is certainly the possibility of both monitoring and controlling very fast chemical reactions. As already shown, this is the basis for using flow conditions in fundamental kinetic studies. In the very early years of the development of injection methods in flow analysis, very fast kinetics of the reaction of sulfur species in a molecular emission cavity detector was utilized for the speciation of sulfur anions [159] (Figure 3a). The differences in the kinetics of either the formation or the dissociation of complexes in a solution were utilized at the same time for multicomponent FIA determinations. This can be illustrated by the simultaneous FIA determination of Ca and Mg based on different reaction rates of the dissociation of cryptand complexes [160] or of Co and Ni based on different reaction rates of the formation of citrate complexes [161]. In the FIA system shown in Figure 3b, the speciation analysis of Fe(II) and Fe(III) was based on the reaction between leucomalachite green and persulfate with 1,10-phenanthroline as an activator [162], thanks to the different catalytic effect of these analytes. Among the fast processes utilized in flow analysis, numerous applications in flow-injection analysis were reported for chemiluminescence detection. It was demonstrated that, for instance, in the case of chemiluminescence resulting from the reaction between urea and hypobromite in an alkaline medium, the maximum emission occurred 2 ms after mixing the reagents, and it was completely extinguished after 400 ms [163]. This can be applied in analytical determination using a double concentric tube mixer, located directly in front of the photomultiplier tube. Such instrumentation was used also much earlier, e.g., to trace the chemiluminescence determination of Co based on the catalytic oxidation of luminol by hydrogen peroxide or sulfide, based on the fluorescein-sensitized oxidation of sulfide by sodium hypochlorite [164]. Among the organic analytes, a fast chemilumine- scence response was utilized, e.g., in the FIA determination of antihypertensive compound known as dihydralazine sulfate, based on the strong enhancement of luminescence produced by the analyte in the reaction of luminol with diperiodatocuprate [165]. It can also be pointed out that the differences in the rate of the emission of chemiluminescence can be utilized for multicomponent flow determinations, which was observed in the determination of two opiate narcotics, namely, morphine and naloxone, based on their oxidation by permanganate [166].

In the much further development of flow synthesis procedures, fast reactions are very widely employed enabling much faster procedures than in conventional batch systems. A convincing example can be, e.g., total synthesis of the fluoroquinoline antibiotic known as ciprofloxacin [167]. This was described as the longest linear sequence of reactions telescoped under flow conditions, which involves six chemical reactions carried out in five flow reactors with 9 min of total residence time. The corresponding patented batch syntheses take up between 24 and >100 h. In yet another example, the synthesis of carboxylic acids via a fast lithiation–carboxylation sequence at room temperature in continuous flow took less than 5 s [168], although a reaction time of only 32 ns was also reported [169]. These are time intervals exploited in fast chemiluminescence detections for analytical purposes.

The field of chemical syntheses involving extremely fast reactions conducted in a highly controlled manner is called "flash chemistry" [170]. Conducting such reactions under flow conditions is especially advantageous and, in some cases, virtually impossible under batch conditions [171]. Only in a continuous-flow system can the reaction time be controlled by the residence time. The formation of unstable intermediates can be suppressed by the time-controlled addition of a quenching reagent.

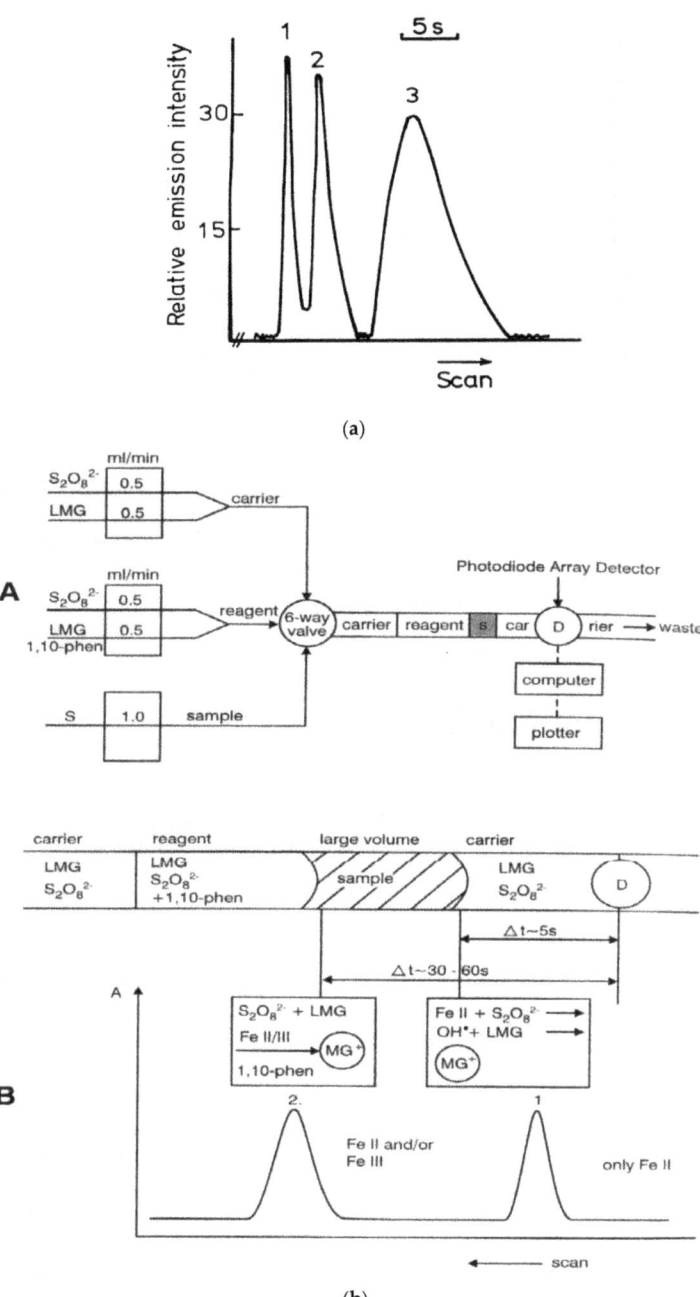

Figure 3. Cont.

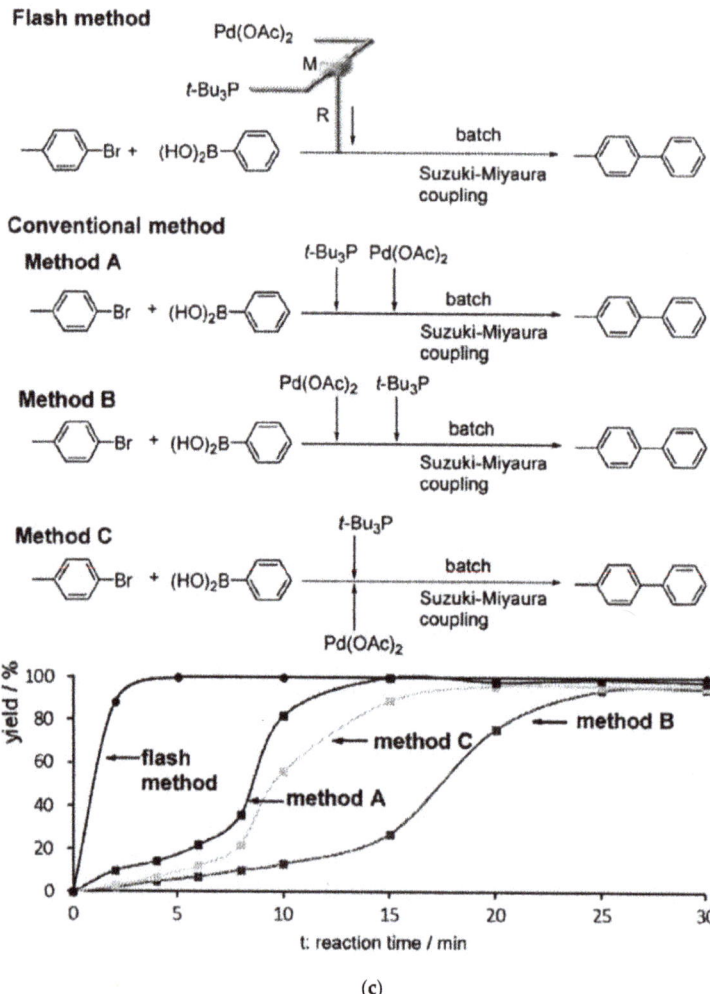

Figure 3. Examples of the kinetic effects for speciation analysis in flow-injection analysis (FIA) systems (a,b) and in continuous-flow synthesis (c): (a) flow-injection signal recorded form a single injection of mixture of sulfide, sulfite, and sulfate into an FIA system with a molecular emission cavity detector [159]; (b) manifold of the FIA system (A) and schematic diagram of the sample/ reagent sequencing (B) in the speciation of iron(II) and iron(III), utilizing kinetic discrimination of analytes in the reaction between leucomalachite green (LMG) and persulfate with 1,10-phenan- throline as an activator [162]; (c) optimization of continuous-flow synthesis procedure for Suzuki–Miyaura coupling of p-bromo-toluene and phenylboronic acid in the presence of KOH catalyzed by [Pd(OAc)$_2$]-t-Bu$_3$P [171].

The flow synthesis approach with strict control over the residence time serves as a powerful method for protecting group-free syntheses. The illustration of this whole flash chemistry concept in a flow system can be the use of a highly reactive short-lived catalyst before its decomposition. Figure 3c shows the use of a Pt(OAc)$_2$ catalyst in the Suzuki–Miyaura coupling of p-bromotoluene and phenylboronic acid, while conducting it in a flow system with 0.65 s of residence time essentially improved the process, as compared to various configurations of the batch procedure. Numerous very fast reactions within flash chemistry are strongly exothermic processes, and conducting them in flow systems is especially favorable due to the excellent heat transfer in the flow reactors, as compared to the batch mode. This can be additionally facilitated by carrying them out under cryogenic conditions. In continuous-flow systems, this can be achieved using a dedicated cryogenic flow reactor operating on compact refrigerator technology [172]. This type of treatment under flowing conditions is virtually not used in analytical flow systems.

3.2. On-Line Processes Supporting Flow Analysis and Flow Synthesis

The possibility of carrying out various physico-chemical processes on-line in a mechanized way, which provides reliable results of determinations and shortens the analysis time, is a special advantage of such a mode of conducting analytical determination since the creation and the development of pioneering constructions of flow analytical systems. Already in the first developed systems for the determination of different analytes in blood samples, this on-line treatment included the mixing of solutions, dialysis in a membrane module, and time-controlled incubation [12,13]. The subsequent decades of progress in this area resulted in the development of modules for carrying out many other processes including multiphase separation operations, thermal treatment, chemical and electrochemical on-line processing, and irradiation at different wavelengths. In analytical systems, these processes are usually performed in additional dedicated flow-through modules which are located between the sample introduction device and the detector. In flow systems for syntheses, there are various sequences of flow-through reactors or columns with immobilized reagents, scavengers, or solid sorbents that enable catch-and-release steps. This is the reason why there are some differences in the used terms; however, more important in this case is the role that a particular operation plays in the whole assembled system, i.e., an analytical or a synthetic one. It also has to be admitted here that some of these operations were selected in this presentation only for the comparison of analytical and synthetic flow systems.

The applications of solid sorbents in both flow analysis and flow synthesis were reported in both types of systems since the early 1980s. In analytical systems, they are used for the separation of analyte(s) from the sample matrix, along with sample clean-up, the preconcentration of trace analytes, and the immobilization of reagents, catalysts, and biocatalysts. Moreover, they are employed mostly as beds of beads packed in columns of different dimensions. In one of the earliest applications, bromide was preconcentrated on an ion-exchange resin in a segmented flow system [173], whereas ammonium ions were preconcentrated on a cationite in the FIA system [174], while a complexing sorbent was used on-line for the preconcentration of trace metal ions prior to their detection with flame AAS [175]. A very early example of using solid sorbents as an immobilizing reagent in analytical flow systems was the application of a cation-exchange resin in the silver form, from which silver ions could be released to remove chloride from the samples to be analyzed [176]. In addition to packed columns, solid sorbents can also be used in analytical flow systems, where they are attached to the internal walls of open tubular reactors [177]. This was also practiced for the immobilization of biocatalysts on the walls of a Teflon tube [178]. Solid sorbents can also be employed in the form of a suspension of magnetic beads with appropriate chemical functiona- lization [179]; they can be placed onto the tip of an automatic injection syringe [180] or as a renewable bed in the mini-columns of LOV valves, enabling the automated exchange of the sorbent bed [181]. The latter mode is especially convenient in the case of using rather instable solid sorbents such as, e.g., extraction resins widely used in determinations of radionuclides [182]. This commonly employed operation was the subject of numerous review papers and even a book [183]. A sample FIA system employing several sorbent columns for on-line solid-phase extraction in the speciation of Cd, Cu, and Zn is shown in Figure 4a [184].

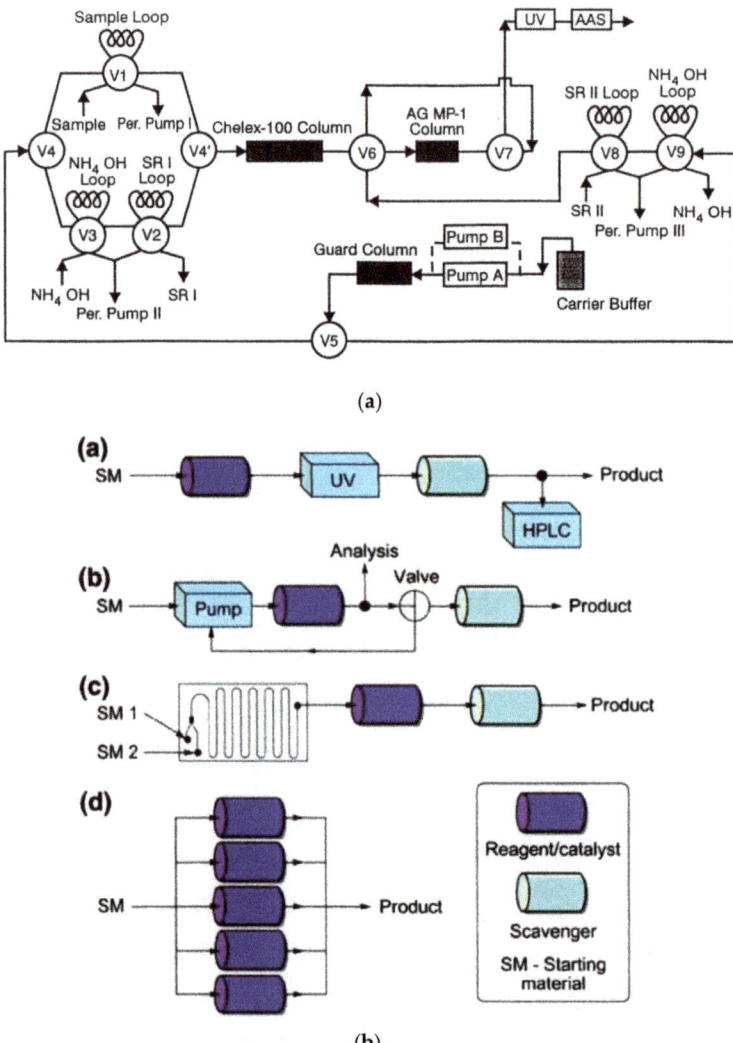

Figure 4. Examples of the application of solid sorbents in flow analysis (a) and flow synthesis (b) systems: (a) schematic diagram of flow-injection system with flame atomic absorption spectrometry detection for speciation of Cd, Cu, and Zn with the use of Chelex-100 chelating resin and anion-exchange AG MP-1 resin [184]; (b) schematic presentation of the main concepts of the arrangement of solid-supported reagents for continuous-flow synthesis [185]. a—linear configuration of series of flow-through column reactors with in-line and on-line analytical monitoring; b—recirculating set-up for processes requiring a longer reaction time; c—set-up combining streams and a flow-through reactor; d—parallel set-up of reactors for producing a larger amount of products.

The role of solid-supported reagents and scavengers in modern organic synthesis, with special emphasis on the generation of combinatorial libraries, was widely discussed in at least two reviews [186,187]. The main concept of the arrangement of solid-supported reagents for continuous-flow synthesis is presented in Figure 4b [185]. This can be a linear sequence of reactors, as described for the flow synthesis of yne-ones [188], a recirculation set-up for transformations which need a longer reaction time than the systems with microfluidics and sorbent-packed columns, or parallel set-up of reactors, for instance, for scaling up the production of a target compound. One example of particularly expanded systems with solid-state supported reagents was the above-mentioned first flow system reported for the total synthesis of a natural alkaloid [7] or a system for the synthesis of γ-aminobutyric acid (GABAA) agonists [189]. Some other examples were also reviewed [129].

Flow analytical systems were designed with the introduction of gases into the carried stream, based on the diffusion of selected components through suitable membranes, since the late 1970s. Usually, the goal of such a procedure is the separation of a given volatile analyte from the sample matrix to the carrier solution. For instance, gas diffusion through a membrane was employed in segmented continuous-flow systems for the determination of CO_2 [190] and ammonia [191] in blood serum. In the continuous-flow determination of free chlorine in waters, the measuring flow system was equipped in a gas separator made of an internal porous Teflon tube and external non-porous Teflon [192]. A similar device developed later on for flow synthesis systems was called a "tube-in-tube reactor" [50]. In modern analytical flow systems, sandwich-type gas diffusion units are frequently used [193]. In flow synthetic systems, hydrogen was introduced to conduct hydrogenation [47], CO_2 was introduced to carry out carboxylation [50], and CO was introduced to perform carbonylation reactions [194]. One can also find some examples of introducing gaseous chlorine, fluorine, oxygen, or ozone. In all these cases, the use of gases instead of various gas surrogates (e.g., aldehydes or metal carbonyls instead of CO) is advantageous from the point of view of product purification, although some limitations of that procedure may occur due to the poor solubility of reactive gases in various solvents [195]. Apart from the above-mentioned diffusion-based gas introduction into an appropriate membrane flow module, another employed possibility is the segmentation of the flowing liquid stream with the given gas or gas introduction in the form of an annular flow, where the liquid flows along the channel walls while the gas flows through the center. The latter mode was successfully used in a micro-fluidic system for conducting gas–liquid–solid hydrogenation [47]. Numerous applications of flow syntheses involving gaseous reactants can be found in many review papers [154,195].

Another unit operation conducted since the early developments of continuous-flow analytical systems with stream segmentation is distillation. This was reported, for instance, in some very early work on the photometric determination of alcohol and acids in beer samples [196] and in the spectrophotometric determination of total and free sulfur dioxide in wine [197]. One can find some other descriptions of such systems, for example, in the American Society for Testing and Materials (ASTM) standards. It seems that on-line distillation in flow synthetic systems may be helpful in removing volatile by-products or in separating the reagents or products from the unwanted materials. Recently, there were certain attempts to use flow distillation with the Hickman distillation apparatus [198] or so-called spinning-band columns [199], but such processes are not employed in the flow syntheses of pharmaceuticals [154].

In the course of multistep syntheses in continuous-flow systems, there may emerge the need for exchanging a solvent for different reasons. This may result from the necessity of maintaining on-line chromatographic control with the use of a particular mobile phase or a change of the reactor from a chemical to electrochemical one [200–202]. Certain flow modules to conduct such an operation were already designed in the early years of the development of flow analytical systems, for instance, as a moving belt interface for the hyphenation of a flow system with HPLC [203] or an ETD module (evaporation to dryness) [204]. Much later, yet another evaporating/condensing system was designed for the same purpose, i.e., switching the solvents in flow synthetic systems [205]. In a glass evaporator with a capillary sprayer, the concentric flow of gas assists the formation of fine spray of a solvent that rapidly evaporates. Then, the concentrated liquid is drawn out, while solvent vapor and carrier gas are directed to the condenser.

Both in analytical laboratories and in organic syntheses, microwave irradiation is used for heating samples or reaction mixtures for several decades now, which is actually also reflected in the construction of flow-through reactors for analytical and synthetic purposes. In analytical systems, microwave heating is applied to digest biological samples since the 1970s, shortening the analysis time from several hours to a few minutes [206]. In some pioneering on-line applications, it was used for the determination of Cu, Fe, and Zn in blood with atomic absorption spectrometry detection [207]. Another early example was the FIA system for the determination of chemical oxygen demand [208] or, more recently, the FIA system for the determination of total organic carbon with atomic emission spectrometry detection [209]. In the majority of syntheses under flow conditions requiring elevated temperatures reported, for instance, for the preparation of pharmaceuticals [154], mostly conventional heat sources like hot plates or oil baths are still used, although they are being more and more commonly replaced by microwave irradiation [41,210–212] or inductive heating [213]. Among different constructions of flow devices used for microwave irradiation in flow synthetic systems, one can find a coating of the flow reactor with a gold film for more efficient absorption of radiation [214], or capillary-based microwave reactor coated capillaries with a thin film of palladium [215]. In yet another approach, a U-shaped glass tube in a microwave-assisted system was filled with PdEnCat^{TM} beads, but it had to be followed by an additional column to remove residual palladium [216]. It seems that some of these designs can also be successfully employed in analytical flow systems. A low-cost microwave unit can also be manufactured from a fluorinated polymer tube and used with commercial microwave ovens [217].

Several interesting approaches were also reported in the application of inductive heating in flow synthetic systems. They include a flow-through microreactor with magnetic nanoparticles [218] and metal oxides such as CrO_2 or NiO_2 as bed material [219]. It was also demonstrated that a copper wire can be heated by induction and may simultaneously serve as a catalyst in the cycloaddition reaction [220].

Similarly to several other attempts to create a flow system for chemical processing, UV irradiation steps were initially employed in flow analytical systems and, later on, widely introduced to flow synthesis. Among the early examples, one can find, for instance, UV irradiation in a segmented flow system for the determination of total cyanide using photometric detection [196]. As for flow-injection systems, one can indicate the determination of chloroorganic compounds using UV digestion and released chloride measurement using potentiometric detection with an ion- selective electrode [221], as well as measurements of dissolved organic carbon after oxidation in a UV reactor [222]. In an FIA system already mentioned for the speciation of nitrogen [118], nitrogen- containing compounds (organic compounds, nitrite, and ammonium ions) were oxidized in a flow-through UV reactor (Figure 5a). Among more recent publications, one can find, for instance, photochemically induced fluorescence, where, due to the UV irradiation of tigecycline (an analyte), a glycylcycline antibiotic, a fluorescent degradation product was produced, which was then determined in the FIA system [223].

Photochemical reactions were given increasing attention in recent years in the area of developing continuous-flow synthetic systems, as well illustrated by several reviews [85,154,211]. This also includes solar photoreactors [224]. The main advantages of using continuous-flow photochemistry in organic synthesis are as follows: short process time, high selectivity, the possi- bility of handling hazardous intermediates in a safe way, and straightforward scaling-up [85]. The advantage of photochemical reactors is the precise control of the energy input and the possibility of placing catalytically active species like TiO_2 doped with Pt on the inner walls in microreactors, enabling the photocatalytic reactions to take place [225]. In the employed photoreactors, both UV and visible radiation are employed using mercury lamps as the source of radiation. Compact fluorescent lamps and white or blue light-emitting diodes (LEDs) are also used here. Numerous photoreactors of that kind are commercially available, and there were a number of developed synthetic routes which use them [85,154]. An example of a four-step synthesis process carried out with the use of a photoreactor was the production of β-amino acids from α-amino acids [226] (Figure 5b). A photocatalytic process involving the use of a photoreactor and irradiation from a white LED was reported for the aerobic oxidation of thiols to disulfides using eosin Y as a metal-free photocatalyst [227]. It can also be noted in this aspect of flow synthesis that numerous innovations achieved in the construction of photoreactors can also be successfully employed in analytical flow systems.

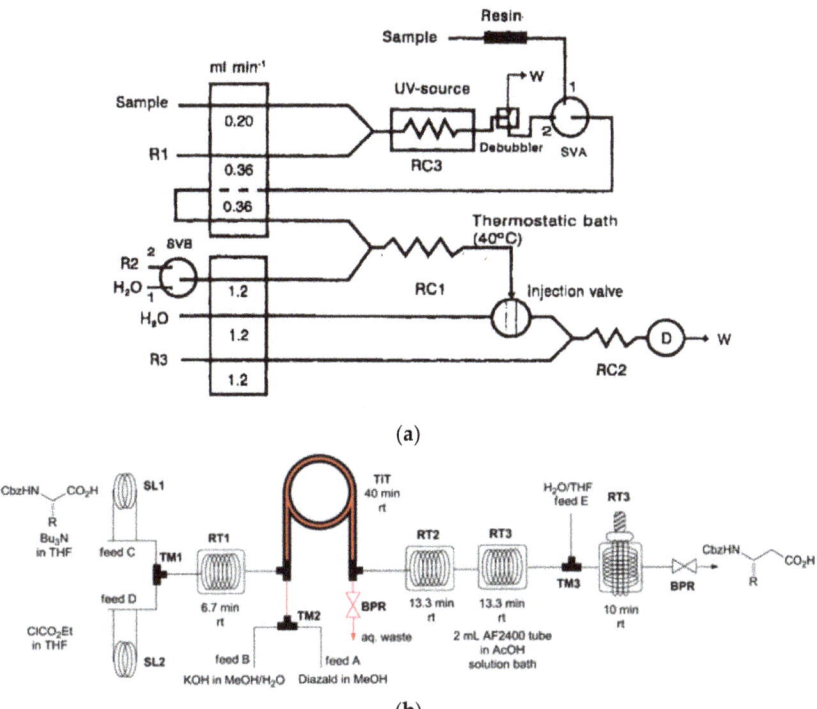

Figure 5. Examples of application of flow-through UV photoreactors in analytical (**a**) and synthetic (**b**) flow systems: (**a**) schematic diagram of flow-injection system with spectrophotometric detection for speciation of nitrogen in wastewater UV oxidation of organic species, nitrite, and ammonium into nitrate [118]; R1—persulfate alkaline solution; R2—reducing solution; R3—chromogenic reagent; resin—Amberlite XAD-7 sorbent; UV source—ultraviolet lamp; injection volume—90 µL; SVA, SVB—selection valves; RC1, RC2—reaction coils; RC3—photo-oxidation coil (3 m × 0.5 mm inner diameter (id)); W—waste; D—detector (540–420 nm); (**b**) schematic diagram of flow system for continuous-flow synthesis of β-amino acids from α-amino acids with photochemically induced Wolff rearrangement [226]; SL—injection valves; TM—T-mixers; RT—residence loops; TiT—tube-in-tube reactor; BPR—back-pressure reactors; PRT—photoreactor.

Similarly to photochemical processes, an increasing interest in flow synthesis was observed in recent years in the use of electrochemical processes, according to several recently published review papers [228–230]. Numerous applications of flow electrosynthesis for pharmaceuticals were also reviewed [154]. In this case, there are not as many similarities in the role and evolution of these processes between analytical and synthetic flow systems, as pointed out for many other aspects above. Basically, electrochemical methods in flow analysis are quite commonly used detection methods, both for inorganic and for organic analytes (in the latter case, mostly employing biosensors and on-line biochemical flow-through reactors). There are, however, some other cases of applying electrochemical processes in the area of flow analysis. This can be, for instance, the generation of very unstable reagents as U(III) [231] or Ag(II) [232] in flow systems, which is similar to flow electrochemistry in flow synthetic systems. In yet another application of an electrochemical reactor in an FIA system, the on-line electrochemical reduction of nitrate to nitrite was used in the spectrophotometric determination of nitrate [233]. Electrochemical dissolution can be used [234] for the flow-injection determination of the composition of metallic alloys. The generation of reagents by electrodialysis for a capillary-scale flow analytical system was also developed for this purpose [235].

The increasing number of applications of electrosynthetic processes was accompanied by a large variety of constructions of flow reactors [229] (especially flow microreactors [228]), a variety of employed electrode materials, and a variety of geometric configurations of reactors. Generally speaking, high surface-area-to-volume ratios of the cell and the proximity of electrodes in the flow cell are two factors that favor obtaining the highest yields. Among the most sophisticated designs of such reactors are the cells with an interdigitated electrode configuration [236] and packed-bed cells with a three-dimensional electrode [237]. Many flow electrochemical cells for flow synthesis are already commercially available, and such examples were shown in a review paper [154] together with some examples of applications involving oxidation, cyclization, dehalogenation, and C–C bond formation processes. One can expect that some electrode processes carried out for synthetic purposes with different flow-cell arrangements might also be also employed in the electroanalytical flow systems.

3.3. Microfluidics in Flow Analysis and Flow Synthesis

A significant technological breakthrough in both the above-discussed fields of flow chemistry was the instrumental down-scaling of flow systems to a microfluidics format. In the case of flow analysis, this was preceded in the early 1980s by a pioneering design of the first integrated miniaturized flow system called micro-conduits [238,239], while, in the case of flow synthetic systems, the beginnings can be attributed to the use of capillary microreactors. Moreover, in this case, the first attempts at developing microfluidic technology were associated with their analytical applications, as already mentioned in this paper. In general, it seems that, in the development of such systems for analytical purposes, in addition to composing the whole hydraulic part from capillary channels, a challenging task is to integrate as many operations (sample introducing, mixing, on-line pretreatment, detection) of the sample on the same chip as possible. In the case of chips for synthetic purposes, they are mostly designed as microreactors for carrying out operations such as a particular reaction or multiphase separation.

While the first design of microfluidic chips was designated for gas chromatography, their first applications in wet analysis took place in the middle of the 1990s, involving capillary electrophoretic determination of metal ions with laser-induced fluorescence detection [240] and the first designs of chips with integrated detectors [28,241]. In the early 2000s, apart from various microfluidic analytical systems, the first very successful applications of microfluidics for flow synthesis were reported at the same time [42,47,242]. Among the early analytical systems, a particularly original application was a single-channel glass microchip for fast screening (using flow-injection mode) or for detailed determination of nitroaromatic explosives (in capillary electrophoretic mode) or organophosphate nerve agents with amperometric detection [243]. A different mode of analysis was possible due to the use of different high-voltage polarization for the same chip, which allowed employing either flow-injection or electrophoretic measurements.

Almost simultaneously, Kitamori's group from the University of Tokyo reported the creation of a microfluidic chip for flow analysis with thermal lens microscopy detection, demonstrating numerous on-line micro-unit operations [244], as well as a microfluidic device for conducting gas–liquid–solid hydrogenation reactions [47].

New analytical applications of microfluidics included, for instance, the creation of a generic microfluidic system for immunoassays with electrochemical detection [245] and a continuous-flow microreactor for cyclical reactions to perform polymerase chain reactions, which can be used for DNA analysis [6]. On the other hand, two microfluidic systems were reported for the synthesis of radiolabeled compounds. One-step synthesis of ^{11}C- or ^{18}F-labeled carboxylic esters was demonstrated in a hydrodynamically driven microreactor [242], while, in yet another approach using an integrated microfluidic chip, a five-step ^{18}F-radiolableled imaging probe for PET was synthesized [42].

The extremely intense development of analytical microfluidic systems brought a variety of designs which involved carrying out various sample treatment operations, as well as various types of detections. These include, for instance, the integration of preconcentration on a molecularly imprinted polymer with sensitive spectrophotometric detection [246] or the hyphenation of a microfluidic system for thermospray sample introduction with flame atomic emission spectroscopic detection [247]. The appropriate design of a single handheld electronic system may offer the possibility of using different configurations of microfluidic systems with different detectors [248], where point-of-care diagnostics seems to be a particularly important and attractive area of application [249].

As already pointed out, microfluidics and microsystems in general are regarded as the most promising devices to enhance the drug discovery processes [40], and, depending on the need for a particular sequence of processes, both microreactors and micro-separation units can be arranged in different combinations [49]. The greatest advantage of microfluidics is the possibility of conducting reactions which cannot be carried out in conventional glassware. This concerns, for instance, using extremely high temperature and pressure, using micro-amounts of substrates, working in sealed systems, or minimizing the contamination of water or oxygen [250]. This was widely illustrated by the vast literature of original research works, including numerous review papers [108,228,251–255].

The objective of both evolution and continuous searching in the construction of microfluidics for flow analysis and flow synthesis is looking for the most appropriate material for their fabrication. The selection of such building materials depends mostly on the target application of a device; however, it also depends on the cost of mass production and the flexibility in preparing the prototypes for research purposes. The capillary micro-structured reactors for flow syntheses are made of silicon–Pyrex, ceramics, stainless steel, or glass [108]. Microfluidic devices for analytical purposes are mainly produced from polydimethylsiloxane (PDMS) due to its low cost, robustness, and simple procedures of fabrication; however, among other polymers employed for their preparation, one can also find thermoset polyester, polyurethane methacrylate, and photocurable resins [256]. Low material consumption and using low-cost materials are the main advantages of the preparation of microfluidics for analytical applications on thin and flexible films [257]. Then, ultra-low-cost microfluidic devices can be produced using polyolefin shrink film with a digital craft cutter involving thermal bonding of the layers [258]. A so-called green alternative to the above-mentioned materials is the use of corn proteins when creating analytical microfluidic devices, where thin zein films with microfluidic chambers and channels are made and bonded to the glass slide with the use of standard lithographic techniques [259].

Since the pioneering work by the Whitesides group on 3D microfluidic devices produced by stacking layers of patterned paper [260], paper-based microfluidic devices gained quite high attention in the analytical chemistry community [261,262]. Able to operate without external pumps owing to strong capillary actions, they are especially appropriate for mass production, healthcare applications, and water analysis [263]. Examples of analytical applications include, for instance, the use for automated staining of malaria parasites prior to microscopic detection [264] and photometric determination of ammonia in freshwater with the device employing a micro-distillation chamber [265].

Teflon-patterned paper was employed in the flow-through synthesis of 100 peptides of 7–14 amino acids [266]. The obtained peptide arrays were used in a cell-based screening to identify the bioactive peptides. In another approach, a flexible cloth-based microfluidic device for analytical purposes was prepared by carving a designated pattern on wax-impregnated papers and attaching it to a cotton cloth via heat treatment [267].

Microfluidics is used for conducting chemical processes in separated droplets both for synthetic and analytical purposes, since the early 2000s. The advantages of such an approach were exploited for the above-discussed analytical and synthetic flow systems with the segmentation of flowing liquid with gas bubbles or immiscible liquids. Droplet-based reactors also operate well on a larger scale, which was demonstrated, e.g., in the reactions carried out in aqueous droplets with catalytically active interfaces to perform Suzuki–Miyaura coupling reactions using a fluorous- tagged palladium catalyst [268]. The main advantage of such a system is preventing fouling by isolating the reaction from the channel walls.

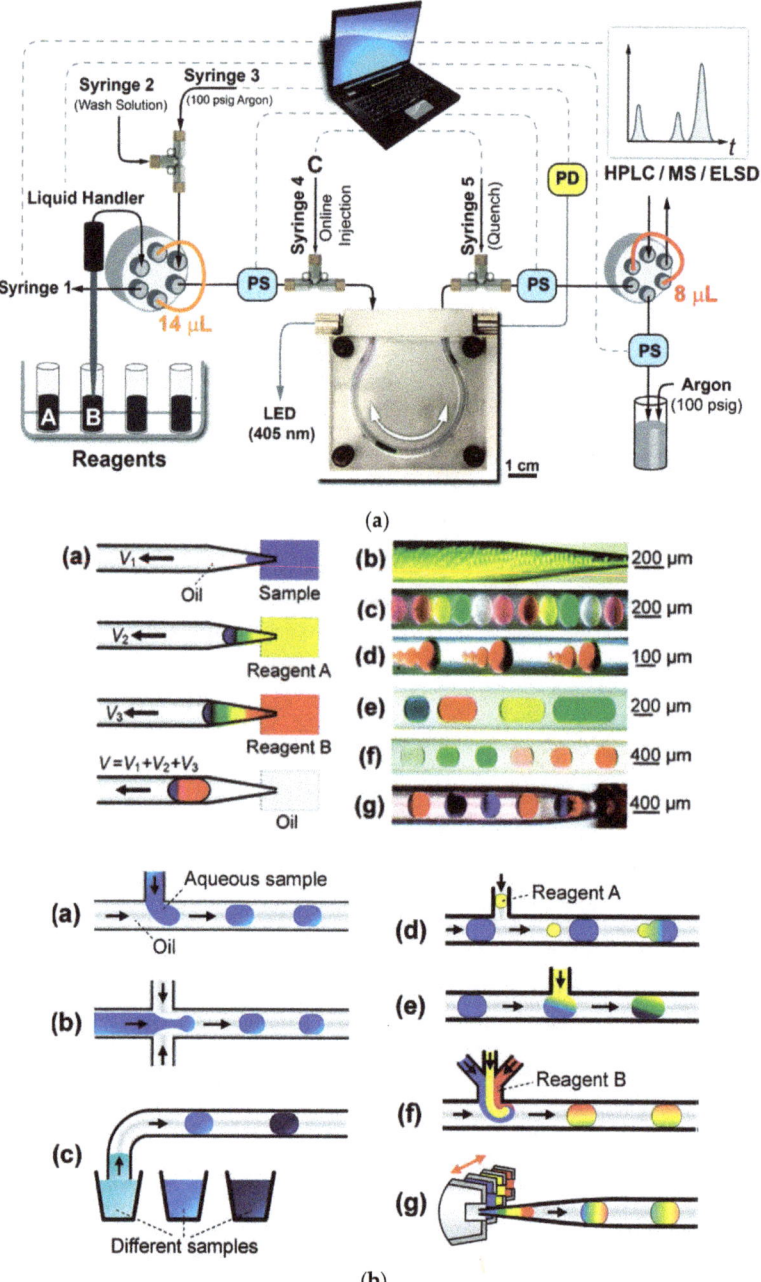

Figure 6. Examples of droplet-based flow systems developed for synthetic (**a**) [269] and analytical (**b**) [270] purposes: (**a**) schematic of the automated droplet-based medicinal chemistry platform [269]. PS—phase sensor; PD—photodetector. Dashed lines indicate personal computer (PC) commu- nication, gray lines correspond to the optical fibers for the light-emitting diode (LED) and the photodetector,

and solid black lines correspond to the fluoropolymer tubing for the delivery and routing of liquid droplets; (**b**) principle and performance of DropLab [270]. (a) Schematic diagram of principle for assembling a three-component droplet using DropLab; (b) continuous generation of 20-pL droplets with a throughput of 4.5 s per droplet by sequentially aspirating a 20-pL fluorescence solution and 80 pL of oil at flow rates of 2 and 8 nL/min, respectively; (c) continuous generation of 1.6-nL droplets containing five different dyes; (d) droplets with different volumes of 20, 40, 160, and 1000 pL generated with 0.6 s of sampling time and increasing flow rates from 2 to 100 nL/min; (e) droplets with different volumes (2.5, 4.5, 5.5, and 8.0 nL) and different dyes; (f) series of 2.5-nL droplets with concentration gradient formed by diluting the green and red dye solutions with volume ratios of 1:19, 1:4, and 19:1 (dye/water); (g) 2.5-nL droplets formed by sequentially introducing red dye, water, and blue dye with different mixing ratios of 1.0:1.5:0, 0.5:1.5:0.5, and 0:1.5:1.0; illustration of various droplet generation modes used in droplet-based microfluidic systems based on T-junction (h), flow-focusing (i), and cartridge (j) techniques; various reagent mixing modes for droplets based on droplet fusion (k), post-mixing (l), and pre-mixing (m) techniques; (n) droplet assembling mode combining droplet generation with reagent mixing.

In another mode, controlled multistep synthesis can be carried out in a three-phase droplet reactor, where inert gas maintains uniform droplet spacing, which was shown in a five-stage quantum-dot synthesis [271]. An even more sophisticated system was invented for in-demand medicinal chemistry with the use of oscillating droplets for multistep synthesis allowing precise control of the reactor temperature and residence time for each step [269] (Figure 6a).

Even earlier than the above-mentioned works, the concept of droplet-based synthetic flow systems was employed in microfluidic formats, providing rapid mixing of reagents and no dispersion, with the suggestion of its application for both chemical analysis and synthesis [272]. In fact, the same research group employed such microfluidic droplet-based systems for the multistep synthesis of nanoparticles, which was performed on a millisecond time scale [273], as well as for the above-mentioned detection of bacteria and determination of their susceptibility to antibiotics [139]. The idea to miniaturize the instrumentation for carrying out polymerase chain reactions [274] is present since the middle of the 1990s, and a nanoliter-sized droplet-based microfluidic set-up was developed for this purpose [275]. With the amplification efficiency comparable to benchtop reactions, the total reaction time was reduced to one-half of that required for benchtop PCR instrumentation. One of the analytical droplet-based microfluidic systems reported in the literature was the automated microfluidic screening assay platform called DropLab. It was developed for various applications including enzyme inhibition assays, protein crystallization screening, and the identification of trace reducible carboxylates [270]. Numerous methods of droplet generation are shown in Figure 6b, together with different modes of the performance of a measuring system. The ionic liquid-based droplet microfluidic set-up for "on-drop" separations and colorimetric pH sensing, involving the creation of two-phase droplet structures in a continuous phase of silicon oil, is yet another example of a wise use of the above-discussed concept [276].

Here, so-called digital microfluidics is also worth mentioning. This concerns the manipulation of discrete volumes of liquids on a surface [277]. The concept of such a device was invented in the late 1990s, employing various mechanisms of the actuation of droplets on the surface such as electro-wetting, surface accounting waves, magnetically controlled droplet movement, dielectrophoresis, or thermocapillary transport. There are numerous analytical applications of this device, for instance, for immunoassays of pathogens [278] in chemical synthesis, [279] or even in synthetic biology [280]. The developed processes with the movement of droplets and non-flow detection are considered to be discrete systems, not flow chemistry procedures; hence, they are not further discussed here.

Pressure-driven microfluidic chips are commonly manufactured using standard soft-litho- graphic techniques. These include, among others, prefabricated microfluidic assembly blocks that can be assembled to form a required microfluidic system for particular needs [281]. Since the middle of the 1990s, a new avenue in micro-constructive manufacturing was given by so-called 3D-printing, i.e., additive layer manufacturing. This was introduced in the early 2000s in the process of designing and fabrication of different instrumental parts and modules for flow chemistry, for both synthetic and analytical systems [282]. Additive manufacturing involves several basic techniques such as stereolithography, multi-jet modeling, selective laser melting, and fused deposition modeling to build a target object layer-by-layer. The main advantages of 3D-printing are its construction cost and production speed. It offers the production of extremely complex geometries; nevertheless, only a limited number of polymeric materials can be used with this technique, and some additional treatment, e.g., electron beam irradiation can be helpful in obtaining satisfactory mechanical properties of 3D-printed elements [283]. The examples of 3D-printed modules for flow synthesis include different flow-through reactors [51,284,285] and microfluidic chips for the synthesis of Ag and Au nanoparticles [286]. A 3D-printed microfluidic system was also produced for the generation of microdroplets, which can be used when creating functionalized microparticles [287].

There were numerous studies on the production of 3D-printed elements for flow analytical systems, such as flow-cells for chemiluminescence [288], as well as UV–Vis and fluorometric measurements [289]. Such devices incorporating disc-based solid-phase extraction were reported for the FIA speciation of iron using spectrophotometric detection [290], as well as fluorometric determination of Cd and Pb in waters using the LOV FIA system [291]. Many more examples can be found in a review on the analytical applications of 3D-printing devices [292].

The application of 3D-printed devices in flow systems combining synthetic and analytical functions seems to be particularly interesting from the point of view of the subject discussed in this paper. A 3D-printed reactor was used for the on-line preparation of Prussian blue nanoparticles in an FIA system with the amperometric detection of hydrogen peroxide. The nanoparticles were further employed for the modification of the working gold electrode, incorporated into a 3D-printed flow cell [293]. Then, in another approach, 3D-printed glass microfluidics was employed with the on-line mass spectrometry monitoring of linezolid synthesis [294].

Concluding this section of the present review, it should be pointed out that, in the field of the novel design and fabrication of microfluidic flow systems, new trends and developments are parallelly introduced and utilized in flow synthetic and analytical systems. In both areas, it is already well documented that these systems provide broad perspectives for many practical applications.

4. Toward the Automation of Chemical Flow Systems

The notions "to automate" or "automation" and their various derivatives are very loosely used when describing chemical instrumentation. The use of mechanical devices to replace, refine, extend, or supplement human effort in operating chemical instrumentation, including flow analytical and synthetic systems, does not mean automation of the instrument, which was clearly indicated already in 1994 in the International Union of Pure and Applied Chemistry (IUPAC) terminological recommendations [295]. Actually, the description covers only a different degree of the mechanization of instrumentation or experimental procedures. The word "automation" means introducing to such a procedure at least one operation which is entirely controlled and carried out by the feedback from the system (usually a computerized one) without any human intervention, which means self-monitoring or self-adjustment.

In the recent literature on flow synthesis, there were certain suggestions of how to diversify different flow synthesis systems, i.e., by differentiating between "automated" and "autonomous" flow synthesis systems [296]. As can be read, *automated* systems require human input to determine the boundaries of operating parameters, thresholds, and protocols, while *autonomous* flow synthesis systems can react to output parameters of the system such as, e.g., reaction yield or purity of the obtained products without human input. The employed control system should properly adjust the input parameters of the process. It seems that these two terms correspond exactly with the terms "mechanization" and "automation" in the above-mentioned IUPAC recommended terminology. This looks like another example of that invisible boundary between the decades of development of flow analysis and the re-discovery of both metrological and instrumental features for synthetic chemistry under flow conditions.

4.1. In-Line Analytical Monitoring in Flow Synthetic Systems

An application of real-time analytical monitoring of the progress of reactions carried out in flow synthesis systems is a crucial factor influencing the yield and the quality, i.e., the purity, of the fabricated product. Several decades of development of flow analysis and liquid chromatography methods resulted in the creation of a large number of flow-through spectrophotometric and electrochemical detectors, reported in the vast literature and commercially available from specialized manufacturers. Flow-through detectors for infrared spectroscopy are rarely used in flow analysis or LC, but they can be used in flow synthesis for the monitoring of selected compounds [297–299]. Raman spectroscopy analysis using a surface-enhanced Raman scatter (SERS) technique under flow conditions was also reported [300]. A number of examples were already presented in the literature on the application of NMR in flow analysis, e.g., in the determination of model drugs [301] or quantitative metabolome analysis of urine [302]. FIA systems can also be designed with mass spectrometry detection, for instance, for rapid determination of pesticides [303] or metabolome studies [304]. These randomly selected examples show the very broad experience gained in the use of the discussed sophisticated instrumental techniques in flow systems, which obviously can be adapted for real-time monitoring of the performance of flow synthesis systems.

One can find numerous examples of using analytical flow measurements for in-line monitoring among a large number of papers on flow synthesis. Most commonly used for this purpose are molecular spectroscopy techniques. For instance, in flow synthesis systems requiring minimal manual intervention, which were designed for the synthesis of a drug known as imatinib, a flow-through UV detector was used to determine when the reaction mixture exited the system in order to fractionate the reactor output for the off-line LC/MS analysis [201]. In the flow synthesis of fluorescent CdS nanoparticles in a microfluidic reactor, an in-line spectrometer to monitor the fluorescence spectra was employed [305], while, in a multi-stage flow synthesis of silica and two types of organic nanoparticles, in-line dynamic light scattering was used for real-time monitoring of the size of the obtained nanoparticles [306]. Numerous applications were reported for the use of infrared spectroscopy in real-time monitoring in flow synthesis systems, mostly using the so-called attenuated total reflection technique (ATR). Such a flow-through cell was attached, e.g., to the outlet of the electrochemical microflow reactor to monitor the on-line formation of the cationic intermediate [307]. An FTIR device with flow cells for ATR measurements using a gold-sealed diamond sensor (ReactIR) can be attached in-line at any part of the flow synthesis system to monitor the reagent consumption or product formation, as well as short-lived intermediates. It was employed, for instance, in the monitoring of fluorination and hydrogenation reactions, as well as the heterocycle saturation reaction for product monitoring; it was also used when screening reaction conditions, monitoring reactive intermediates in Curtius rearrangement, and monitoring hazardous wastes [308]. The last case was also separately presented showing the continuous monitoring of hazardous azide contaminants for the round-the-clock operation of a flow system in a fully automated fashion [309]. A multichannel Raman system equipped with ball-probe immersion optics was reported for monitoring each of the reagent lines and the product line in a flow system for the esterification of benzoic acid.

The hyphenation of flow synthesis with an appropriate analytical technique can be very helpful for quick optimization of the flow parameters and screening of the combinations of reagents. This was well illustrated by the so-called µSYNTAS system (microchip-based synthesis with total analysis system), which integrated the microreactor with a time-of-flight mass spectrometer via an electro-spray unit [310]. It was employed for the efficient continuous-flow generation of compound libraries on the microscale with real-time identification of the reaction components. In yet another example of applying such hyphenation, the microflow synthesis system was coupled to a miniaturized electrospray ionization (ESI) mass spectrometer to monitor reactive intermediates, screen starting materials, and optimize reaction parameters. This was employed for the generation of benzyne and its subsequent reaction with furan, showing the ability of such a hyphenated system to be adapted for safe operation due to the monitoring of hazardous intermediates and the generation of the desired target products [311].

An especially expanded flow synthesis platform, integrated in-line with a UPLC–MS set-up and made of commercially available components, was recently reported for both nanomole-scale reaction screening and micromole-scale synthesis [312]. It is characterized by enabling the preparation and analysis of up to 1500 reaction segments in a 24-h period. In a robotized synthesis–purification–sample-management platform, the preparative LC/MS system was integrated with the flow synthesis set-up for the generation of pharmaceutically active substances, where a 1000:1 split of a preparative system was directed toward another mass spectrometer for analysis [313]. Direct hyphenation of the HPLC set-up to a flow synthesis system can also be used for the same purpose, which was demonstrated, e.g., for a thermal isomerization reaction carried out in a flow reactor [314]. For the efficient generation of structure–activity relationship data, the flow synthesis system was implemented with a microfluidic biosensor chip employing fluorimetric detection [315]. It was developed for the preparation of the most active inhibitors of β-secretase (BACA1), considered as a key target in the research on Alzheimer's disease.

4.2. Automated Flow Analytical and Synthetic Systems

In spite of the above-mentioned terminological recommendations [295], calling the flow analytical or flow synthesis systems automated is a misnomer. However, in the vast literature on flow chemistry in the fullest sense of the word, one can find several examples of really automated set-ups.

Titrations belong to the category of classic, yet still widely used, analytical quantitative procedures routinely employed in modern analytical chemistry, which are also adapted in flow systems. A very original concept which tells us how to carry out titrations is a so-called "binary search strategy". It is based on a computer-controlled adjustment of the volumes of segments of a titrant and a titrand in order to reach stoichiometry, corresponding to the end point of titration. This is carried out in a computer-controlled system with automatic control and adjustment of the volume of segments using solenoid valves. Such systems were developed for acid–base titrations with photometric detection [316] and for the determination of copper complexation capacity, e.g., of milk, based on complexometric titration with chemiluminescence detection [317]. Two other examples of analytical automated flow systems concern the FIA set-ups with spectrophotometric detection developed for the speciation of Fe(III) and Fe(II) in water [318], as well as the simultaneous determination of thorium and uranium in environmental samples with the use of separation employing an extraction resin [319]. Different measuring procedures were programmed in these two both entirely computer-controlled measuring systems, and their selection for particular samples to be analyzed was made automatically based on a series of preliminary test measurements without human intervention. The main differences between the advanced programmed procedures are, e.g., the selection of an additional preconcentration step of analyte(s) or the dilution of the analyzed sample.

Although, in a very recent paper [296], one can learn that "future advances may include the hyphenation of online reactor analytics that would allow a read-out of the reaction progress to alter the reaction parameters using feedback algorithms", and that it can be achieved in the upcoming years, some examples of this approach can already be found in the literature on flow synthesis. For instance, in the designed microfluidic platform for multidimensional screening, chemical reactions were evaluated employing multiple variables [320]. UV on-line monitoring was utilized in such a system in a feedback loop to control both the selection and the injection of a reagent, setting temperature, and sample collection, while the analysis of the reaction screen was carried out using UPLC/MS in the samples collected into 96-well plates. Then, an infrared flow-through detector was employed in another multistep segmented flow processing system for the synthesis of pyrazoles to control the pump for the accurate addition of reagents in combination with a LabVIEW software application [321].

It was also mentioned that the output from the infrared cell can also be potentially used, e.g., to control a fraction collector in a fully automated process.

The creation of a fully computerized autonomous flow chemistry platform for the studies of rhodium-catalyzed hydroformylation reactions was reported with the use of photodetectors, phase sensors, and mass flow controllers, but no information on the employed feedback loops was provided [322]. In some recently published papers, two extremely advanced and fully computerized systems for the flow synthesis of organic compounds were described for the synthesis of pharmaceuticals [52,312]. Both systems are reported as automated platforms, but no details on the use of the feedback loops for self-adjusting of any operations in their functioning are available.

4.3. Robotics, Digital Transmission, and Camera-Enabled Techniques in Flow Chemistry Instrumentation

A particular improvement in terms of the mechanization of complex, multi-modular instrumental systems, as well as their functioning in chemical laboratories, is the application of robots in their design. They offer significant technical enhancement in multistep procedures. Moreover, decision feedback loops can also be involved in their functioning, providing additional elements of automation to the operation of the whole instrumental set-up. A robot can be defined as a programmable, multifunctional machine which is capable of carrying out different operations using materials, parts, and specialized tools. In their second generation, they were equipped with numerous sensors, which make them automatic devices. As manipulators in the laboratory environment, they are designed in Cartesian, cylindrical, and spherical configurations. Their applications in chemical analytical laboratories for sample preparation were observed since the early 1980s [323]. In the same decade, their first applications for flow-injection analysis were reported for the preparation of oil samples prior to analysis using inductively coupled plasma (ICP) emission spectrometry [324]. A full description of a robotized sample preparation station was published in later works on the flow-injection photometric determinations of total polyphenols in olive oil [325] and starch determination in food products [326]. Figure 7a shows a scheme of a robotic station with a commercial, cylindrical-type robot called Zymate II Plus for sample preparation, involving sample weighting, microwave digesting, dissolving, and solvent extraction, as well as centrifugation [326]. The progress in the analytical applications of robots observed in next decades showed their essential advantages in the improvements achieved in small laboratories, in industrial research and development laboratories, and in the production of chemicals [327].

It seems obvious that the application of laboratory robots may also improve multistep and multi-instrumental systems for conducting organic syntheses in flow systems, especially when a given procedure requires manipulation with numerous reagents and different modules of the system. The discussed works were published in the most recent decade and most of them reported developments with the contribution of widely known pharmaceutical companies (Pfizer, GlaxoSmithKline, Abbvie). A Kawasaki six-axis robot was used for liquid handling in a flow system for the monophasic synthesis of drug-like compounds via a plug-flow approach already mentioned in this paper [141]. Using a nucleophilic aromatic substitution reaction and diazo transfer, it was demonstrated that the developed system for performing chemical reactions on a microliter scale can be scaled up to provide a much larger quantity of the required product by working with larger plugs. In another example of a robotized flow synthesis system, a Mitsubishi six-axis robot was also employed to hand off samples between different instruments involved in the whole process [313]. These stations included a segmented flow-chemistry reactor, a preparative LC/MS purification set-up, pumps for dilution, a labeling station, and a centrifugal evaporator. The process ended with MS and NMR analysis; the multistep reaction-time was only about 30 min due to the possibility of carrying out various parallel operations, which significantly shortened the overall production time for the library. A combinatorial synthesis robot incorporating a continuous hydrothermal reactor was developed in the area of direct solid-state chemistry and employed in the flow synthesis of Zn–Ce oxides as crystalline nano powders [328] and iron-doped lanthanum nickelates, which are difficult to prepare in a single step [329].

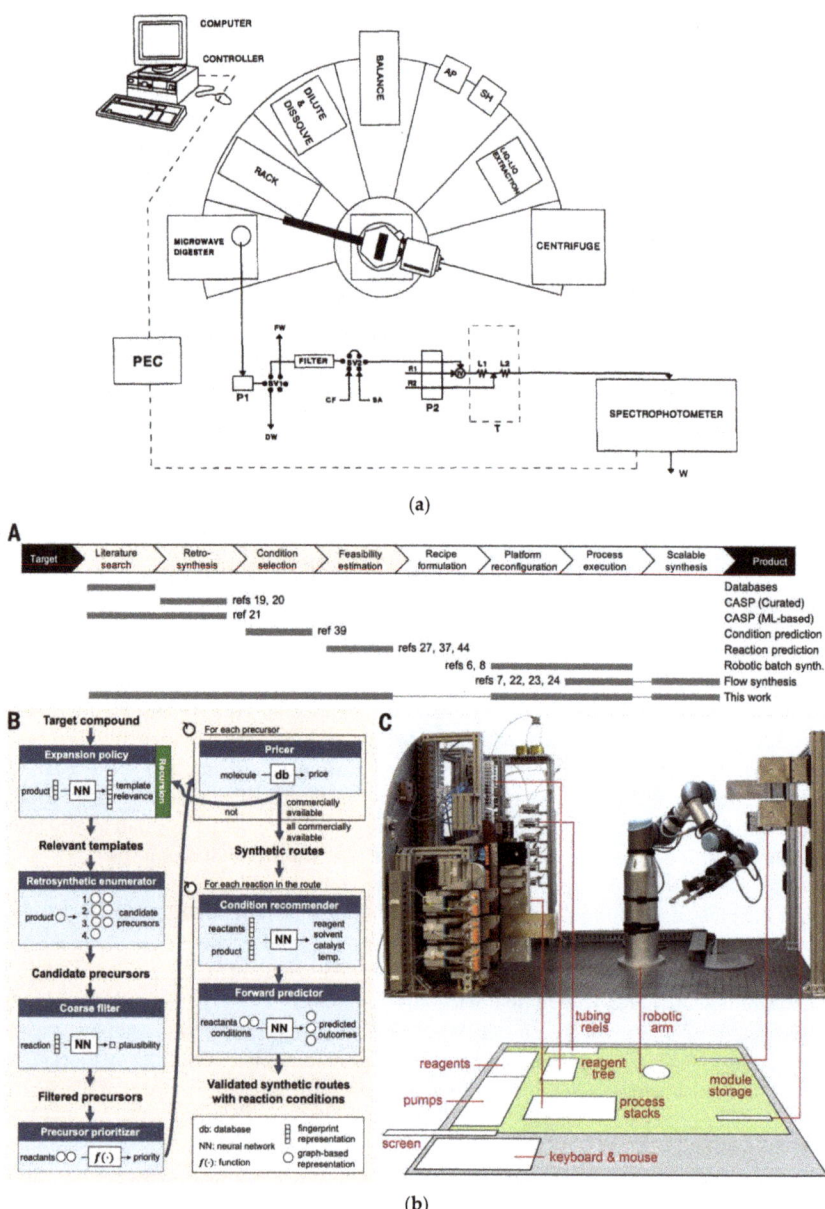

Figure 7. Examples of robotic systems developed for flow-injection analysis (a) [326] and flow synthesis (b) [52]: (a) FIA system developed for colorimetric determination of starch in food based on the determination of sugars using the neocuproine method; PEC—power and event controller; Ap—all-purpose hand; SH—syringe hand; P—peristaltic pump; FW—filter waste; DW—digester waste; CF—clean-up filter solution; SA—solution for assaying sugars; SV—switching valve; Rl—neocuproine

channel; R2—NaOH channel; IV—injection valve; L—reactor; T—thermostat; W—waste; (**b**) robotic flow system for flow synthesis of organic compounds based on artificial intelligence planning; A—workflow for on-demand synthesis of a targeted organic compound; B—software modules combining cheminformatics and machine learning to design and validate synthetic pathways; C—photograph of the robotic flow chemistry platform, with a projected floorplan of the 6 × 4 ft working table.

The most advanced robotic platform for the above-mentioned flow synthesis [52] was recently reported. It employs artificial intelligence planning and generalizes millions of known reactions, via which the prediction of a particular route for execution in a modular continuous-flow platform can be made (Figure 7b).

The whole constructed robotic platform consists of two towers containing eight universal process bays, connecting fluidic, electrical, and pneumatic lines using the UR3 Universal Robot®. The selected process modules are moved from the storage location and configured into a given manifold. The workflow for retrosynthetic planning firstly identifies the reactor for a given target compound; then, the generation of precursors and a proposal of the reaction conditions take place. In the developed instrumentation, ATR–FTIR was only occasionally employed in-line; however, in further modifications of that set-up, in-line HPLC and frit will be implemented for the optimization of the robotic platform by designing an analytic tower in parallel with the process towers. The synthesis planning and its execution using this robotically reconfigurable flow chemistry platform was illustrated by the synthesis of 15 drug and drug-like molecules.

Although the applications of robots for sample processing in flow analysis were already presented in the early 1990s, this method of enhancement of flow analytical systems did not gain much support in subsequent decades. This can be ascribed to the quite common perception of the instrumentation for flow analysis as a cost-effective way to improve laboratory operations; hence, adding robots to such set-ups could be too costly. A completely different story is the case of flow synthesis systems which are most beneficial for the pharmaceutical chemistry. The robotization of the searching process for new active pharmaceutical ingredients and the optimization of their syntheses has to be considered as an exceptionally correct and valuable choice in the further development of instrumentation for such needs.

An increasing trend in the construction of new devices and measuring systems, as well as in chemical analysis, is the hyphenation of conventional devices with mobile consumer electronics (e.g., smartphones or web cameras), as well as their use for the transmission of measurement data of global teleinformatic networks [330,331]. Mobile phones and smartphones are more and more commonly used as signal transducers for optical sensors and biosensors, as well as for the same purpose in the development of microfluidic devices with spectrophotometric or luminescence detections in recent years. For instance, a microfluidic-based smartphone dongle with colorimetric detection was developed for simultaneous assays of hemoglobin and human immunodeficiency virus (HIV) antibodies [332], while a wearable, cotton–thread–paper-based microfluidic device was also designed for colorimetric enzymatic glucose sensing in sweat [333]. A fluorescence immunoassay of *Escherichia coli* was adapted to a microfluidic format with smartphone signal transducing for the determination of that pathogen in urine [334], and electrochemiluminescence detection was employed in a paper-based microfluidic system with wax screen-printing channels and carbon ink-based screen-printed electrodes for the determination of hydrogen peroxide, which can also be adapted for point-of-care detections of glucose [335].

Separate digital web cameras can also be employed for that purpose, replacing conventional spectrophotometers or luminescence detectors. This was illustrated, for instance, by the FIA lab-on-chip miniaturized system developed with the use of a photometric detector for acidity detection [336]. The fluorometric detection employing a web camera was reported in a conventional flow-batch system for the determination of a drug known as acetylcysteine with the use of Cd–Te quantum dots [337]. Smartphones or webcams can also be used in flow instrumentation to control the same functions, in addition to transducing a detection signal. This was reported, for instance, in a microfluidic set-up, where, in addition to the read-out biosensor, a smartphone was a part of the microfluidic liquid handling system combining elastomeric on-chip valves and compact pneumatic liquid pumping [338].

This can be applied in a bead-based fluoroimmunoassay, PCR-based analysis, flow cytometry, and nucleic acid sequencing. A similar concept was introduced into a smartphone electrochemical set-up developed for the determination of selected heavy metals in milk and fruit juices [339] (Figure 8a). In this case, the smartphone controls a programmable, prototype solid-state microwave flow digestion through a digital interface circuit, and it also performs fluidic operations (using a pump/valve) and differential pulse stripping voltammetry.

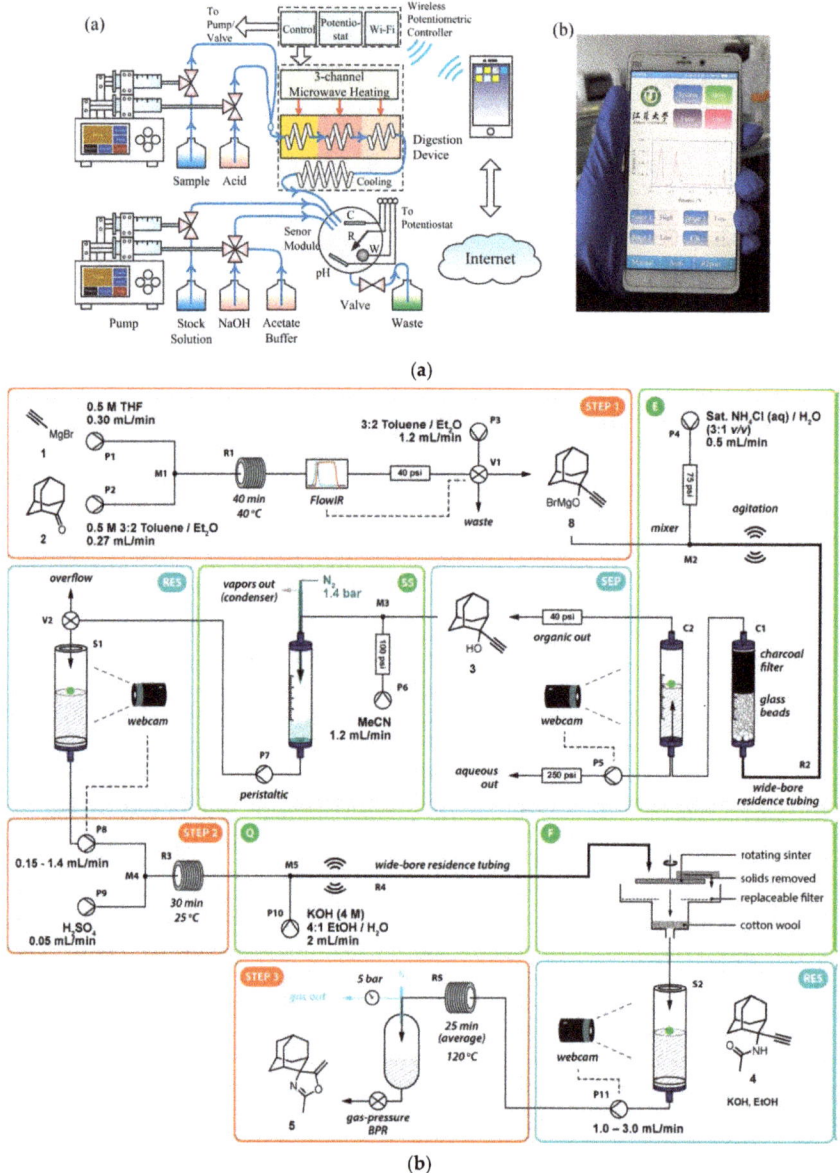

Figure 8. Examples of applications of webcam cameras in flow systems for analytical (**a**) [339] and synthetic (**b**) [340] applications: (**a**) manifold of measuring flow system involving programmable solid-state microwave flow digestion (a) and typical user interface (b) of the smart-phone-based electrochemical

platform developed for determination of heavy metals in liquid foods; (**b**) seven-operation integrated flow platform for the synthesis of 2-aminoadamantane-2-carboxylic acid employing webcam cameras (P—pump; V—valve; M—mixer; R—reactor; C—column; S—reservoir). The output from the initial Grignard step is subjected to in-line quenching and then computer-controlled liquid–liquid phase separation. This solution undergoes a solvent switch, and the output is stored in a reservoir before being used for the Ritter reaction stage. The acidic output is quenched with a base, and the resulting salts are removed using a continuous filter. The filtrate is stored in a second reservoir before finally being heated to undergo cyclization.

The use of web cameras was also reported in controlling certain fluidic operations in flow systems for the synthesis of organic compounds [341]. They involve, among others, the monitoring of chemical events in microdroplets, observing the reaction within the microwave cavity, the formation of aggregate deposits in the glass chip, recording crystallization processes, and the visualization of the liquid/liquid extraction process with the use of a plastic float, e.g., to adjust the flow rate of pumped phases. An exemplary application of three webcams in a self-controlling, telescoped seven-step synthesis was shown in a process including Grignard, Ritter, and cyclization reactions, where extractions, solvent-switching, filtration, and quenching steps were involved [340] (Figure 8b). The functioning of such a complex system can be controlled by only one operator.

In scientific and technical literature since the 1960s, there can be found various attempts at using telecommunication for medical purposes in order to facilitate long-distance doctor–patient interactions, as well as for environmental purposes, such as the remote monitoring of instrumen- tation and collected data transmission. Based on the tremendous progress made over the recent decades in personal computing and mobile telephony, it is not surprising to also find some examples of the application of modern telemetry in flow chemistry instrumentation. For instance, employing a computer sound card and a transmission wireless microphone, wireless recording of outputs from a spectrophotometer used for detection in the FIA system was reported for a distance up to 30 m [342]. Obviously, all the systems developed with data acquisition using mobile phones can be easily adapted for the same purpose. One more pioneering development in this field to be mentioned here was the application of camera phones for signal acquisition from paper-based microfluidic devices for low-cost off-site diagnosis [343]. This was demonstrated for carrying out the colorimetric determination of glucose and protein in urine. Then, in another approach employing a dedicated circuit plugged into the Universal Serial Bus (USB) port of a mobile phone, amperometric detection could also be carried out in a microfluidic system. This was reported for the determination of a protein, which is a biomarker of the malaria parasite, using a poly(dimethylsiloxane) microfluidic chip [344].

One more important aspect of the current developments of flow chemistry is worth mentioning here. In the last two decades, there was an increase in the associations of flow chemistry with the fast development of nanotechnology. In the field of flow analysis, this can be seen as an increasing interest in the application of two- and three-dimensional nanostructures for the enhancement of detection techniques, in separation processes carried out under flow conditions and in the creation of nanofluidic and microfluidic analytical systems [345]. In the development of new methods for spectral detections, the use of quantum dots proved to be particularly fruitful [346], while, in the development of electrochemical detections and separation methods, other inorganic nanoparticles with different functionalization were used [347]. Then, the development of especially efficient and fast methods for the synthesis of various nanoparticles was the subject of numerous works in the field of flow synthesis. These initially involved inorganic nanoparticles [348–350], where metallic nanoparticles found numerous catalytic applications in microflow reactors for organic syntheses [348]. Various nanoparticles and microparticles were synthesized under flow conditions [351], and functionalized nanoparticles [352] were designated for biomedical applications. Using microfluidic systems for this purpose enables precise control of the processes, as well as their significant shortening via appropriate modulation of the most critical stages. This leads to an improvement in the reproducibility of synthetic processes and a better control of the morphology and the size distribution of nanoparticles [352].

5. Conclusions and Perspectives

Flow chemistry is doubtlessly a very significant field in modern chemistry, with respect to conducting scientific research and various kinds of applications, moving from traditional approaches into inorganic, organic, and physical chemistry and chemical technology. In addition to the specialized books listed below, there were tens if not hundreds of reviews published, so far, on the different areas of flow chemistry discussed above, of which many were cited. Very recent reviews in the field of flow synthesis include a general review on flow syntheses as green methods of chemical syntheses [353], a presentation of equipment and separation units [199], an outline of the use of flow methods for taming hazardous chemistry [128], a description of the applications of flow micro- reactors and electrolysis cells in electrosynthetic processes [228,229], and a review on the pharmaceutical applications of flow synthesis [154]. In the field of flow analysis, general reviews on its recent advances [14] and 60-year development, based on publications in the analytical journal *Talanta* [354], can be mentioned. Furthermore, there were also reviews on the spectroanalytical applications of multi-syringe flow-injection methods [355], as well as on the application of fluidized particles in flow analysis methods [356], microfluidic paper-based analytical devices [262], or the use of flow analysis for the determination of radionuclides in nuclear industry [357]. Among the recent reviews on the technological applications of flow methods, much attention was focused on photochemical methods [85] and solar photochemistry [224].

The invention of flow chemistry in the middle of the previous century has to be attributed to the creation of the concept of flow analysis, which, at that time, was a real breakthrough in the world of chemical analytical methods. It was employed, first of all, in medical diagnostics, very quickly expanding into the areas of environmental analysis, food chemistry, and industrial processes. The last years of the previous century brought a very essential transformation of its instrumental format to microfluidic systems, and a parallel phenomenon took place in the rapid increase in the interest in conducting chemical syntheses under flow conditions in organic chemistry.

The first apogee of the interest in flow analysis and its real routine applications was in the 1960s/1970s. Then, a certain renaissance took place between 1980 and 1990 along with the invention of injection methodologies. However, there were no spectacular commercial successes in this field, which is why chromatographic and atomic absorption techniques are more widely employed in routine analytical laboratories since then. Then, a new impetus was again provided in the last two decades with the creation of microsystems with wide prospects in medical applications, particularly for the design of devices for patient personal use. In the case of flow synthesis, the strongest impetus for its development came, without a doubt, from the pharmaceutical industry, where the search for new drugs is a top priority, and where microflow synthetic methodology trumps any other kind of instrumentation.

The main goal of this review article was to demonstrate how unjustified and far from reality it is to limit the term *flow chemistry* to flow synthesis, as can be found in thousands of published research papers. In fact, this notion concerns publications even in the most prestigious chemical scientific journals or books published by worldwide leading scientific publishers. Virtually all the metho- dologies of conducting chemical reactions or utilizing different unit operations are the same in flow analysis and flow synthesis, whereas their earlier applications are usually reported in flow analysis. The only difference between these two main fields of flow chemistry is the wide application of cryogenic conditions in conducting numerous reactions in organic syntheses, which are basically not used in flow analysis. It is truly bewildering how hermetic the progress is in the development of instrumental systems and methodologies in analytical and synthetic flow systems. Considering these two fields, twice as many books on flow analysis were published so far (Table 1), as compared to the books on flow synthesis (Table 2). However, in books on flow analysis published after 2000, one cannot find any remark on flow syntheses, although there is vast literature on carrying out different on-line reactions in flow analytical systems for sample treatment or the derivatization of analyte(s). On the other hand, none of the books on flow chemistry or flow synthesis mention the decades of development of flow analytical systems. One can find only some remarks on the use of analytical devices as additional accessories installed within flow synthetic systems for real-time monitoring of the progress of conducted reactions.

Table 1. Books published in English on flow analysis.

Title	Author(s)	Publisher
Continuous Analysis of Chemical Process Systems	S. Siggia	Wiley, 1959
Continuous-Flow Analysis. Theory and Practice	W.B. Furman	Marcel Dekker, 1976
Automatic Chemical Analysis	J.K. Foreman, P.B. Stockwell	Ellis Horwood, 1975
Automated Stream Analysis for Process Control	D.P. Manka	Academic Press, 1982
Flow-Injection Analysis	J. Ruzicka, E.H. Hansen	Wiley, 1st Edition 1981, 2nd Edition, 1988
Flow-Injection Analysis, Principles, and Applications	M. Valcarcel, M.D. Luque de Castro	Ellis Horwood, 1987
Flow-Injection Atomic Spectroscopy	J.L. Burguera (Ed.)	Marcel Dekker, 1989
Flow-Injection Analysis. A Practical Guide	B. Karlberg, G.E. Pacey	Elsevier, 1989
Flow-Injection Separation and Preconcentration	Z.-L. Fang	VCH Verlag, 1993
Flow-Injection Atomic Spectrometry	Z.-L. Fang	Wiley, 1994
Flow-Injection Analysis of Pharmaceuticals: Automation in the Laboratory	J.M. Calatayud	Taylor and Francis, 1997
Flow Analysis with Spectrometric Detectors	A. Sanz-Medel (Ed.),	Elsevier, 1999
Flow-Injection Analysis. Instrumentation and Applications	M. Trojanowicz	World Scientific, 2000
Advances in Flow Analysis	M. Trojanowicz (Ed.)	Wiley-VCH, 2008
Advances in Flow-Injection Analysis and Related Techniques	S.D. Kolev, I.D. McKelvie (Eds.)	Elsevier, 2008
Flow-Injection Analysis of Marine Samples	M.C. Yebra-Biurrun	Nova Science Publishers, 2009
Flow Analysis with Spectrophotometric and Luminometric Detection	E.A.G. Zagatto, C.C. Oliveira, A. Townshend, P. Worsfold	Elsevier, 2011
Flow Analysis: A Practical Guide	V. Cerda, L. Ferrer, J. Avivar, A. Cerda	Elsevier, 2014
Flow-Injection Analysis of Food Additives	C. Ruiz-Capillas, L.M.L. Nollet (Eds.)	CRC Press, 2015.
Flow and Capillary Electrophoretic Analysis	P. Kościelniak, M. Trojanowicz (Eds.)	Nova Science Publishers, 2017

Table 2. Books published on flow synthesis.

Title	Author(s)	Publisher
Chemical reactions and processes under flow conditions	S.V.Luis, E. Garcia-Varga	RSC, 2010
Micro reaction technology in organic synthesis	C. Viles, P. Watts	CRC Press, 2011
Flow chemistry. Fundamentals	F. Darvas, V. Hessel, G. Dorman	De Gruyter, 2014
Organometallic flow chemistry	T. Noël (Ed.)	Springer, 2016
Continuous-flow chemistry in the research laboratory	T. Glasnov	Springer, 2016
Sustainable flow chemistry: Methods and Applications	L. Vaccaro (Ed.)	Wiley, 2016
Flow chemistry for the synthesis of heterocycles	U.K. Sharma, E.V. Van der Eycken (Eds.)	Springer, 2018
Science of synthesis: Flow chemistry in organic synthesis	T.F. Jamison, G. Koch (Eds.)	Thieme,2018
Flow chemistry: Integrated approaches for practical applications	S.V. Luis, E. Garcia-Verdugo	RSC, 2019

With regard to the briefly overviewed developments of these methodologies and instrumentation, it seems to be fully justified to use the term flow chemistry to represent all other chemical processes carried out under flow conditions of the reacting mixture, a sample to be analyzed, and other media chemically transformed under flow conditions. Certain symptoms of a slight evolution toward such a situation can be noted in some recent papers on microfluidics. the outstanding dynamics of the development of these technologies is reflected by the long-standing publication of dedicated scientific journals and books (already about 50 to date), as well as the presence of numerous manufacturers of such systems on the market. In an increasing number of research papers reporting new achievements, one can find remarks on their potential applications, both for analytical and for synthetic purposes, which is now the reality.

There are several numbers worth mentioning at the end of this review. They are related to bibliographical or scientometric aspects of the discussed field of scientific research, based on the Institute for Scientific Information (ISI) Web of Science in all database modes, viewed on 25 January 2020. It shows 1505 publications when the key phrase "flow chemistry" is entered, with a rapid increase after 2007 and about 210 papers per year published in the last three years. Quite similar results can be found when searching for the "flow synthesis" phrase, with 1139 publications, and a fast increase in the number of publications since 2005, with about 160 papers per year published in the last three years. A search for the "flow analysis" phrase shows 21,804 publications, but this cannot be considered as adequate since it also includes papers on fluid mechanics, economy, or marketing issues. More appropriate seem to be the data obtained for the "flow-injection analysis" phrase, which, in fact, can be considered as the main stream of the development of flow analysis in recent decades. The search for this entry shows 13,649 published papers, with about 250 papers published per year in the last three years. Such statistics clearly show the large activity in all considered fields; however, it must be pointed out that the search for the "microfluidic" phrase gives 72,585 papers published since 1989, with about 7000 papers per year published in the last three years. This seems to express the most important trend in this field, showing an equal importance of the future development of both flow analysis and flow synthesis.

Conflicts of Interest: The authors declare no conflict of interest.

References

1. Driscoll, R.J.; Moon, L.F. Pressure recovery in chemical lasers. *AIAA J.* **1977**, *15*, 665–673. [CrossRef]
2. Liu, Y.Y.; Billone, M.C.; Fischer, A.K.; Tam, S.W.; Clemmer, R.G.; Hollenberg, G.W. Solid tritium breeder materials-Li_2O and $LiAlO_2$: A data base review. *Fision Technol.* **1985**, *8*, 1970–1984. [CrossRef]
3. Joseph, H.C.; Origley, E.D. Role of suspended sediment in irrigation return flow chemistry, Southern Alberta. *Water Resour. Res.* **1986**, *22*, 643–654. [CrossRef]
4. Gupta, R.N. Aerothermodynamic analysis of stardust sample return capsule with coupled radiation and ablation. *J. Spacecraft Rock.* **2000**, *37*, 507–514. [CrossRef]
5. Wang, X.D.; Cardwell, T.J.; Cattrall, R.W.; Jenkins, G.E. Pulsed flow chemistry. A new approach to the generation of concentration profiles in flow analysis. *Anal. Commun.* **1998**, *33*, 97–101. [CrossRef]
6. West, J.; Karamata, B.; Lillis, B.; Gleeson, J.P.; Alderman, J.; Collins, J.K.; Lane, W.; Mathewson, A.; Berney, H. Application of magneto-hydrodynamic actuation to continuous flow chemistry. *Lab. Chip.* **2002**, *2*, 224–230. [CrossRef]
7. Baxendale, I.R.; Deeley, J.; Griffiths, C.M.; Ley, S.V.; Saaby, S.; Tranmer, G.K. A flow process for the multi-step synthesis of the alkaloid natural product oxomaritidine: A new paradigm for molecular assembly. *Chem. Commun.* **2006**, *24*, 2566–2568. [CrossRef]
8. Bayer, E.; Jung, E.; Halasz, I.; Sebastian, I. A new support for polypeptide synthesis in columns. *Tetrahedron Lett.* **1970**, 4503–4505. [CrossRef]
9. Cwiet, M.S. О новой категорииадсорбционных явлений ио применении их к био-химическому анализу. *Proc. Warsaw Soc. Nat. Sci. Biol. Sec.* **1903**, *14*, 20–39.
10. Tswett, M. Adsorptionsanalyse und chromatographishe Methode. *Anwendung auf die Chemie des Chlorophylls. Ber. Dtsch. Botan. Ges.* **1906**, *24*, 385–392.
11. Reed, L. Notes on capillary separation of substances in solution. *Proc. Chem. Soc.* **1893**, *9*, 123–126.

12. Skeggs, L.T., Jr. An automatic method for colorimetric analysis. *Clin. Chem.* **1956**, *2*, 241. [CrossRef]
13. Skeggs, L.T., Jr. An automatic method for colorimetric analysis. *Am. J. Clin. Path.* **1957**, *28*, 311–322. [CrossRef] [PubMed]
14. Trojanowicz, M.; Kołacińska, K. Flow injection methods in chemical analysis. *Analyst* **2016**, *141*, 2085–2139. [CrossRef] [PubMed]
15. Furman, W.B. *Continuous Flow Analysis. Theory and Practice*; Marcel Dekker, Inc.: New York, NY, USA, 1976.
16. Blaedel, W.J.; Hicks, G.P. Continuous analysis by measurement of the rate of enzyme catalyzed reactions. *Anal. Chem.* **1962**, *34*, 388–394. [CrossRef]
17. Nagy, G.; Fegher, Z.; Pungor, E. Application of silicone rubber-based graphite electrodes for continuous flow measurements: Part II. Voltammetric study of active substances injected into electrolyte streams. *Anal. Chim. Acta* **1970**, *52*, 47–54. [CrossRef]
18. Ruzicka, J.; Hansen, E.H. Flow injection analysis. 1. New concept of fast continuous-flow analysis. *Anal. Chim. Acta* **1975**, *78*, 145–157.
19. Ruzicka, J.; Stewart, J.W.B. Flow injection analysis. 2. Ultrafast determination of phosphorus in plant material by continuous-flow spectrophotometry. *Anal. Chim. Acta* **1975**, *79*, 79–91.
20. Stewart, J.W.B.; Ruzicka, J.; Bergamin Filho, H.; Zagatto, E.A.G. Flow injection analysis. 3. Comparison of conti- nuous-flow spectrophotometry and potentiometry for rapid-determination of total nitrogen content in plant digests. *Anal. Chim. Acta* **1976**, *81*, 371–386. [CrossRef]
21. Stewart, K.K.; Beecher, G.R.; Hare, P.E. Automated high-speed analyses of discrete samples – Use of nonsegmented, continuous-flow systems. *Fed. Proc.* **1974**, *33*, 1439.
22. Stewart, K.K.; Beecher, G.R.; Hare, P.E. Rapid analysis of discrete samples: The use of nonsegmented, continuous flow. *Anal. Biochem.* **1976**, *70*, 167–173. [CrossRef]
23. Trojanowicz, M. Commercially available instrumentation for FIA. In *Flow Injection Analysis. Instrumentation and Applications*; World Scientific: Singapore, 2000.
24. Ruzicka, J.; Marshall, G.D. Sequential injection: A new concept for chemical sensors, process analysis and laboratory assays. *Anal. Chim. Acta* **1990**, *237*, 329–343. [CrossRef]
25. Ruzicka, J. Lab-on-valve: Universal microflow analyzer based on sequential and bead injection. *Analyst* **2000**, *125*, 1053–1060. [CrossRef]
26. Terry, S.S.; Jerman, J.H.; Angell, J.B. Gas-chromatogaphic air analyzer fabricated on a silicon wafer. *IEEE Trans. Electron. Dev.* **1979**, *26*, 1880–1886. [CrossRef]
27. Johnson, B.; Lofas, S.; Lindquist, G. Immobilization of proteins to a carboxymethyldextrin-modified gold surface for biospecific interaction analysis in surface plasmon resonance sensors. *Anal. Biochem.* **1991**, *198*, 268–277. [CrossRef]
28. Verpoorte, E.M.J.; Van Der Schoot, B.H.; Jeanneret, S.; Manz, A.; Widmer, H.M.; De Rooij, N.F. Three-dimensional micro flow manifolds for miniaturized chemical analysis systems. *J. Micromech. Microeng.* **1994**, *4*, 246–256. [CrossRef]
29. Karnatz, F.A.; Whitmore, F.C. Dehydration of diethylcarbinol. *J. Am. Chem. Soc.* **1932**, *54*, 3461. [CrossRef]
30. Walker, R.E. Chemical reaction and diffusion in a catalytic tubular reactor. *Phys. Fluids* **1961**, *4*, 1211–1216.
31. Pines, H.; Csicsery, S.M. Alumina: Catalyst and support. XVI. Aromatization and dehydroisomerization of branched C_6-C_8 hydrocarbons over "nonacidic" chromia-alumina catalyst. *J. Catal.* **1962**, *1*, 313–328. [CrossRef]
32. Davis, B.R.; Scott, D.S. Rate of isomerization of cyclopropane in a flow reactor. *I&EC Fundamentals* **1964**, *3*, 20–23.
33. Bodi, L.J. A flow synthesis of gallium phosphide and some properties of gallium phosphide powder layers. *J. Electrochem. Soc.* **1962**, *109*, 497–501. [CrossRef]
34. Carroll, D. Continuous-flow salt gradient dialysis for the preparation of polynucleotide-polypeptide complexes. *Anal. Biochem.* **1971**, *44*, 496–502. [CrossRef]
35. Dye, J.L.; Lok, M.T.; Tehan, F.J.; Ceraso, J.M.; Voorhees, K.J. Flow synthesis. A substitute for the high-dilution steps in cryptate synthesis. *J. Org. Chem.* **1973**, *38*, 1773–1775. [CrossRef]
36. Lukas, T.J.; Prystowsky, M.B.; Erickson, B.W. Solid-phase peptide synthesis under continuous-flow conditions. *Proc. Natl. Acad. Sci. USA* **1981**, *78*, 2791–2795. [CrossRef] [PubMed]
37. Andrews, R.P. Automated continuous flow peptide synthesis. *Nature* **1986**, *319*, 429–430. [CrossRef]

38. Warner, B.D.; Warner, M.E.; Karns, G.A.; Lu, L. Construction and evaluation of an instrument for the automated synthesis of oligodeoxyribonucleotides. *DNA* **1984**, *3*, 401–411. [CrossRef]
39. Vagner, J.; Kocna, P.; Krchnak, V. Continuous-flow synthesis of alpha-gliadin peptides in an ultrasonic field and assay of their inhibition of intestinal sucrase activity. *Peptide Res.* **1991**, *4*, 284–288.
40. Wong Hawkes, S.Y.V.; Chapela, M.J.V.; Montembault, M. Leveraging the advantages offered by microfluidics to enhance the drug discovery process. *QSAR Comd. Sci.* **2005**, *24*, 712–721. [CrossRef]
41. Baxendale, I.R.; Pitts, M.R. Microwave flow chemistry: The next evolutionary step in synthetic chemistry? *Chim. Oggi* **2006**, *24*, 41–45. [CrossRef]
42. Lee, C.-C.; Sui, G.; Elizarov, A.; Shu, C.J.; Shin, Y.-S.; Dooley, A.N.; Huang, J.; Daridon, A.; Wyatt, P.; Stout, D.; et al. Multistep synthesis of a radiolabeled imaging probe using integrated microfluidics. *Science* **2005**, *310*, 1793–1796. [CrossRef]
43. Richards, P. MIT and Novartis in new partnership aimed at transforming pharmaceutical manufacturing. Available online: http://news.mit.edu/2007/novartis-0928 (accessed on 15 March 2020).
44. Chemat, F.; Esveld, D.C.; Poux, M.; Di-Martino, J.L. The role of selective heating in the microwave activation of hetero-geneous catmnsis reactions using a continuous microwave reactor. *J. Microwave Power Electrom. Energ.* **1998**, *33*, 88–94. [CrossRef]
45. Nikbin, N.; Watts, P. Solid-supported flow synthesis in microreactors using electroosmotic flow. *Org. Proc. Res. Dev.* **2004**, *8*, 942–944. [CrossRef]
46. Marre, S.; Park, J.; Rempel, J.; Guan, J.; Bawendi, M.G.; Jensen, K.F. Supercritical continuous-microflow synthesis of narrow size distribution quantum dots. *Adv. Mater.* **2008**, *20*, 4830–4834. [CrossRef]
47. Kobayashi, J.; Mori, Y.; Okamoto, K.; Akiyama, R.; Ueno, M.; Kitamori, T.; Kobayashi, S. A microfluidic device for conducting gas-liquid-solid hydrogenation reactions. *Science* **2004**, *304*, 1305–1308.
48. Duraiswamy, S.; Khan, S.A. Droplet-based microfluidic synthesis of anisotropic metal nanocrystals. *Small* **2005**, *5*, 2828–2834. [CrossRef] [PubMed]
49. Sahoo, H.R.; Krali, J.G.; Jensen, K.F. Multistep continuous-flow microchemical synthesis involving multiple reactions and separations. *Angew. Chem. Int. Ed.* **2007**, *46*, 5704–5708. [CrossRef] [PubMed]
50. Polyzos, A.; O'Brien, M.; Petersen, T.P.; Baxendale, I.R.; Ley, S.V. The continuous-flow synthesis of carboxylic acids using CO_2 in a tube-in-tube gas permeable membrane reactor. *Angew. Chem. Ind. Ed.* **2011**, *50*, 1190–1193. [CrossRef]
51. Okafor, O.; Weilhard, A.; Fernandes, J.A.; Karjalainen, E.; Goodridge, R.; Sans, V. Advanced reactor engineering with 3D printing for the continuous-flow synthesis of silver nanoparticles. *React. Chem. Eng.* **2017**, *2*, 129–136. [CrossRef]
52. Coley, C.W.; Thomas, D.A.; Lummiss, J.A.M.; Jaworski, J.N.; Breen, C.P.; Schultz, V.; Hart, T.; Fishman, J.S.; Rogers, L.; Gao, H.; et al. A robotic platform for flow synthesis of organic compounds informed by AI planning. *Science* **2019**, *365*, eaax1566. [CrossRef]
53. Wynkoop, R.; Wilhelm, R. Kinetics in tubular flow reactor – hydrogenation of ethylene over copper-magnesia catalyst. *Chem. Eng. Progr.* **1950**, *46*, 300–310.
54. Chance, B. The kinetics of the enzyme-substrate compound of peroxidase. *J. Biol. Chem.* **1943**, *151*, 553–577.
55. Vreeman, H.J.; van Rooijen, P.J.; Visser, S. Segmented flow analysis as applied to kinetic studies of peptide bond hydrolysis. *Anal. Biochem.* **1977**, *77*, 251–264. [CrossRef]
56. Young, H.H., Jr.; Hammett, L.P. Kinetic measurements in a stirred flow reactor; the alkaline bromination of acetone. *J. Am. Chem. Soc.* **1950**, *72*, 280–283. [CrossRef]
57. Keles, H.; Susanne, F.; Livingstone, H.; Hunter, S.; Wade, C.; Bourdon, R.; Rutter, A. Development of a Robust and Reusable Microreactor Employing Laser Based Mid-IR Chemical Imaging for the Automated Quantification of Reaction Kinetics. *Org. Process. Res. Dev.* **2017**, *21*, 1761–1768. [CrossRef]
58. Seibt, S.; With, S.; Bernet, A.; Schmidt, H.; Förster, S. Hydrogelation kinetics measured in a microfluidic device with in-situ X-ray and fluorescence detection. *Langmuir* **2018**, *34*, 5535–5544. [CrossRef] [PubMed]
59. Schwolow, S.; Braun, F.; Rädle, M.; Kockman, N.; Röder, T. Fast and efficient acquisition of kinetic data in microreactors using in-line Raman analysis. *Org. Proc. Res. Dev.* **2015**, *19*, 1286–1292. [CrossRef]
60. Butkolvsjaya, N.I.; Setser, D.W. Infrared chemiluminescence study of the reaction of hydroxyl radicals with formamide and the secondary unimolar reaction of chemically activated carbamic acid. *J. Phys. Chem. A* **2018**, *122*, 3735–3746. [CrossRef]

61. Ma, J.; Zhou, H.; Yan, S.; Song, W. Kinetics studies and mechanistic considerations on the reactions of superoxide radical ions with dissolved organic matter. *Water Res.* **2019**, *149*, 56–64. [CrossRef]
62. Chance, B. Rapid and sensitive spectrophotometry. I. The accelerated and stopped-flow methods for the measurement of the reaction kinetics and spectra of unstable compounds in the visible region of the spectrum. *Rev. Sci. Instr.* **1951**, *22*, 619–627. [CrossRef]
63. Mukai, K.; Nagai, K.; Ouchi, A.; Izumisawa, K.; Naganka, S. Finding of remarkable synergistic effect on the aroxyl radical-scavenging rate under the coexistence of α-tocopherol and catechnis. *Int. J. Chem. Kinet.* **2019**, *51*, 643–656. [CrossRef]
64. Delgadillo, R.F.; Mueser, T.C.; Zaleta-Rivera, K.; Carnes, K.A.; Gonzalez-Valdez, J.; Parkhurst, L.J. Detailed characterization of the solution kinetics and thermodynamics of biotin, biocytin and HABA binding to avidin and streptavidin. *PLOS ONE* **2019**, 0204194. [CrossRef] [PubMed]
65. Walklate, J.; Geeves, M.A. Temperature manifold for a stopped-flow machine to allow measurements from −10 to +40 °C. *Anal. Biochem.* **2015**, *476*, 11–16. [CrossRef] [PubMed]
66. Bajuszova, Z.; Naif, H.; Ali, Z.; McGinnis, J.; Islam, M. Cavity enhanced liquid-phase stopped-flow kinetics. *Analyst* **2018**, *143*, 493–502. [CrossRef]
67. Bayley, P.; Anson, M. Stopped-flow circular dichroism: A rapid kinetic study of the binding of a sulphonamide drug to bovine carbonic anhydrase. *Biochem. Biophys. Res. Commun.* **1975**, *62*, 717–722. [CrossRef]
68. Makabe, K.; Nakamura, T.; Dhar, D.; Ikura, T.; Koine, S.; Kuwajima, K. An overlapping region between the two terminal holding units of the outer surface protein A (Ospa) controls its holding behavior. *J. Mol. Biol.* **2018**, *430*, 1799–1813. [CrossRef]
69. Reback, M.L.; Roske, C.W.; Bitterwolf, T.E.; Griffiths, P.R.; Manning, C.J. Stopped-flow ultra-rapid-scanning Fourier transform infrared spectroscopy on the millisecond time scale. *Appl. Spectrosc.* **2010**, *64*, 907–911. [CrossRef] [PubMed]
70. Christianson, M.D.; Tan, E.H.P.; Landis, C.R. Stopped-flow NMR: Determining the kinetics of [rac-(C_2H_4(1-indenyl)$_2$) ZrMe][MeB(C_6F_5)$_3$]-catalyzed polymerization of 1-hexene by direct observation. *J. Am. Chem. Soc.* **2010**, *132*, 11461–11463. [CrossRef]
71. Sirs, J.A. Electrometric stopped flow measurements of rapid reactions in solution. Part 1 – Conductivity measurements. *Trans. Faraday Soc.* **1958**, *54*, 201–206. [CrossRef]
72. Chauhan, S.; Hosseinzadeh, P.; Lu, Y.; Blackburn, N.J. Stopped-flow studies of the reduce-tion of the copper centers suggest a bifuricated electron transfer pathway in peptidyl-glycine monooxygenase. *Biochemistry* **2016**, *55*, 2008–2021. [CrossRef]
73. Peacock, D.G.; Richardson, J.F. *Chemical Engineering, Vol.3: Chemical and Biochemical Reactors and Process Control*; Elsevier: Amsterdam, The Netherlands, 2012.
74. Ravi, R.; Vinu, R.; Gummadi, S.N. *Coulson and Richardson's Chemical Engineering, Vol.3A: Chemical and Biochemical Reactors and Reaction Engineering*; Butterworth-Heinemann: Oxford, UK, 2017.
75. Reynolds, J.H. An immobilized α-galactose continuous flow reactor. *Biotechnol. Bioeng.* **1974**, *16*, 135–147. [CrossRef]
76. Basso, A.; Moreira, R.; Jose, H.J. Effect of operational conditions on photocatalytic ethylene degradation applied to control tomato ripening. *J. Photochem. Phtobiol. A: Chemistry* **2018**, *367*, 294–301. [CrossRef]
77. Maia, M.; Soares, T.R.; Mota, A.I.; Rosende, M.; Magalhaes, L.; Miró, M.; Segundo, M.A. Dynamic flow-through approach to evaluate readily bioaccessible antioxidants in solid food samples. *Talanta* **2017**, *166*, 162–168. [CrossRef] [PubMed]
78. Jiang, K.; Dickinson, J.E.; Galvin, K.P. The kinetics of fast flotation using the reflux flotation cell. *Chem. Eng. Sci.* **2010**, *196*, 463–477. [CrossRef]
79. Maliutina, K.; Tahmasebi, A.; Yu, J.; Saltykov, S.N. Comparative study on flash purolysis characteristics of microalgal and lignocellulosic biomass in entrained-flow reactor. *Energy Conv. Manag.* **2017**, *151*, 426–438. [CrossRef]
80. Elliott, D.C.; Schmidt, A.J.; Hart, T.R.; Billing, J.M. Conversion of a wet waste feedstock to biocrude by hydrothermal processing in a continuous-flow reactor: Grape pomace. *Biomass Conv. Bioref.* **2017**, *7*, 455–465. [CrossRef]
81. Cheng, F.; Jarvis, J.M.; Yu, J.; Jena, U.; Nirmalakhandan, N.; Schaub, T.M.; Brewer, C.E. Bio-crude oil from hydro-thermal liquefaction of wastewater microalgae in a pilot-scale continuous flow reactor. *Bioresource Technol.* **2019**, *294*, 122184. [CrossRef] [PubMed]

82. García-Martínez, Y.; Chirinos, J.; Bengoa, C.; Stuber, F.; Font, J.; Fortuny, A.; Fabregat, A. Nitrate removal in an innovative up-flow stirred packed-bed bioreactor. *Chem. Eng. Proc. Process. Intens.* **2017**, *121*, 57–64. [CrossRef]
83. Lin, Y.-H. Adsorption and biodegradation of 2-chlorophenol by mixed culture using activated carbon as a supporting medium-reactor performance and model verification. *Appl. Water Sci.* **2017**, *7*, 3741–3757. [CrossRef]
84. Meng, F.; Huang, W.; Liu, D.; Zhao, Y.; Huang, W.; Lei, Z.; Zhang, Z. Application of aerobic granules-continuous flow reactor for saline wastewater treatment: Granular stability, lipid production and symbiotic relationship between bacteria and algae. *Bioresource Technol.* **2020**, *295*, 122291. [CrossRef]
85. Cambie, D.; Bottecchia, C.; Straathof, N.J.W.; Hessel, V.; Noël, T. Applications of continuous-flow photochemistry in organic synthesis, material science, and water treatment. *Chem. Rev.* **2016**, *116*, 10276–10341. [CrossRef]
86. Petrella, A.; Spasiano, D.; Cosma, P.; Rizzi, V.; Race, M. Evaluation of the hydraulic and hydrodynamic parameters influencing photo-catalytic degradation of bio-persistent pollutants in a pilot plant. *Chem. Eng. Commun.* **2019**, *206*, 1286–1296. [CrossRef]
87. Yaday, A.; Verma, N. Carbon bead-supported copper-dispersed carbon nanofibers: An efficient catalyst for wet air oxidation of industrial wastewater in a recycle flow reactor. *J. Ind. Eng. Chem.* **2018**, *67*, 448–460.
88. Hu, Y.; Boyer, T.H. Removal of multiple drinking water contaminants by combined ion exchange resin in a completely mixed flow reactor. *J. Wat. Supp: Res. Technol.* **2018**, *67*, 659–672. [CrossRef]
89. Naddeo, V.; Ricco, D.; Scannapieco, D.; Belgiorono, V. Degradation of antibiotics in wastewater during sonolysis, ozonation, and their simultaneous application: Operating condition effects and process evaluation. *Int. J. Photoenergy* **2012**, 624270. [CrossRef]
90. Xu, L.; Ma, X.; Niu, J.; Chen, J.; Zhou, C. Removal of trace naproxen from aqueous solution using a laboratory-scale reactive flow-through membrane electrode. *J. Hazard. Mater.* **2019**, *379*, 120692. [CrossRef] [PubMed]
91. Brito, C.N.; Ferreira, M.B.; de Moura Santos, E.C.M.; Leon, J.J.L.; Ganiyu, S.O.; Martinez-Huitle, C.A. Electro-chemical degradation of Azo-dye Acid Violet 7 using BDD anode: Effect of flow reactor configuration on cell hydrodynamics and dye removal efficiency. *J. Appl. Electrochem.* **2018**, *48*, 1321–1330. [CrossRef]
92. Urtiaga, A.; Fernandez-Gonzalez, C.; Gomez-Lavin, S.; Ortiz, I. Kinetics of the electrochemical mineralization of perfluorooctanoic acid on ultrananocrystalline boron doped conductive diamond electrodes. *Chemosphere* **2015**, *129*, 20–26. [CrossRef]
93. Pikaev, A.K.; Podzorova, E.A.; Bakhtin, O.M. Combined electron beam and ozone treatment of wastewater in the aerosol flow. *Radiat. Phys. Chem.* **1997**, *49*, 155–157. [CrossRef]
94. Trojanowicz, M. Flow chemistry vs. flow analysis. *Talanta* **2016**, *146*, 621–640. [CrossRef]
95. Ruzicka, J. Flow chemistry and flow analysis. *J. Flow Chem.* **2015**, *5*, 55. [CrossRef]
96. Feng, S.; Yuan, D.; Huang, Y.; Lin, K.; Zhu, Y.; Ma, J. A catalytic spectrophotometric method for determination of nanomolar manganese in seawater using reverse flow injection analysis and a long path length liquid waveguide capillary cell. *Talanta* **2018**, *178*, 577–582. [CrossRef] [PubMed]
97. Valcarcel, M.; Luque de Castro, M.D. Integration of reaction (retention) and spectroscopic detection in continuous-flow systems. *Analyst* **1990**, *115*, 699–703. [CrossRef]
98. Trojanowicz, M.; Worsfold, P.J.; Clinch, J.R. Solid-state photometric detectors for flow injection analysis. *Trends Anal.Chem.* **1988**, *7*, 301–305. [CrossRef]
99. Trojanowicz, M.; Szpunar-Łobińska, J. Simultaneous flow-injection determination of aluminium and zinc using LED photometric detection. *Anal. Chim. Acta* **1990**, *230*, 125–130. [CrossRef]
100. Fiedoruk, M.; Mieczkowska, E.; Koncki, R.; Tymecki, Ł. A bimodal optoelectronic flow-through detector for phosphate determination. *Talanta* **2014**, *128*, 211–214. [CrossRef] [PubMed]
101. Strzelak, K.; Koncki, R. Nephelometry and turbidimetry with paired emitter detector diodes and their application for determination of total urinary protein. *Anal. Chim. Acta* **2013**, *788*, 68–73. [CrossRef]
102. Trojanowicz, M. Recent Developments in Electrochemical Flow Detections – A Review. Part I. Flow Analysis and Capillary Electrophoresis. *Anal. Chim. Acta* **2009**, *653*, 36–58. [CrossRef]
103. Bavol, D.; Economou, A.; Zima, J.; Barek, J.; Dejmkova, H. Simultaneous determination of *tert*-butyl-hydroquinone, propyl gallate, and butylated hydroxyanisole by flow-injection analysis with multiple- pulse amperometric detection. *Talanta* **2018**, *178*, 231–236. [CrossRef]

104. Wilson, D.; del Valle, M.; Alegret, S.; Valderrama, C.; Florido, A. Potentiometric electronic tongue-flow injection analysis system for the monitoring of heavy metal biosorption processes. *Talanta* **2012**, *91*, 285–292. [CrossRef]
105. Rodriguez-Mozaz, S.; Reder, S.; Lopez de Alda, M.; Gauglitz, G.; Barcelo, D. Simultaneous multi-analyte deter- mination of estrone, isoproturon and atrazine in natural waters by River ANAlyser (RIANA), an optical immunosensor. *Biosens. Bioelectron.* **2004**, *19*, 633–640. [CrossRef]
106. Cantillo, D.; de Frutos, O.; Rincon, J.A.; Mateos, C.; Kappe, C.O. A scalable procedure for light-benzylic brominations in continuous flow. *J. Org. Chem.* **2014**, *79*, 223–229. [CrossRef] [PubMed]
107. Öhrngren, P.; Fardost, A.; Russo, F.; Schanmche, J.-S.; Fagrell, M.; Larhed, M. Evaluation of a nonresonant microwave applicator for continuous-flow chemistry applications. *Org. Process. Res. Dev.* **2012**, *16*, 1053–1063. [CrossRef]
108. Jensen, K.F.; Reizman, B.J.; Newmann, S.G. Tools for chemical synthesis in micro-systems. *Lab. Chip* **2014**, *14*, 3206–3212. [CrossRef] [PubMed]
109. Brandt, J.C.; Wirth, T. Controlling hazardous chemicals in microreactors: Synthesis with iodine azide. *Beilstein J. Org. Chem.* **2009**, *5*, 30. [CrossRef]
110. Szymborski, T.; Jankowski, P.; Garstecki, P. Teflon microreactors for organic syntheses. *Sens. Actuat. B Chemical* **2018**, *255*, 2274–2281. [CrossRef]
111. Hornung, C.H.; Hallmark, B.; Baumann, M.; Baxendale, I.R.; Ley, S.V.; Hester, P.; Clayton, P.; Mackley, M.R. Multiple microcapillary reactor for organic synthesis. *Ind. Eng. Chem. Res.* **2010**, *49*, 4576–4582. [CrossRef]
112. Sachse, A.; Galarneau, A.; Coq, B.; Fajula, F. Monolithic flow microreactors improve fine chemical synthesis. *New J. Chem.* **2011**, *35*, 259–264. [CrossRef]
113. Gunther, F.A.; Ott, D.E. Rapid automated determination of biphenyl in citrus fruit rind. *Analyst* **1966**, *91*, 475–481. [CrossRef]
114. Sonne, K.; Dasgupta, P.K. Simultaneous photometric flow injection determination of sulfide, polysulfide, sulfite, thiosulfate, and sulfate. *Anal. Chem.* **1991**, *63*, 427–432. [CrossRef]
115. Kika, F.S. Low pressure separations using automated flow and sequential injection analysis coupled to monolithic columns. *J. Chromatogr. Sci.* **2009**, *47*, 648–655. [CrossRef]
116. Soldi, L.; Ferstl, W.; Loebbecke, S.; Maggi, R.; Malmassari, C.; Sartori, G.; Yada, S. Use of immobilized organic base catalysts for continuous-flow fine chemical synthesis. *J. Catal.* **2008**, *258*, 289–295. [CrossRef]
117. Tian, K.; Dasgupta, P.K. Simultaneous flow-injection measurement of hydroxide, chloride, hypochlorite and chlorate in chlor-alkali cell effluents. *Talanta* **2000**, *52*, 623–630. [CrossRef]
118. Cerda, A.; Oms, M.T.; Forteza, R.; Cerda, V. Speciation of nitrogen in wastewater by flow injection. *Analyst* **1996**, *121*, 13–17. [CrossRef]
119. Lopez Pasquali, C.E.; Fernandez Hernando, P.; Durand Alegria, J.S. Spectrophotometric simultaneous determination of nitrite, nitrate and ammonium in soils by flow injection analysis. *Anal. Chim. Acta* **2007**, *600*, 177–182. [CrossRef]
120. Zhang, Y.; Miro, M.; Kolev, S.D. A novel hybrid flow platform for on-line simultaneous dynamic fractionation and evaluation of mercury liability in environmental solids. *Talanta* **2018**, *178*, 622–628. [CrossRef]
121. Wieczorek, M.; Kościelniak, P.; Świt, P.; Paluch, J.; Kozak, J. Solenoid micropump-based flow system for generalized calibration strategy. *Talanta* **2015**, *133*, 21–26. [CrossRef]
122. Urbanyi, T.; O'Connell, A. Simultaneous automated determination of hydralazine hydro-chloride, hydrochloro- thiazide, and reserpine in single tablet formulations. *Anal. Chem.* **1972**, *44*, 565–570. [CrossRef]
123. Luque de Castro, M.D.; Valcarcel, M. Flow injection methods based on multidetection. *Trends Anal. Chem.* **1986**, *5*, 71–74. [CrossRef]
124. Saurina, J. Flow-injection analysis for multi-component determinations of drugs based on chemometric approaches. *Trends Anal. Chem.* **2010**, *29*, 1027–1037. [CrossRef]
125. Miro, M.; Cerda, V.; Estela, J.M. Multisyringe flow injection analysis. *Trends Anal. Chem.* **2002**, *21*, 199–210. [CrossRef]
126. Sharma, Y.; Nikam, A.V.; Kulkarni, A.K. Telescoped sequence of exothermic and endothermic reactions in multistep flow synthesis. *Org. Proc. Res. Dev.* **2019**, *23*, 170–176. [CrossRef]

127. Usutani, H.; Tomida, Y.; Nagaki, A.; Okamoto, H.; Nokami, T.; Yoshida, J. Generation and reactions of o-bromo- phenyllithium without benzyne formation using a microreactor. *J. Am. Chem. Soc.* **2007**, *129*, 3046–3047. [CrossRef] [PubMed]
128. Movsisyan, M.; Delbeke, E.I.P.; Berton, J.K.E.T.; Battilocchio, C.; Ley, S.V.; Stevens, C.V. Taming hazardous chemistry by continuous flow technology. *Chem. Soc. Rev.* **2016**, *45*, 4892–4928. [CrossRef] [PubMed]
129. Webb, D.; Jamison, T.F. Continuous flow multi-step organic synthesis. *Chem. Sci.* **2010**, *1*, 675–680. [CrossRef]
130. Wegner, J.J.; Ceylan, S.; Kirschning, A. Flow chemistry – a key enabling technology for (multistep) organic synthesis. *Adv. Synth. Catal.* **2012**, *354*, 17–57. [CrossRef]
131. Britton, J.; Raston, C.L. Multi-step continuous-flow synthesis. *Chem. Soc. Rev.* **2017**, *46*, 1250–1271. [CrossRef]
132. Baxendale, I.R.; Ley, S.V.; Smith, C.D.; Tranmer, G.K. A flow reactor process for the synthesis of peptides utilizing immobilized reagents, scavengers and catch and release protocols. *Chem. Commun.* **2006**, 4835–4837. [CrossRef]
133. Adamo, A.; Beingessner, R.L.; Behnam, M.; Chen, J.; Jamison, T.F.; Jensen, K.F.; Monbaliu, J.-C.; Myerson, A.S.; Revalor, E.M.; Snead, D.R.; et al. On-demand continuous-flow production of pharmaceuticals in a compact, reconfigurable system. *Science* **2016**, *352*, 61–67. [CrossRef]
134. Abolhasani, M.; Jensen, K.F. Oscillatory multiphase flow strategy for chemistry and biology. *Lab. Chip* **2016**, *16*, 2775–2784. [CrossRef]
135. Barlow, I.M.; Harrison, S.P.; Hogg, G.L. Evaluation of the Technicon CHEM-1. *Clin. Chem.* **1988**, *34*, 2340–2344. [CrossRef]
136. Alexander, P.W.; Thalib, A. Nonsegmented rapid-flow analysis with ultraviolet/visible spectrophotometric determi- nation for short sampling times. *Anal. Chem.* **1983**, *55*, 497–501. [CrossRef]
137. Pasquini, C.; de Oliveira, W.A. Monosegmented system for continuous flow analysis. Spectrophotometric determination of chromium(VI), ammonia, and phosphorus. *Anal. Chem.* **1985**, *57*, 2575–2579. [CrossRef]
138. Tian, L.; Sun, X.; Xu, Y.; Zhi, Z. Segmental flow-injection analysis: Device and applications. *Anal. Chim. Acta* **1990**, *238*, 183–190.
139. Boedicker, J.Q.; Li, L.; Kline, T.R.; Ismagilov, R.F. Detecting bacteria and determining their susceptibility to antibiotics by stochastic confinement in nanoliter droplets using plug-based microfluidics. *Lab. Chip* **2008**, *8*, 1265–1272. [CrossRef] [PubMed]
140. Onal, Y.; Lucas, M.; Claus, P. Application of a capillary microreactor for selective hydrogenation of α, β-unsaturated aldehydes in aqueous multiphase catalysis. *Chem. Eng. Technol.* **2005**, *28*, 972–978. [CrossRef]
141. Wheeler, R.C.; Benali, O.; Deal, M.; Farrant, E.; MacDonald, S.J.F.; Warrington, B.H. Mesoscale flow chemistry: A plug-flow approach to reaction optimization. *Org. Proc. Res. Dev.* **2007**, *11*, 704–710. [CrossRef]
142. Kreutz, J.E.; Shukhaev, A.; Du, W.; Druskin, S.; Daugulis, O.; Ismagilov, R.F. Evolution of catalysts directed by genetic algorithms in a plug-based microfluidic device tested with oxidation of methane by oxygen. *J. Am. Chem. Soc.* **2010**, *132*, 3128–3132. [CrossRef]
143. Hochlowski, J.E.; Searle, P.A.; Tu, N.P.; Pan, J.Y.; Spanton, S.G.; Djuric, S.W. An integrated synthesis-purification system to accelerate the generation of compounds in pharmaceutical discovery. *J. Flow Chem.* **2011**, *2*, 56–61. [CrossRef]
144. Dasgupta, P.K.; Liu, S. Electroosmosis: A reliable fluid propulsion system for flow injection analysis. *Anal. Chem.* **1994**, *66*, 1792–1798. [CrossRef]
145. Haswell, S.J. Development and operating characteristics of micro flow injection analysis systems based on electro- osmotic flow. *Analyst* **1997**, *122*, 1R–10R. [CrossRef]
146. Fletcher, P.D.; Haswell, S.J.; Pombo-Villar, E.; Warrington, B.H.; Watts, P.; Wong, S.Y.; Zhang, X. Micro reactors: principles and applications in organic synthesis. *Tetrahedron* **2002**, *58*, 4735–4757. [CrossRef]
147. Wiles, C.; Watts, P.; Haswell, S.J. The use of electroosmotic flow as a pumping mechanism for semi-preparative scale continuous flow synthesis. *Chem. Commun.* **2007**, 966–968. [CrossRef] [PubMed]
148. Sweileh, J.A.; Dasgupta, P.K. Applications of in situ detection with an automated micro batch analyzer. *Anal. Chim. Acta* **1988**, *214*, 107–120. [CrossRef]
149. Honorato, R.S.; Araujo, M.C.U.; Lima, R.A.C.; Zagatto, E.A.G.; Lapa, R.A.S.; Costa Lima, J.L.F. A flow-batch titrator exploiting a one-dimensional optimization algorithm for end point search. *Anal. Chem. Acta* **1999**, *396*, 91–97. [CrossRef]
150. Honorato, R.C.; Carneiro, J.M.T.; Zagatto, E.A.G. Spectrophotometric flow-batch determination of aluminum in plant tissues exploiting a feedback mechanism. *Anal. Chim. Acta* **2001**, *441*, 309–315. [CrossRef]

151. Almeida, L.F.; Vale, M.G.; Dessuy, M.B.; Da Silva, M.M.; Lima, R.S.; Dos Santos, V.B.; Diniz, P.H.G.D.; Araujo, M. A flow-batch analyzer with piston propulsion applied to automatic preparation of calibration solutions for Mn determination in mineral waters by ET AAS. *Talanta* **2007**, *73*, 906–912. [CrossRef]
152. Barreto, I.S.; Lima, M.B.; Andrade, S.I.E.; Araujo, M.C.U.; Almeida, L.F. Using a flow-batch analyzer for photometric determination of Fe(III) in edible and lubricating oils without external pretreatment. *Anal. Meth.* **2013**, *5*, 1040–1045. [CrossRef]
153. Fernandez, C.J.; Domini, C.E.; Grűnhut, M.; Lista, A.G. A soft material for chromium speciation in water samples using a chemiluminescence automatic system. *Chemosphere* **2018**, *196*, 361–367. [CrossRef]
154. Bogdan, A.R.; Dombrowski, A.W. Emerging trends in flow chemistry and applications to the pharma- ceutical industry. *J. Med. Chem.* **2019**, *62*, 6422–6468. [CrossRef]
155. Hafner, A.; Filipponi, P.; Piccioni, L.; Meisenbach, M.; Schenkel, B.; Venturoni, F.; Sedelmeier, J. A simple scale-up strategy for organolithium chemistry in flow mode: From feasibility to kilogram quantities. *Org. Proc. Res. Dev.* **2016**, *20*, 1833–1837. [CrossRef]
156. Hafner, A.; Mancino, V.; Meisenbach, M.; Schenkel, B.; Sedelmeier, J. Dichloromethyllithium: Synthesis and application in continuous flow mode. *Org. Lett.* **2017**, *19*, 786–789. [CrossRef] [PubMed]
157. Dallinger, D.; Kappe, C.O. Lab-scale production of anhydrous diazomethane using membrane separation technology. *Nat. Protoc.* **2017**, *12*, 2138–2147. [CrossRef] [PubMed]
158. Fitzpatrick, D.E.; Ley, S.V. Engineering chemistry; integrating batch and flow reactions on a single, automated reactor platform. *React. Chem. Eng.* **2016**, *1*, 629–635. [CrossRef]
159. Burguera, J.L.; Burguera, M. Determination of sulphur anions by flow injection with a molecular emission cavity detector. *Anal. Chim. Acta* **1984**, *157*, 177–181. [CrossRef]
160. Kagenov, H.; Jensen, A. Kinetic determination of magnesium and calcium by stopped-flow injection analysis. *Anal. Chim. Acta* **1983**, *145*, 125–133. [CrossRef]
161. Betteridge, D.; Fields, B. Two point kinetic simultaneous determination of cobalt(II) and nickel(II) in aqueous solution using flow injection analysis (FIA). *Fresenius Z. Anal. Chem.* **1983**, *314*, 386–390. [CrossRef]
162. Műller, H.V.; Műller, V.; Hansen, E.H. Simultaneous differential rate determination of iron(II) and iron(III) by flow- injection analysis. *Anal. Chim. Acta* **1990**, *230*, 113–123. [CrossRef]
163. Hu, X.; Takenaka, N.; Kitano, M.; Bandow, H.; Maeda, Y. Determination of trace amounts of urea by using flow injection with chemiluminescence detection. *Analyst.* **1994**, *119*, 1829–1833. [CrossRef]
164. Burguera, J.L.; Townshend, A.; Greenfield, S. Flow injection analysis for monitoring chemiluminescent reactions. *Anal. Chim. Acta* **1980**, *114*, 209–214. [CrossRef]
165. Yang, C.; Zhang, Z.; Wang, J. Flow-injection chemiluminescence determination of dihydralazine sulfate in serum using luminol and diperiodatocuprate (III) system. *Spectrochim. Acta A Mol. Biomol. Spectr.* **2010**, *75*, 77–82. [CrossRef]
166. Pulgarin, J.A.M.; Bermejo, L.F.G.; Gallego, J.M.L.; Garcia, M.N.S. Simultaneous stopped-flow determination of morphine and naloxone by time-resolved chemiluminescence. *Talanta* **2008**, *74*, 1539–1546. [CrossRef] [PubMed]
167. Lin, H.; Dai, C.; Jamison, T.F.; Jensen, K.F. A rapid total synthesis of ciprofloxacin hydrochloride in continuous flow. *Angew. Chem. Int. Ed.* **2017**, *56*, 8870–8873. [CrossRef] [PubMed]
168. Pieber, B.; Glasnov, T.; Kappe, C.O. Flash carboxylation: Fast lithiation-carboxylation sequence at room temperature in continuous flow. *RSC Adv.* **2014**, *4*, 13430–13433. [CrossRef]
169. Nieuwland, P.J.; Koch, K.; van Harskamp, N.; Wehrens, R.; van Hest, J.C.M.; Rutjes, F.J.T. Flash chemistry extensively optimized: High-temperature Swern-Moffatt oxidation in an automated microreactor platform. *Chem. Asian J.* **2010**, *5*, 799–805. [CrossRef] [PubMed]
170. Yoshida, J. Flash chemistry using electrochemical method and microsystems. *Chem. Commun.* **2005**, 4509–4516. [CrossRef]
171. Yoshida, J.; Takahashi, Y.; Nagaki, A. Flash chemistry: Flow chemistry that cannot be done in batch. *Chem. Commun.* **2013**, *49*, 9896–9904. [CrossRef]
172. Browne, D.L.; Baumann, M.; Harji, B.H.; Baxendale, I.R.; Ley, S.V. A new enabling technology for convenient laboratory scale continuous flow processing at low temperatures. *Org. Lett.* **2011**, *13*, 3312–3315. [CrossRef]
173. Basel, C.L.; Defreese, J.D.; Whittemore, D.O. Interferences in automated Phenol Red method for determination of bromide in water. *Anal. Chem.* **1982**, *54*, 2090–2094. [CrossRef]

174. Bergamin, F.H.; Reis, B.F.; Jacintho, A.O.; Zagatto, E.A.G. Ion exchange in flow injection analysis: Determination of ammonium ions at the µg L^{-1} level in natural waters with pulsed Nessler reagent. *Anal. Chim. Acta* **1980**, *117*, 81–89. [CrossRef]

175. Olsen, S.; Pessenda, L.C.R.; Ruzicka, J.; Hansen, E.H. Combination of flow injection analysis with flame atomic-absorption spectrophotometry: Determination of trace amounts of heavy metals in polluted seawater. *Analyst* **1983**, *108*, 905–917. [CrossRef]

176. Faizullah, A.T.; Townshend, A. Applications of ion-exchange minicolumns in a flow-injection system for the spectrophotometric determination of anions. *Anal. Chem. Acta* **1986**, *179*, 233–244. [CrossRef]

177. Miro, M.; Gomez, E.; Estela, J.M.; Casa, M.; Cerda, V. Sequential injection ^{90}Sr determination in environmental samples using a wetting-film extraction method. *Anal. Chem.* **2002**, *74*, 826–833. [CrossRef] [PubMed]

178. Chirillo, R.; Caenaro, G.; Pavan, B.; Pin, A. The use of immobilized enzyme reactors in continuous-flow analyzers for the determination of glucose, urea, and uric acid. *Clin. Chem.* **1979**, *25*, 1744–1748. [CrossRef] [PubMed]

179. Yamini, Y.; Safari, M. Modified magnetic nanoparticles with catechol as a selective sorbent for magnetic solid phase extraction of ultra-trace amounts of heavy metals in water and fruit samples followed by flow injection ICP-OES. *Talanta* **2018**, *143*, 503–511. [CrossRef]

180. Trojanowicz, M.; Koźmiński, P.; Dias, H.; Brett, C.M.A. Batch injection stripping voltammetry (tube-less flow-injection analysis) of trace metals with on-line sample pretreatment. *Talanta* **2005**, *68*, 394–400. [CrossRef] [PubMed]

181. Quintana, J.B.; Miro, M.; Estela, J.M.; Cerda, V. Automated on-line renewable solid-phase extraction-liquid chromatography exploiting multisyringe flow injection-bead injection lab-on-valve analysis. *Anal. Chem.* **2006**, *78*, 2832–2840. [CrossRef]

182. Kołacińska, K.; Chajduk, E.; Dudek, J.; Samczyński, Z.; Łokas, E.; Bojanowska-Czajka, A.; Trojanowicz, M. Automation of sample processing for ICP-MS determination of ^{90}Sr radionuclide at ppq level for nuclear technology and environmental purposes. *Talanta* **2017**, *169*, 216–226. [CrossRef]

183. Fang, Z.-L. *Flow-injection Separation and Preconcentration*; VCH Verlag: Weinheim, Germany, 1993.

184. Liu, Y.; Ingle, J.D., Jr. Automated two-column ion exchange system for determination of the speciation of trace metals in natural waters. *Anal. Chem.* **1989**, *51*, 525–529. [CrossRef]

185. Baxendale, I.R.; Hayward, J.J.; Lanners, S.; Ley, S.V.; Smith, C.D. Heterogeneous reaction. In *Microreactors in Organic Synthesis and Catalysis*; Wiley-VCH: Weinheim, Germany, 2008.

186. Ley, S.V.; Baxendale, I.R.; Bream, R.N.; Jackson, P.S.; Leach, A.G.; Longbottom, D.A.; Nesi, M.; Scott, J.S.; Storer, R.I.; Taylor, S.J. Multi-step organic synthesis using solid-supported reagents and scavengers: A new paradigm in chemical library generation. *J. Chem. Soc., Perkin Trans.* **2000**, *1*, 3815–4195. [CrossRef]

187. Ley, S.V.; Baxendale, I.R. New tools and concepts for modern organic synthesis. *Nature Rev.* **2002**, *1*, 573–586. [CrossRef]

188. Baxendale, I.R.; Schou, S.C.; Sedelmaier, J.; Ley, S.V. Multi-step synthesis by using modular flow reactors: The preparation of yne-ones and their use in heterocyclic synthesis. *Chem. Eur. J.* **2010**, *16*, 89–94. [CrossRef] [PubMed]

189. Guetzoyan, L.; Nikbin, N.; Baxendale, I.R.; Ley, S.V. Flow chemistry synthesis of zolpidem, alpidem and other ABA$_A$ agonists and their biological evaluation and through the use of in-line frontal affinity chromatography. *Chem. Sci.* **2013**, *4*, 764–769. [CrossRef]

190. Baadenhuijsen, H.; Seuren-Jacobs, H.E.H. Determination of total CO_2 in plasma by automated flow-injection analysis. *Clin. Chem.* **1979**, *25*, 443–445. [CrossRef] [PubMed]

191. Svenson, G.; Anfält, T. Rapid determination of ammonia in whole blood and plasma using flow injection analysis. *Clin. Chim. Acta* **1982**, *119*, 7–14. [CrossRef]

192. Aoki, T.; Munemori, M. Continuous flow determination of free chlorine in water. *Anal. Chem.* **1989**, *55*, 209–212. [CrossRef]

193. Rodrigues, S.S.M.; Oleksiak, Z.; Ribeiro, D.S.M.; Poboży, E.; Trojanowicz, M.; Prior, J.A.; Santos, J.L. Selective determination of sulphide based on photoluminescence quenching of MPA-capped CdTe nanocrystals by exploiting a gas-diffusion multi-pumping flow manifold. *Anal. Meth.* **2014**, *6*, 7956–7966. [CrossRef]

194. Gross, U.; Koos, P.; O'Brien, M.; Polyzos, A.; Ley, S.V. A general continuous flow method for palladium catalysed carbonylation reactions using single and multiple tube-in-tube gas-liquid microreactors. *Eur. J. Org. Chem.* **2014**, 6418–6430. [CrossRef]

195. Mallia, C.J.; Baxendale, I.R. The use of gases in flow synthesis. *Org. Proc. Res. Dev.* **2016**, *20*, 327–360. [CrossRef]
196. Sawyer, R.; Dixon, E.J. The automatic determination of original gravity of beer. Part II. The determination of alcohol and gravity lost. *Analyst* **1968**, *93*, 680–687. [CrossRef]
197. Maquieira, A.; Casamayor, F.; Puchades, R.; Sagrado, S. Determination of total and free sulphur dioxide in wine with a continuous-flow microdistillation system. *Anal. Chim. Acta* **1993**, *283*, 401–407. [CrossRef]
198. Baumann, M. Integrating reactive distillation with continuous flow processing. *React. Chem. Eng.* **2019**, *4*, 368–371. [CrossRef]
199. Bittorf, L.; Reichmann, F.; Schmalenberg, M.; Soboll, S.; Kockmann, N. Equipment and separation units for flow chemistry applications and process development. *Chem. Eng. Technol.* **2019**, *42*, 1985–1995. [CrossRef]
200. Kabeshov, M.A.; Musio, B.; Murray, P.R.D.; Browne, D.L.; Ley, S.V. Expedient preparation of nazlinine and a small library of indole alkaloids using flow electrochemistry as an enabling technology. *Org. Lett.* **2014**, *16*, 4618–4621. [CrossRef] [PubMed]
201. Hopkin, M.D.; Baxendale, I.R.; Ley, S.V. An expeditious synthesis of imatinib and analogues utilising flow chemi- stry methods. *Org. Biomol. Chem.* **2013**, *11*, 1822–1839. [CrossRef]
202. Cole, K.P.; Groh, J.M.; Johnson, M.D.; Burcham, C.L.; Campbell, B.M. Kilogram-scale prexasertib monolactate monohydrate synthesis under continuous-flow CGMP conditions. *Science* **2017**, *356*, 1144–1150. [CrossRef]
203. Dolan, J.W.; Gant, J.R.; Tanaka, N.; Giese, R.W.; Karger, B.L. Continuous-flow automated HPLC analysis of fat- soluble vitamins in tablets. *J. Chromatogr. Sci.* **1978**, *16*, 616–622. [CrossRef]
204. Snyder, L.R. Continuous-flow analysis: Present and future. *Anal. Chim. Acta* **1980**, *114*, 3–18. [CrossRef]
205. Deadman, B.J.; Battilocchio, C.; Slivinski, E.; Ley, S.V. A prototype device for evaporation in batch and flow chemical processes. *Green Chem.* **2013**, *15*, 2050–2055. [CrossRef]
206. Luque-Garcia, J.L.; Luque de Castro, M.D. Where is microwave-based analytical equipment for solid sample pretreatment. *Trends Anal. Chem.* **2003**, *22*, 90–98. [CrossRef]
207. Burguera, M.; Burguera, J.L.; Alarcon, O.M. Flow-injection and microwave-oven sample decomposition for determination of copper, zinc and iron in whole-blood by atomic-absorption spectrometry. *Anal. Chim. Acta* **1986**, *179*, 351–357. [CrossRef]
208. Balconi, M.L.; Borgarello, M.; Ferraroli, R.; Realini, F. Chemical Oxygen Demand determination in well and river waters by flow-injection analysis using a microwave oven during the oxidation step. *Anal. Chim. Acta* **1992**, *261*, 295–299. [CrossRef]
209. Han, B.; Jiang, X.; Hou, X.; Zheng, C. Miniaturized dielectric barrier discharge carbon atomic emission spectrometry with online microwave-assisted oxidation for determination of Total Organic Carbon. *Anal. Chem.* **2014**, *86*, 6214–6219. [CrossRef] [PubMed]
210. Ley, S.V.; Baxendale, I.R. The changing face of organic synthesis. *Chimia* **2008**, *62*, 162–168. [CrossRef]
211. Wegner, J.; Ceylan, S.; Kirschning, A. Ten key issues in modern flow chemistry. *Chem. Commun.* **2011**, *47*, 4583–4592. [CrossRef]
212. Noël, T.; Buchwald, S.L. Cross-coupling in flow. *Chem. Soc. Rev.* **2011**, *40*, 5010–5029. [CrossRef]
213. Hartwig, J.; Ceylan, S.; Kupracz, L.; Coutable, L.; Kirschning, A. Heating under high-frequency inductive conditions: Application to the continuous synthesis of the Neurolepticum Olanzapine (Zyprexa). *Angew. Chem. Int. Ed.* **2013**, *52*, 9813–9817. [CrossRef]
214. He, P.; Haswell, S.J.; Fletcher, P.D.I. Microwave heating of heterogeneously catalysed Suzuki reactions in a micro reactor. *Lab. Chip* **2004**, *4*, 38–41. [CrossRef]
215. Comer, E.; Organ, M.G. A microreactor for microwave-assisted capillary (continuous flow) organic synthesis. *J. Am. Chem. Soc.* **2005**, *127*, 8160–8167. [CrossRef]
216. Ramarao, C.; Ley, S.V.; Smith, S.C.; Shirley, I.M.; DeAlmeida, N. Encapsulation of palladium in polyurea micro- capsules. *Chem. Commun.* **2002**, 1132–1133. [CrossRef]
217. Smith, C.J.; Iglesias-Siguenza, F.J.; Baxendale, I.R.; Ley, S.V. Flow and batch mode focused microwave synthesis of 5-amino-4-cyanopyrazoles and their further conversion to 4-aminopyrazolopyrimidines. *Org. Biomol. Chem.* **2007**, *5*, 2758–2761. [CrossRef]
218. Ceylan, S.; Friese, C.; Lammel, C.; Mazac, K.; Kieschning, A. Inductive heating for organic synthesis by using functionalized magnetic nanoparticles inside microreactors. *Angew. Chem. Int. Ed.* **2008**, *120*, 8950–8953. [CrossRef] [PubMed]

219. Wegner, J.; Ceylan, S.; Friese, C.; Kirschning, A. Inductively heated oxides inside micro-reactors - Facile oxidations under flow conditions. *Eur. J. Org. Chem.* **2010**, 4372–4375.
220. Ceylan, S.; Klande, T.; Vogt, C.; Friese, C.; Kirschning, A. Chemical synthesis with inductively heated copper flow reactors. *Synlett.* **2010**, 2009–2013.
221. Ilcheva, L.; Cammann, K. A simple, selective and sensitive liquid-chromatographic of flow-injection detector for chloroorganic compounds based on ion-selective electrodes. *Fresenius Z. Anal. Chem.* **1986**, *325*, 11–14. [CrossRef]
222. Edwards, R.T.; McKelvie, I.D.; Ferret, P.C.; Hart, B.T.; Bapat, J.B.; Koshy, K. Sensitive flow-injection technique for the determination of dissolved organic carbon in natural and wastewaters. *Anal. Chim. Acta* **1992**, *261*, 287–294. [CrossRef]
223. Molina-Garcia, L.; Llorent-Martinez, E.J.; Ortega-Barrales, P.; Fernandez de Cordoba, M.L.; Ruiz-Medina, A. Photo-chemically induced fluorescence determination of tigecycline by a stopped-flow multicom- mutated flow-analysis assembly. *Anal. Lett.* **2011**, *44*, 127–136. [CrossRef]
224. Cambie, D.; Noël, T. Solar photochemistry in flow. *Topics Curr. Chem.* **2018**, *376*, 45. [CrossRef]
225. Takei, G.; Kitamori, T.; Kim, H. Photocatalytic redox-combined synthesis of L-pipecolininc acid with a titania-modified microchannel chip. *Catal. Commun.* **2005**, *6*, 357–360. [CrossRef]
226. Pinho, V.D.; Gutmann, B.; Kappe, C.O. Continuous flow synthesis of β-amino acids from α-amino acids via Ardndt-Eistert homologation. *RSC Adv.* **2014**, *4*, 37419–37472. [CrossRef]
227. Talla, A.; Driessen, B.; Straathof, N.J.W.; Milroy, L.-G.; Brunsveld, L.; Hessel, V.; Noël, T. Metal-free photocatalytic aerobic oxidation of thiols to disulfides in batch and continuous-flow. *Adv. Synth. Catal.* **2015**, *357*, 2180–2186. [CrossRef]
228. Atobe, M.; Tateno, H.; Matsumara, Y. Applications of flow microreactors in electrosynthetic processes. *Chem. Rev.* **2018**, *118*, 4541–4572. [CrossRef] [PubMed]
229. Pletcher, D.; Green, R.A.; Brown, R.C.D. Flow electrolysis cells for the synthetic organic chemistry laboratory. *Chem. Rev.* **2018**, *118*, 4573–4591. [CrossRef] [PubMed]
230. Noël, T.; Cao, Y.; Laudadio, G. The fundamentals behind the use of flow reactors in electrochemistry. *Acc. Chem. Res.* **2019**, *52*, 2858–2869. [CrossRef] [PubMed]
231. Schothorst, R.C.; Van Son, M.; Den Boef, G. The application of strongly reducing agents in flow injection analysis: Part 4. Uranium(III). *Anal. Chim. Acta* **1984**, *162*, 1–8. [CrossRef]
232. Schothorst, R.S.; Den Boef, G. The application of strongly oxidizing agents in flow injection analysis: Part 1. Silver(II). *Anal. Chim. Acta* **1985**, *169*, 99–107. [CrossRef]
233. Nakata, R. Spectrophotometric determination of nitrite produced by flow-electrolysis of nitrate. *Fresenius Z. Anal. Chem.* **1984**, *317*, 115–117. [CrossRef]
234. Bergamin, F.H.; Krug, F.J.; Zagatto, E.A.G.; Arruda, E.C.; Coutinho, C.A. On-line electrolytic dissolution of alloys in flow-injection analysis: Part 1. Principles and applications in the determination of soluble aluminum in steels. *Anal. Chim. Acta* **1986**, *190*, 177–184. [CrossRef]
235. Mishra, S.K.; Dasgupta, P.K. Electrodialytic reagent introduction in flow systems. *Anal. Chem.* **2010**, *82*, 3981–3984. [CrossRef]
236. Belmont, C.; Girault, H.H. Coplanar interdigitated band electrodes for electrosynthesis - Part II: Methoxy-lation of furan. *J. Appl. Electrochem.* **1994**, *24*, 719–724. [CrossRef]
237. Pintauro, P.N.; Johnson, D.K.; Park, K.; Baizer, M.M.; Nobe, K. The paired electrochemical synthesis of sorbitol and gluconic acid in undivided flow cells. I. *J. Appl. Electrochem.* **1984**, *14*, 209–220. [CrossRef]
238. Ruzicka, J.; Hansen, E.H. Integrated microconduits for flow injection analysis. *Anal. Chim. Acta* **1984**, *161*, 1–25. [CrossRef]
239. Ruzicka, J. Flow injection analysis. *From test tube to integrated microconduits. Anal. Chem.* **1983**, *56*, 1040A–1053A.
240. Jacobson, S.C.; Moore, A.W.; Ramsey, J.M. Fused quartz substrates for microchip electrophoresis. *Anal. Chem.* **1995**, *67*, 2059–2063. [CrossRef]
241. Tantra, R.; Manz, A. Integrated potentiometric detector for use in chip-based flow cells. *Anal. Chem.* **2000**, *72*, 2875–2878. [CrossRef] [PubMed]
242. Lu, S.; Watts, P.; Chin, F.T.; Hong, J.; Musachio, J.L.; Briard, E.; Pike, V.W. Syntheses of ^{11}C- and ^{18}F-labeled carboxylic esters within a hydrodynamically-driven micro-reactor. *Lab. Chip* **2004**, *4*, 523–525. [CrossRef] [PubMed]

243. Wang, J.; Pumera, M.; Chatrathi, M.P.; Escarpa, A.; Musameh, M.; Collins, G.; Mulchandani, A.; Lin, Y.; Olsen, K. Single-channel microchip for fast screening and detailed identification of nitroaromatic explosives or organophosphate nerve agents. *Anal. Chem.* **2002**, *74*, 1187–1191. [CrossRef]
244. Tokeshi, M.; Minagawa, T.; Uchiyama, K.; Hibara, A.; Sato, K.; Hisamoto, H.; Kitamori, T. Continuous-flow chemical processing on a microchip by combining microunit operations and a multiphase flow network. *Anal. Chem.* **2002**, *74*, 1565–1571. [CrossRef]
245. Choi, J.-W.; Oh, K.W.; Thomas, J.H.; Heineman, W.R.; Halsall, H.B.; Nevin, J.H.; Helmicki, A.J.; Henderson, H.T.; Ahn, C.H. An integrated microfluidic biochemical detection system for protein analysis with magnetic bead-based sampling capabilities. *Lab. Chip* **2002**, *2*, 27–30. [CrossRef]
246. Hong, C.; Chang, P.; Lin, C.; Hong, C. A disposable microfluidic biochip with on-chip molecularly imprinted biosensors for optical detection of anesthetic propofol. *Biosens. Bioelectron.* **2010**, *25*, 2058–2064. [CrossRef]
247. Kiss, A.; Gaspar, A. Fabrication of a microfluidic flame atomic emission spectrometer: A flame-on-a-chip. *Anal. Chem.* **2018**, *90*, 5995–6000. [CrossRef]
248. Neuzil, P.; Campos, C.D.M.; Wong, C.C.; Soon, J.B.W.; Reboud, J.; Manz, A. From chip-in-a-lab to lab-on-a-chip: Towards a single handheld electronic system for multiple application-specific lab-on-a- chip. *Lab. Chip* **2014**, *14*, 2168–2176. [CrossRef] [PubMed]
249. Weaver, W.; Kittur, H.; Dhar, M.; Di Carlo, D. Research highlights: Microfluidic point-of-care diagnostics. *Lab. Chip* **2014**, *14*, 1962–1965. [CrossRef]
250. Murphy, E.R.; Martinelli, J.R.; Zaborenko, N.; Buchwald, S.L.; Jensen, K.F. Accelerating reactions with microreactors at elevated temperatures and pressures: Profiling amino-carbonylation reactions. *Angew. Chem. Int. Ed.* **2007**, *46*, 1734–1737. [CrossRef] [PubMed]
251. Abou-Hassan, A.; Sandre, O.; Cabuil, O.V. Microfluidics in inorganic chemistry. *Angew. Chem. Int. Ed.* **2010**, *49*, 6268–6286. [CrossRef] [PubMed]
252. Watts, P.; Wiles, C. Micro reactors, flow reactors and continuous flow synthesis. *J. Chem. Res.* **2012**, 181–193. [CrossRef]
253. Marre, S.; Roig, Y.; Aymonier, C. Supercritical microfluidics: Opportunities in flow-through chemistry and materials science. *J. Supercrit. Fluids* **2012**, *66*, 251–264. [CrossRef]
254. Badilescu, S.; Packirisamy, M. Microfluidics-nano-integration for synthesis and sensing. *Polymers* **2012**, *4*, 1278–1310. [CrossRef]
255. Scheler, O.; Postek, W.; Garstecki, P. Recent developments of microfluidics as a tool for biotechnology and microbiology. *Curr. Opinion Biotechnol.* **2019**, *55*, 60–67. [CrossRef]
256. Sollier, E.; Murray, C.; Maoddi, P.; Di Carlo, D. Rapid prototyping polymers for microfluidic devices and high pressure injections. *Lab. Chip* **2011**, *11*, 3752–3765. [CrossRef]
257. Focke, M.; Kosse, D.; Müller, C.; Reinecke, H.; Zengerle, R.; von Stetten, F. Lab-on-a-Foil: Microfluidics on thin and flexible films. *Lab. Chip* **2010**, *10*, 1365–1366. [CrossRef]
258. Taylor, D.; Dyer, D.; Lew, V.; Khine, M. Shrink film patterning by craft cutter: Complete plastic chips with high resolution/high-aspect ratio channel. *Lab. Chip* **2010**, *10*, 2472–2475. [CrossRef] [PubMed]
259. Luecha, J.; Hsiao, A.; Brodsky, S.; Liu, G.L.; Kokini, J.L. Green microfluidic devices made of corn proteins. *Lab. Chip* **2011**, *11*, 3419–3425. [CrossRef] [PubMed]
260. Martinez, A.W.; Phillips, S.T.; Whitesides, G.M. Three-dimensional microfluidic devices fabricated in layered paper and tape. *Proc. Nat. Acad. Sci. USA* **2008**, *105*, 19606–19611. [CrossRef] [PubMed]
261. He, Y.; Wu, Y.; Fu, J.; Wu, W. Fabrication of paper-based microfluidic analysis devices: A review. *RSC Adv.* **2015**, *5*, 78109–78127. [CrossRef]
262. Akyazi, T.; Basabe-Desmonds, L.; Benito-Lopez, F. Review on microfluidic paper-based analytical devices towards commercialisation. *Anal. Chim. Acta* **2018**, *1001*, 1–17. [CrossRef]
263. Almeida, M.I.G.S.; Jayawardane, B.M.; Kolev, S.D.; McKelvie, I.D. Developments of microfluidic paper-based analytical devices (μPADs) for water analysis: A review. *Talanta* **2018**, *177*, 176–190. [CrossRef]
264. Horning, M.P.; Delahunt, C.B.; Singh, S.R.; Garing, S.H.; Nichols, K.P. A paper microfluidic cartridge for automated staining of malaria parasites with an optically transparent microscopy window. *Lab. Chip* **2014**, *14*, 2040–2046. [CrossRef]
265. Peters, J.J.; Almeida, M.I.G.S.; Sraj, L.O.; McKelvie, I.D.; Kolev, S.D. Development of a micro-distillation microfluidic paper-based analytical device as a screening tool for total ammonia monitoring in freshwaters. *Anal. Chim. Acta* **2019**, *1079*, 120–128. [CrossRef]

266. Deiss, F.; Matochko, W.L.; Govindasamy, N.; Lin, E.Y.; Derda, R. Flow-through synthesis on Teflon- patterned paper to produce peptide arrays for cell-based assays. *Angew. Chem. Int. Ed.* **2014**, *53*, 6374–6377. [CrossRef]
267. Nilghaz, A.; Wicaksono, D.H.B.; Gustiono, D.; Majid, F.A.A.; Supriyanto, E.; Kadir, M.R.A. Flexible microfluidic cloth-based analytical devices using a low-cost wax patterning technique. *Lab. Chip* **2012**, *12*, 209–218. [CrossRef]
268. Theberge, A.B.; Whyte, G.; Frenzel, M.; Fidalgo, L.M.; Wootton, R.C.R.; Huck, T.S. Suzuki-Miyaura coupling reactions in aqueous microdroplets with catalytically active fluorous interfaces. *Chem. Commun.* **2009**, 6225–6227. [CrossRef] [PubMed]
269. Hwang, Y.-J.; Coley, C.W.; Abolhasani, M.; Marzinzik, A.L.; Koch, G.; Spanka, C.; Lehmann, H.; Jensen, K.F. A segmented flow platform for on-demand medicinal chemistry and compound synthesis in oscillating droplets. *Chem. Commun.* **2017**, *53*, 6649–6652. [CrossRef]
270. Du, W.; Sun, M.; Gu, S.; Zhu, Y.; Fang, Q. Automated microfluidic screening assay platform based on DropLab. *Anal. Chem.* **2010**, *82*, 9941–9947. [CrossRef] [PubMed]
271. Nightingale, A.M.; Phillips, T.W.; Bannock, J.H.; de Mello, J.C. Controlled multistep synthesis in a three-phase droplet reactor. *Nature Commun.* **2014**, *5*, 3777. [CrossRef]
272. Song, H.; Tice, J.D.; Isamgilov, R.F. A microfluidic system for controlling reaction networks in time. *Angew. Chem. Int. Ed.* **2003**, *42*, 768–772. [CrossRef] [PubMed]
273. Shestooalov, B.; Tice, J.D.; Ismagilov, R.F. Multi-step synthesis of nanoparticles performed on millisecond time scale in a microfluidic droplet-based system. *Lab. Chip* **2004**, *4*, 316–321. [CrossRef] [PubMed]
274. Wilding, P.; Shoffner, M.A.; Kricka, L.J. PCR in a silicon microstructure. *Clin. Chem.* **1994**, *40*, 1815–1818.
275. Wang, F.; Burns, M.A. Performance of nanoliter-sized droplet-based microfluidic PCR. *Biomed. Microdevices* **2009**, *11*, 1071–1080. [CrossRef]
276. Barikbin, Z.; Rahman, M.T.; Parthiban, P.; Rane, A.S.; Jain, V.; Duraiswamy, S.; Lee, S.S.; Khan, S.A. Ionic liquid-based compound droplet microfluidics for "on-drop" separations and sensing. *Lab. Chip* **2010**, *10*, 2458–2463. [CrossRef]
277. Fair, R.B. Digital microfluidics: Is a true lab-on-a-chip possible? *Microfluid. Nanofluid.* **2007**, *3*, 245–281. [CrossRef]
278. Coudron, L.; McDonnell, M.B.; Munro, I.; McCluskey, D.K.; Johnston, I.; Tan, C.K.; Tracey, M.C. Fully integrated digital microfluidics platform for automated immunoassay; A versatile tool for rapid, specific detection of a wide range of pathogens. *Biosens. Bioelectron.* **2019**, *128*, 52–60. [CrossRef] [PubMed]
279. Wu, B.; van der Ecken, S.; Swyer, I.; Li, C.; Jenne, A.; Vincent, F.; Schmidig, D.; Kuehn, T.; Beck, A.; Busse, F.; et al. Rapid chemical reaction monitoring by digital micro- fluidics-NMR: Proof of principle towards an automated synthetic discovery. *Angew. Chem. Int. Ed.* **2019**, *58*, 15372–15376. [CrossRef] [PubMed]
280. Gach, P.C.; Iwai, K.; Kim, P.W.; Hillson, N.J.; Singh, A.K. Droplet microfluidics for synthetic biology. *Lab. Chip* **2017**, *17*, 3388–3400. [CrossRef] [PubMed]
281. Rhee, M.; Burns, M.A. Microfluidic assembly blocs. *Lab. Chip* **2008**, *8*, 1365–1373. [CrossRef] [PubMed]
282. Capel, A.J.; Edmondson, S.; Christie, S.; Goodridge, R.D.; Bibb, R.J.; Thurstans, M. Design and additive manufacture for flow chemistry. *Lab. Chip* **2013**, *13*, 4583–4590. [CrossRef] [PubMed]
283. Hong, S.Y.; Kim, Y.C.; Wang, M.; Kim, H.-I.; Byun, Y.; Nam, J.-D.; Chou, T.-W.; Ajayan, P.M.; Ci, L.; Suhr, J. Experimental investigation of mechanical properties of UV-curable 3D printing materials. *Polym.* **2018**, *143*, 88–94. [CrossRef]
284. Dragone, V.; Sans, V.; Rosnes, M.H.; Kitson, P.J.; Cronin, L. 3D-printed devices for continuous-flow organic chemistry. *Beilstein J. Org. Chem.* **2013**, *9*, 951–959.
285. Peng, M.; Mittmann, E.; Wenger, L.; Hubbuch, J.; Engqvist, M.K.M.; Niemeyer, C.M.; Rabe, K.S. 3D-printed phenacrylate decarboxylase flow reactors for the chemoenzymatic synthesis of 4-hydroxystilbene. *Chem. - A Eur. J.* **2019**, *25*, 15998–16001. [CrossRef]
286. Bressan, L.P.; Robles-Najar, J.; Adamo, C.B.; Quero, R.F.; Costa, B.M.C.; de Jesus, D.P.; da Silva, J.A. 3D-printed microfludic device for the synthesis of silver and gold nanoparticles. *Microchem. J.* **2019**, *146*, 1083–1089. [CrossRef]
287. Zhang, J.M.; Aguirre-Pablo, A.A.; Li, E.Q.; Buttner, U.; Thoroddsen, S.T. Droplet generation in cross-flow for cost-effective 3D-printed "plug-and-play" microfluidic devices. *RSC Adv.* **2016**, *6*, 81120–81129. [CrossRef]

288. Spilstead, K.; Learey, J.J.; Doeven, E.H.; Barbante, G.J.; Mohr, S.; Barnett, N.W.; Terry, J.M.; Hall, R.M.; Francis, P.S. 3D-printed and CNC milled flow cells for chemiluminescence detection. *Talanta* **2014**, *126*, 110–115. [CrossRef] [PubMed]
289. Michalec, M.; Tymecki, Ł. 3D printed flow-through cuvette insert for UV-Vis spectrophotometric and fluorescence measurements. *Talanta* **2016**, *190*, 423–428. [CrossRef] [PubMed]
290. Calderilla, C.; Maya, F.; Cerda, V.; Leal, L.O. 3D printed device including disk-based solid-phase extraction for the automated speciation of iron using the multisyringe flow injection analysis technique. *Talanta* **2017**, *175*, 463–469. [CrossRef] [PubMed]
291. Mattio, E.; Robert-Peillard, F.; Vassalo, L.; Branger, C.; Margaillan, A.; Brach-Papa, C.; Knoery, J.; Boudenne, J.-L.; Coulomb, B. 3D-printed lab-on-valve for fluorescent determination of cadmium and lead in water. *Talanta* **2018**, *183*, 201–208. [CrossRef] [PubMed]
292. Gross, B.; Lockwood, S.Y.; Spence, D.M. Recent advances in analytical chemistry by 3D printing. *Anal. Chem.* **2017**, *89*, 57–70. [CrossRef]
293. Bishop, G.; Satterwhite, J.E.; Bhakta, S.; Kadimisetty, K.; Gillette, K.M.; Chen, E.; Rusling, J.F. 3D-printed fluidic devices for nanoparticle preparation and flow-injection amperometry using integrated Prussian blue nanoparticle-modified electrodes. *Anal. Chem.* **2015**, *87*, 5437–5443. [CrossRef]
294. Gal-Or, E.; Gershoni, Y.; Scotti, G.; Nilsson, S.M.E.; Saarinen, J.; Jokinen, V.; Strachan, C.J.; Gennäs, G.B.A.; Yli-Kauhaluoma, J.T.; Kotiaho, T. Chemical analysis using 3D printed glass microfluidics. *Anal. Meth.* **2019**, *11*, 1802–1810. [CrossRef]
295. Kingstone, H.M.; Kingstone, M.L. Nomenclature in laboratory robotics and automation. *Pure Appl. Chem.* **1994**, *66*, 609–630. [CrossRef]
296. Empel, C.; Koenigs, R.M. Artificial-intelligence-driven organic synthesis – En route towards autonomous synthesis? *Angew. Chem. Int. Ed.* **2019**, *58*, 2–5. [CrossRef]
297. Schindler, R.; Lendl, B. FTIR spectroscopy as detection principle in aqueous flow analysis. *Anal. Commun.* **1999**, *36*, 123–126. [CrossRef]
298. Gallignani, M.; Ayala, C.; del Rodario Brunetto, M.; Burguera, J.L.; Burguera, M. A simple strategy for determining ethanol in all types of alcoholic beverages based on its on-line liquid-liquid extraction with chloroform, using a flow injection system and Fourier transform infrared spectrometric detection in the mid-IR. *Talanta* **2005**, *68*, 470–479. [CrossRef] [PubMed]
299. Lendl, B.; Frank, J.; Schindler, R.; Muller, A.; Beck, M.; Faist, J. Mid-infrared quantum cascade lasers for flow injection analysis. *Anal. Chem.* **2000**, *72*, 1645–1648. [CrossRef] [PubMed]
300. Thygese, L.G.; Jorgensen, K.; Møller, B.L.; Engelsen, S.B. Raman spectroscopoic analysis of cynaogenic glucosides in plants: Development of a flow injection surface-enhanced Raman scatter (SERS) method for determination of cyanide. *Appl. Spectrosc.* **2004**, *58*, 212–217. [CrossRef] [PubMed]
301. Louden, D.; Handley, A.; Taylor, S.; Lenz, E.; Miller, S.; Wilson, I.D.; Sage, A. Flow injection spectroscopic analysis of model drugs using on-line UV-diode array, FT-infrared and ^{1}H-nuclear magnetic resonance spectroscopy and time-of-flight mass spectrometry. *Anal.* **2000**, *125*, 927–931. [CrossRef] [PubMed]
302. Da Silva, L.; Godejohann, M.; Martin, F.-P.; Collino, S.; Bürkle, A.; Moreno-Villanueva, M.; Bernhardt, J.; Toussaint, O.; Grubeck-Loebenstein, B.; Gonos, E.S.; et al. High-resolution quantitative metabolome analysis of urine by automated flow injection NMR. *Anal. Chem.* **2013**, *85*, 5801–5809. [CrossRef] [PubMed]
303. Mol, H.G.J.; van Dam, R.C.J. Rapid detection of pesticides not amenable to multi-residue methods by flow injection-tandem mass spectrometry. *Anal. Bioanal. Chem.* **2014**, *406*, 6817–6825. [CrossRef]
304. Chen, G.-Y.; Liao, H.-W.; Tseng, Y.J.; Tsai, I.-L.; Kuo, C. A matrix-induced ion suppression method to normalize concentration in urinary metabolomics studies flow injection analysis electrospray ionization mass spectrometry. *Anal. Chim. Acta* **2015**, *864*, 21–29. [CrossRef]
305. Krishnadasan, S.; Brown, R.J.C.; deMello, A.J.; deMello, J.C. Intelligent routes to the controlled synthesis of nanoparticles. *Lab Chip* **2007**, *7*, 1434–1441. [CrossRef]
306. Von Bomhard, S.; Schramm, J.; Bleul, R.; Thiermann, R.; Höbel, P.; Krtschil, U.; Löb, P.; Maskos, M. Modular manufacturing platform for continuous synthesis and analysis of versatile nanomaterials. *Chem. Eng. Technol.* **2019**, *42*, 2085–2094. [CrossRef]
307. Suga, S.; Okajima, M.; Fujiwara, K.; Yoshida, J. "Cation flow" method: A new approach to conventional and combinatorial organic syntheses using electrochemical microflow systems. *J. Am. Chem. Soc.* **2001**, *123*, 7941–7942. [CrossRef]

308. Carter, C.F.; Lange, H.; Ley, S.V.; Baxendale, I.R.; Wittkamp, B.; Goode, J.G.; Gaunt, N.L. ReactIR flow cell: A new analytical tool for continuous flow chemical processing. *Org. Proc. Res. Dev.* **2010**, *14*, 393–403. [CrossRef]
309. Smith, C.J.; Nikbin, N.; Ley, S.V.; Lange, H.; Baxendale, I.R. A fully automated, multistep flow synthesis of 5-amino-4-cynao-1,2,3-triazoles. *Org. Biomol. Chem.* **2011**, *9*, 1938–1947. [CrossRef] [PubMed]
310. Roberto, M.F.; Dearing, T.I.; Martin, S.; Marquardt, B.J. Integration of continuous flow reactors and online Raman spectroscopy for process optimization. *J. Pharm. Innov.* **2012**, *7*, 69–75. [CrossRef]
311. Browne, D.L.; Wright, S.; Deadman, B.; Dunnage, S.; Baxendale, I.R.; Turner, R.M.; Ley, S.V. Continuous flow reaction monitoring using an on-line miniature mass spectrometer. *Rapid Commun. Mass Spectrom.* **2012**, *26*, 1999–2010. [CrossRef]
312. Perera, D.; Tucker, J.W.; Brahmbhatt, S.; Helal, C.J.; Chong, A.; Farrell, W.; Richardson, P.; Sach, N.W. A platform for automated nanomole-scale reaction screening and micromole-scale synthesis in flow. *Science* **2018**, *359*, 429–434. [CrossRef]
313. Sutherland, J.D.; Tu, N.P.; Nemcek, T.A.; Searle, P.A.; Hochlowski, J.E.; Djuric, S.W.; Pan, J.Y. An automated synthesis-purification-sample-management platform for the accelerated generation of pharmaceutical candidates. *J. Lab. Autom.* **2014**, *19*, 176–182. [CrossRef]
314. Welch, C.J.; Gong, X.; Cuff, J.; Dolman, S.; Nyrop, J.; Lin, F.; Rogers, H. Online analysis of flowing streams using microflow HPLC. *Org. Proc. Res. Dev.* **2009**, *13*, 1022–1025. [CrossRef]
315. Werner, M.; Kuratli, C.; Martin, R.E.; Hochstrasser, R.; Wechsler, D.; Enderle, T.; Alanine, A.I.; Vogel, H. Seamless integration of dose-response screening and flow chemistry: Efficient generation of structure-activity relationship data of β-secretase (BACE1) inhibitors. *Angew. Chem. Int. Ed.* **2014**, *53*, 1704–1708. [CrossRef]
316. Korn, M.; Gouveia, L.F.B.P.; de Oliveira, E.; Reis, B.F. Binary search in flow titration employing photometric end-point detection. *Anal. Chim. Acta* **1995**, *313*, 177–184. [CrossRef]
317. Lima, M.J.A.; Reis, B.F.; Zagatto, E.A.G.; Kamogawa, M.Y. An automatic titration setup for the chemiluminometric determination of the copper complexation capacity in opaque solutions. *Talanta* **2020**, *209*, 120530. [CrossRef]
318. Pons, C.; Forteza, R.; Cerda, V. Expert multi-syringe flow-injection system for the determination and speciation analysis of iron using chelating disks in water samples. *Anal. Chim. Acta* **2004**, *524*, 79–88. [CrossRef]
319. Avivar, J.; Ferrer, L.; Casas, M.; Cerda, V. Smart thorium and uranium determination exploiting renewable solid-phase extraction applied to environmental samples in a wide concentration range. *Anal. Bioanal. Chem.* **2011**, *400*, 3585–3594. [CrossRef] [PubMed]
320. Goodell, J.R.; McMullen, J.P.; Zaborenko, N.; Maloney, J.R.; Ho, C.-X.; Jensen, K.F.; Porco, J.A.; Beeler, A.B. Development of an automated microfluidic reaction platform for multidimensional screening: Reaction discovery employing bicycle [3.2.1]octanoid scaffolds. *J. Org. Chem.* **2009**, *74*, 6169–6180. [CrossRef] [PubMed]
321. Lange, H.; Carter, C.F.; Hopkin, M.D.; Burke, A.; Goode, J.G.; Baxendale, I.R.; Ley, S.V. A breakthrough method for the accurate addition of reagents in multi-step segmented flow processing. *Chem. Sci.* **2011**, *2*, 765–769. [CrossRef]
322. Zhu, C.; Raghuvanshi, K.; Mason, D.; Rodgers, J.; Janka, M.E.; Abolhasani, M.; Coley, C.W. Flow chemistry-enabled studies of rhodium- catalyzed hydroformylation reactions. *Chem. Commun.* **2018**, *54*, 8567–8570. [CrossRef] [PubMed]
323. Owens, G.D.; Eckstein, R.J. Robotic sample preparation station. *Anal. Chem.* **1982**, *54*, 2347–2351. [CrossRef]
324. Granchi, M.P.; Biggerstaff, J.A.; Hillard, L.H.; Grey, P. Use of a robot and flow injection for automated sample preparation and analysis of used oils by ICP emission spectrometry. *Spectrochim. Acta* **1987**, *42B*, 169–180. [CrossRef]
325. Garcia-Mesa, J.A.; Luque de Castro, M.D.; Valcarcel, M. Coupled robot-flow injection analysis system for fully automated determination of total polyphenols in olive oil. *Anal. Chem.* **1993**, *65*, 3540–3542. [CrossRef]
326. Velasco-Arjona, A.; Luque de Castro, M.D. A robotic-flow injection approach to the fully automated determination of starch in food. *Anal. Chim. Acta* **1996**, *333*, 205–213. [CrossRef]
327. Prabhu, G.R.D.; Urban, P.L. The dawn of unmanned analytical laboratories. *Trends Anal. Chem.* **2017**, *88*, 41–52. [CrossRef]

328. Kellici, S.; Gong, K.; Lin, T.; Brown, S.; Clark, R.J.H.; Vickers, M.; Cockcroft, J.K.; Middelkoop, V.; Barnes, P.; Perkins, J.M.; et al. High-throughput continuous hydrothermal flow synthesis of Zn-Ce oxides: Unprecedented solubility of Zn in the nanoparticle fluorite lattice. *Phil Trans. R. Soc. A* **2010**, *368*, 4331–4349. [CrossRef] [PubMed]
329. Alexander, S.J.; Lin, T.; Brett, D.; Evans, J.R.; Cibin, G.; Dent, A.; Sankar, G.; Darr, J.A. A combinatorial nanoprecursor route for direct solid state chemistry: Discovery and electronic properties of new iron- doped lanthanum nickelates up to La4Ni2FeO10−δ. *Solid State Ionics* **2012**, *225*, 176–181. [CrossRef]
330. Urban, P.L. Universal electronics for miniature and automated chemical assays. *Analyst* **2015**, *140*, 963–975. [CrossRef] [PubMed]
331. Trojanowicz, M. Mobile-phone based chemical analysis – instrumental innovations and smartphone apps. *Modern Chem. Appl.* **2017**, *5*, 1000220. [CrossRef]
332. Guo, T.; Patnaik, R.; Kuhlmann, K.; Rai, A.J.; Sia, S.K. Smartphone dongle for simultaneous measurement of hemoglobin concentration and detection of HIV antibodies. *Lab. Chip* **2015**, *15*, 3514–3520. [CrossRef] [PubMed]
333. Xiao, G.; He, J.; Chen, X.; Qiao, Y.; Wang, F.; Xia, Q.; Yu, L.; Lu, Z.S. A wearable, cotton thread/paper-based microfluidic device coupled with smartphone for sweat glucose sensing. *Cellulose* **2019**, *26*, 4553–4562. [CrossRef]
334. Alves, I.P.; Reis, N.M. Microfluidic smartphone quantitation of *Escherichia coli* in synthetic urine. *Biosens. Bioelectron.* **2019**, *145*, 111624. [CrossRef]
335. Chen, L.; Zhang, C.; Xing, D. Paper-based bipolar electrode-electrochemiluminescence (BPE-ECL) device with battery energy supply and smartphone read-out: A handheld ECL system for biochemical analysis at the point-of-care level. *Sens. Actuat. B* **2016**, *237*, 308–317. [CrossRef]
336. Wongwilai, W.; Lapanantnoppakhun, S.; Grudpan, S.; Grudpan, K. Webcam camera as a detector for a simple lab-on-chip time based approach. *Talanta* **2010**, *81*, 1137–1141. [CrossRef]
337. Lima, M.B.; Barreto, I.S.; Andrade, S.I.E.; Almeida, L.F.; Araujo, M.C.U. Using webcam CdTe quantum dots and flow-batch system for automatic spectrofluorimetric determination of N-acetyl-L-cysteine in pharmaceutical formulations. *J. Braz. Chem. Soc.* **2014**, *25*, 1638–1646.
338. Li, B.; Li, L.; Guan, A.; Dong, Q.; Ruan, K.; Hu, R.; Li, Z. A smartphone controlled handheld microfluidic liquid handling system. *Lab. Chip* **2014**, *14*, 4085–4092. [CrossRef] [PubMed]
339. Zhang, W.; Liu, C.; Liu, F.; Zou, X.; Xu, Y.; Xu, X. A smart-phone-based electrochemical platform with programmable solid-state-microwave flow digestion for determination of heavy metals in liquid food. *Food Chem.* **2020**, *303*, 125378. [CrossRef] [PubMed]
340. Ingham, R.J.; Battilocchio, C.; Fitzpatrick, D.E.; Hawkinsk, J.M.; Ley, S.V. A systems approach towards an intelligent and self-controlling platform for integrated continuous reaction sequences. *Angew. Chem. Int. Ed.* **2015**, *54*, 144–148. [CrossRef] [PubMed]
341. Ley, S.V.; Ingham, R.J.; O'Brien, M.; Brownbe, D.L. Camera-enabled techniques for organic synthesis. *Beilstein J. Org. Chem.* **2013**, *9*, 1061–1072. [CrossRef] [PubMed]
342. Nacapricha, D.; Amornthammarong, N.; Sereenonchai, K.; Anujarawat, P.; Wilairat, P. Low cost telemetry with PC sound card for chemical analysis applications. *Talanta* **2007**, *71*, 605–609. [CrossRef] [PubMed]
343. Martinez, A.W.; Phillips, S.T.; Carrilho, E.; Thomas III, S.W.; Sindi, H.; Whitesides, G.M. Simple telemedicine for developing regions: Camera phones and paper-based microfluidic devices for real-time, off-site diagnosis. *Anal. Chem.* **2008**, *80*, 3699–3707. [CrossRef]
344. Lillehoj, P.B.; Huang, M.; Truong, N.; Ho, C. Rapid electrochemical detection on a mobile phone. *Lab. Chip* **2013**, *13*, 2950–2955. [CrossRef]
345. Trojanowicz, M. Nanostructures in flow analysis. *J. Flow-Inj. Anal.* **2008**, *25*, 5–13.
346. Frigerio, C.; Ribeiro, D.S.; Rodrigues, S.S.M.; Abreu, V.L.; Barbosa, J.; Prior, J.A.V.; Marques, K.; Santos, J.L. Application of quantum dots as analytical tools in automated chemical analysis: A review. *Anal. Chim. Acta* **2012**, *735*, 9–22. [CrossRef]
347. Passos, M.L.C.; Pinto, P.C.A.G.; Santos, J.L.M.; Saraiva, M.L.M.F.S.; Araujo, A.R.T.S. Nanoparticle-based assays in automated flow systems: A review. *Anal. Chim. Acta* **2015**, *889*, 22–34. [CrossRef]
348. Shahbazali, E.; Hessel, V.; Noël, Y.T.; Wang, Q. Metallic nanoparticles made in flow and their catalytic applications in micro-flow reactors for organic synthesis. *Phys. Sci. Rev.* **2016**, 20150016. [CrossRef]

349. Darr, J.A.; Zhang, J.; Makwana, N.M.; Weng, X. Continuous hydrothermal synthesis of inorganic nanoparticles: Applications and future directions. *Chem. Rev.* **2017**, *117*, 11125–11238. [CrossRef] [PubMed]
350. Sui, J.; Yan, J.; Liu, D.; Wang, K.; Luo, G. Continuous synthesis of nanocrystals via flow chemistry technology. *Small* **2019**, e1902828. [CrossRef] [PubMed]
351. Hassan, N.; Oyarzan-Ampuero, F.; Larta, P.; Guerrero, S.; Cabuil, V. Flow chemistry to control the synthesis of nano and microparticles for biomedical applications. *Curr. Topics Med. Chem.* **2014**, *14*, 676–689. [CrossRef] [PubMed]
352. Ma, J.; Lee, S.M.; Yi, C.; Li, C. Controllable synthesis of functional nanoparticles by microfluidic platforms for biomedical applications. *Lab. Chip* **2017**, *17*, 209–226. [CrossRef] [PubMed]
353. Rogers, L.; Jensen, K.F. Continuous manufacturing – the Green Chemistry promise? *Green Chem.* **2019**, *21*, 3481–3498. [CrossRef]
354. Rocha, F.R.P.; Zagatto, E.A.G. Flow analysis during the 60 years of *Talanta*. *Talanta* **2010**, *206*, 120186. [CrossRef]
355. Cerda, V.; Ferrer, L.; Portugal, L.A.; de Souza, C.T.; Ferreira, S.L.C. Multisyringe flow injection analysis in spectro- analytical techniques – A review. *Trends Anal. Chem.* **2018**, *98*, 1–18. [CrossRef]
356. Dias, T.R.; Melchert, W.R.; Kamogawa, M.Y.; Rocha, F.R.P.; Zagatto, E.A.G. Fluidized particles in flow analysis: Potentialities, limitations and applications. *Talanta* **2018**, *184*, 325–331. [CrossRef]
357. Trojanowicz, M.; Kołacińska, K.; Grate, J.W. A review of flow analysis methods for determination of radionuclides in nuclear wastes and nuclear reactor coolants. *Talanta* **2018**, *183*, 70–82. [CrossRef]

© 2020 by the author. Licensee MDPI, Basel, Switzerland. This article is an open access article distributed under the terms and conditions of the Creative Commons Attribution (CC BY) license (http://creativecommons.org/licenses/by/4.0/).

Article

Development of a Simple Reversible-Flow Method for Preparation of Micron-Size Chitosan-Cu(II) Catalyst Particles and Their Testing of Activity

Apichai Intanin [1], Prawpan Inpota [1], Threeraphat Chutimasakul [1], Jonggol Tantirungrotechai [1], Prapin Wilairat [2] and Rattikan Chantiwas [1,*]

[1] Department of Chemistry and Center of Excellence for Innovation in Chemistry, Faculty of Science, Mahidol University, Rama VI Rd., Bangkok 10400, Thailand; apichai.itn@gmail.com (A.I.); praw.inpota@gmail.com (P.I.); threeraphat.chu@gmail.com (T.C.); jonggol.jar@mahidol.ac.th (J.T.)
[2] National Doping Control Centre, Mahidol University, Rama VI Rd., Bangkok 10400, Thailand; prapin.wil@mahidol.ac.th
* Correspondence: rattikan.cha@mahidol.ac.th or rattikan.cha@mahidol.edu; Tel.: +66-2-201-5199; Fax: +66-23-54-7151

Academic Editor: Pawel Koscielniak
Received: 30 March 2020; Accepted: 12 April 2020; Published: 14 April 2020

Abstract: A simple flow system employing a reversible-flow syringe pump was employed to synthesize uniform micron-size particles of chitosan-Cu(II) (CS-Cu(II)) catalyst. A solution of chitosan and Cu(II) salt was drawn into a holding coil via a 3-way switching valve and then slowly pumped to drip into an alkaline solution to form of hydrogel droplets. The droplets were washed and dried to obtain the catalyst particles. Manual addition into the alkaline solution or employment of flow system with a vibrating rod, through which the end of the flow line is inserted, was investigated for comparison. A sampling method was selected to obtain representative samples of the population of the synthesized particles for size measurement using optical microscopy. The mean sizes of the particles were 880 ± 70 µm, 780 ± 20 µm, and 180 ± 30 µm for the manual and flow methods, without and with the vibrating rod, respectively. Performance of the flow methods, in terms of rate of droplet production and particle size distribution, are discussed. Samples of 180 µm size CS-Cu(II) particles were tested for catalytic reduction of 0.5 mM p-nitrophenol to p-aminophenol by 100-fold excess borohydride. The conversion was 98% after 20 min, whereas without the catalyst there was only 14% conversion.

Keywords: flow method; chitosan; catalyst particles; micron-size; sampling study; p-nitrophenol reduction

1. Introduction

Heterogeneous metal catalyst has been widely used in polymer and chemical production and in petroleum, pharmaceutical, and food industries [1,2]. They provide advantages such as ease of removal and recovery for recycling [3]. Recently, methods have been developed for preparation of heterogeneous catalyst particles with different sizes (ranging from nanometers to millimeters) and various shapes (spherical, rod, and fiber) [4–7]. Chitosan (CS) is a natural biopolymer. It can be obtained from alkaline deacetylation of chitin contained in the exoskeletons of crabs, shrimps, or insects. Chitosan is composed of polysaccharides and heteropolymers of glucosamine and acetylglucosamine, with pKa values in the range of 6.3–6.7 [4]. Chitosan has many applications because it is hydrophilic, non-toxic, biodegradable, eco-friendly, and cost-effective. It is thermally stable and can be chemically modified. It also has many metal-binding functional groups with a high affinity for various metal ions, such as Pb(II) [8], Ru(III) [8], Cu(II) [9], and Au(III) [10–12].

The preparation of catalyst particles can be carried out to give a range of sizes (millimeter, micrometer, and nanometer sizes [13]) for different applications [14]. In general, nano-size catalyst particles provide a greater catalytic activity for the same mass of catalyst when compared to micron-size or milli-size particles. However, reusing nanometer catalysts requires more procedure since it needs either evaporation using solvent or ultrafiltration. Hence, its reusability is a major issue especially for industrial applications in terms of cost-effectiveness or environmental impact. The millimeter- or micrometer-size catalysts can be separated simply by gravity settling for reuse. The micron- to milli-size particles are prepared by different methods, such as spray drying [15], jet cutting [14,16], and manual addition [17,18].

Manual dropping method is the conventional method for the preparation of millimeter-size particles [17,18]. The procedure produces hydrogel drop by drop. The limitations of manual dropping, spray drying, and jet cutting method are difficulty in scaling-up, low production rate, and difficulty in controlling size and shape, especially with the manual dropping method [17–22]. The methods (i.e., spraying, jet cutting) can be carried out in continuous flow procedure. Automation control is possible for spraying method, although it is complicated to set up [23]. Jet cutting method is suitable for a wide range of viscous fluidic medium and is designed for industrial scale production of monodisperse beads for pharmaceutical or biotechnology applications [14].

Micron-size particles exhibit more catalytic activity than millimeter-size catalysts. Simple preparation method can be used, such as manual dropping method, to produce these particles. A further advantage of using micron-size particle is the convenience of reusability without the need for separation via settling or filtration. Thus, this work aims to produce uniform micron-size CS-Cu(II) catalyst particles with narrow size distribution employing a simple flow method. It is efficient to produce the uniform with consistent size of synthesized catalyst due to the designs of flow instruments are promising with various platforms for flow synthesis [24]. Chitosan is employed as the substrate for synthesizing the heterogeneous metal catalysts. Copper is one of the most widely used transition metals for preparation of either homogenous or heterogeneous catalyst [25]. It has high activity with a wide range of industrial applications [26]. A simple reversible-flow system incorporating a vibrating rod was developed and compared with the flow method without the vibrator and also with the manual dropping method. A sampling procedure was employed to provide representative samples of the CS-Cu(II) particles for size analysis using images recorded under an optical microscope and analyzed with ImageJ program. The synthesized CS-Cu(II) catalyst particles were evaluated for their catalytic activity by using the well-known reaction of reduction of p-nitrophenol by aqueous borohydride solution to give the aminophenol product. p-nitrophenol has been used as a chemical in various industrial manufacturing processes, such as analgesic and antipyretic drugs, photographic developer, corrosion inhibitor, and dyes. However, it is hazardous and causes serious hypoxia. It has major effects on blood, liver, and the central nervous system [27]. Because of the extensive use of this compound, it is selected as the model compound for testing the catalytic property of the synthesized material. This development of a simple and convenient set-up of a flow method for preparation of heterogeneous catalyst of reproducible size on a small scale is convenient and efficient. It is less time consuming and more cost-effective for synthesis of various possible hydrogel formulations. This would lead to faster testing of the activity of the catalysts with different chemical reactions.

2. Results and Discussion

2.1. Evaluation of Sampling Procedure for Measurement of Particle Size of CS-Cu(II) Catalyst Particles

The study of sampling procedure was carried out by employing the reversible flow system for synthesis of CS-Cu(II) catalyst particles, see Figure 1A. The sampling procedure (see Section 3.3 and Figure 2A) provided particles for determination of the size distribution using ImageJ software analysis of the recorded images taken on an optical stereo microscope. As shown in Figure 2B, the mean particle size of 30 particles in each sector (Q1–Q4) for all 3 aliquots (Number 1–3) is in the range of 390–490 µm.

The average of the mean particle size of the sectors ($n = 12$) is 440 ± 50 µm. Thus, the sampling procedure is effective for obtaining representative particles for particle size measurements.

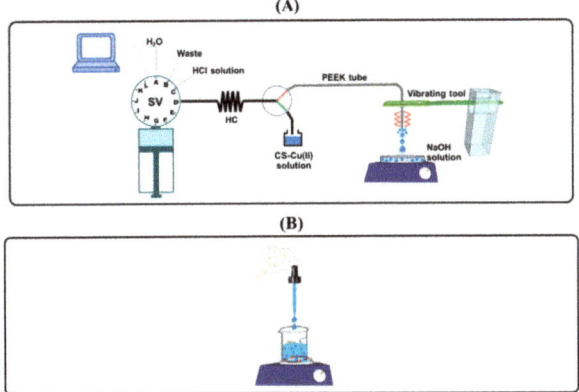

Figure 1. (**A**) Schematic of the flow system for synthesis of CS-Cu(II) catalyst particles with the PEEK tubing inserted through the vibrating rod. The reagent droplets at the tip of the vibrating tubing falls into a solution of NaOH. SV: selection valve, HC: holding coil. (**B**) Manual dropping method using a Pasteur pipette with a rubber teat.

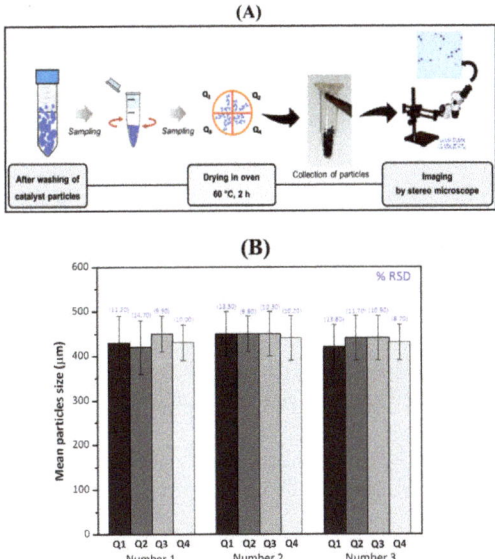

Figure 2. (**A**) Schematic of the sampling procedure for particle size measurements of the CS-Cu(II) particles. Bulk catalyst particles (~10.8% w/v) are vortexed before pipetting 0.3 mL of suspension into a microtube. Then, 0.3 mL of suspension is pipetted onto a filter paper, dried and divided into 4 quarters before drying in an oven at 60 °C (2 h). Particles in each quarter collected for recording of image on an optical microscope and size measurement using ImageJ software. (**B**) Bar graphs of mean particle size and % RSD for each quarter (Q1–Q4) and 3 sampling aliquots (Number 1–Number 3).

2.2. Investigation of Flow Parameters

Figure 1A shows the schematic of the flow system with the vibrating tool for the preparation of the CS-Cu(II) catalyst particles, see Table 1 for details of operational flow procedure. Various flow parameters can affect the particle shape and size distribution, i.e., the chitosan content, the NaOH concentration, and the flow rate of the CS-Cu(II) reactant solution. These parameters were studied with the objective of obtaining uniform size and shape of the catalyst particles.

Table 1. Operation procedure of the flow system shown in Figure 1A for preparation of micron-size chitosan (CS)-Cu(II) catalyst particles.

Step	Port	Flow Rate ($\mu L\ s^{-1}$)	Operation (Volume/μL)	Description
Aspirate H$_2$O carrier into syringe pump (repeat steps 1–2 twice)				
1	1	50	Aspirate (5000 µL)	Syringe pump valve in; aspirate H$_2$O.
2	4	50	Dispense (5000 µL)	Syringe pump valve out; dispense H$_2$O.
Aspirate air, CS-Cu(II) catalyst solution into holding coil				
3	4	10	Aspirate (200 µL)	Switch 3-port switching valve; dispense air plug.
4	4	10	Aspirate (4000 µL)	Aspirate plug of CS-Cu(II) catalyst solution.
Dispense CS-Cu(II) catalyst solution, air to the tip of tubing				
5	4	10	Dispense (4200 µL)	Switch 3-port switching valve; dispense CS-Cu(II) catalyst solution and air plugs.
Cleaning of holding coil by 0.2 M HCl solution and H$_2$O (repeat steps 6–9 twice)				
6	3	50	Aspirate (5000 µL)	Syringe pump valve in; aspirate 0.2 M HCl solution plug.
7	4	50	Empty	Syringe pump valve out; dispense 0.2 M HCl solution plug.
8	1	50	Aspirate (5000 µL)	Syringe pump valve in; aspirate H$_2$O plug.
9	1	50	Empty	Syringe pump valve out; dispense H$_2$O plug.

2.2.1. Effect of Chitosan Content

The chitosan concentrations were varied from 0.5% to 2.0% (w/v) (see Section 3.1 for preparation of solution). Catalyst particles were formed with increasing size (230 ± 70 µm to 340 ± 40 µm) when the chitosan concentration was increased from 1.0% to 1.5% (w/v) (see Figure 3A). However, at 0.5% (w/v), chitosan did not form hydrogel droplets because of its low viscosity. Conversely, the high viscosity of the 2.0% (w/v) solution limited the flow of the chitosan solution. Hence, histograms of the size distribution for the 0.5% and 2.0% (w/v) chitosan solutions were not constructed. The concentration of chitosan solution of 1.5% (w/v) was selected since it gave particles with more spherical shape and narrower size distribution than the 1.0% (w/v) chitosan sample (see Figure 3A).

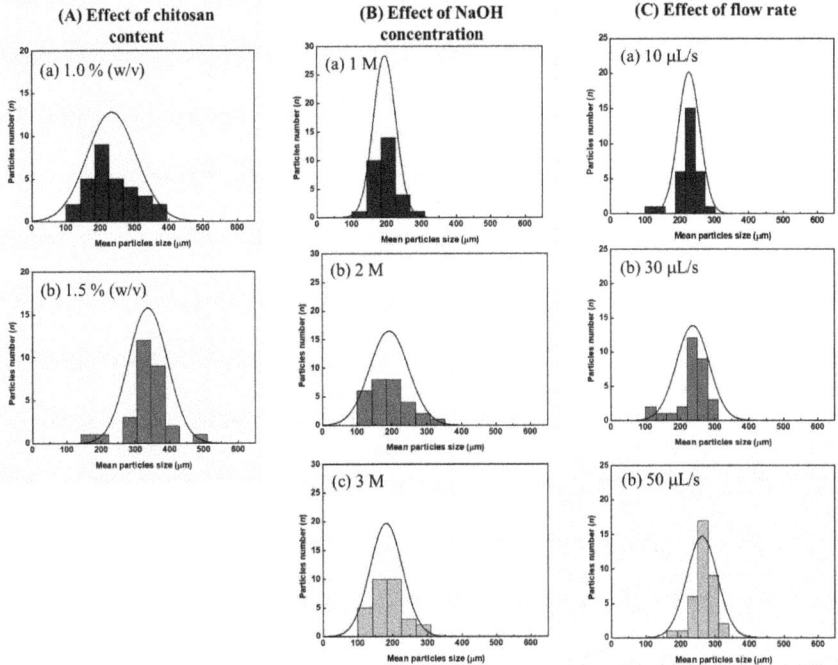

Figure 3. Histograms of the particle size distribution of CS-Cu(II) catalyst particles produced under various flow conditions. (**A**) Effect of chitosan content: 1.0 1.5% (*w/v*). Chitosan at 0.5% and 2.0% did not produce hydrogel droplets due to the low and high viscosity of the reagent solutions, respectively. (**B**) Effect of NaOH concentration: (a) 1 M, (b) 2 M, and (c) 3 M NaOH is used for solidification of chitosan-Cu(II) catalyst particles. (**C**) Effect of flow rate of the CS-Cu(II) reagent: (a) 10, (b) 20, and (c) 30 µL s^{-1}, respectively.

2.2.2. Effect of NaOH Concentration

Suitable NaOH concentration was investigated since the concentration of NaOH of the receptacle solution (Figure 1A) affects the rate of chitosan solidification. An optimum solidification rate is needed to obtain a narrow particle size distribution. Chitosan is soluble in hydrochloric acid solution due to the protonation of amino groups in the polymeric chain. In alkaline solution, the protons on the polymeric chain are dissociated and the neutral polymer chain can aggregate together. Concentrations of the NaOH solutions selected were 1 M, 2 M, and 3 M, respectively. Figure 3B shows the results obtained for the different NaOH concentrations. The use of 1 M and 2 M NaOH solutions gave similar results for the mean size of the particles, viz. 260 ± 50 µm and 240 ± 50 µm, respectively. Particle size distributions were narrower for the 1 M and 2 M NaOH solutions than for the 3 M solution. Therefore, 2 M NaOH was selected as the operating NaOH concentration.

2.2.3. Effect of Flow Rate of the CS-Cu(II) Reactant Solution

The flow rate of the CS-Cu(II) solution was varied from 10 µL s^{-1} to 50 µL s^{-1}. The flow rate of 10 µL s^{-1} is the smallest rate possible for the syringe pump. The resulting size distribution of the synthesized CS-Cu(II) catalyst particles are shown in Figure 3C. The catalyst particles have mean sizes of 230 ± 30 µm, 240 ± 50 µm, and 260 ± 40 µm, when the flow rates are 10 µL s^{-1}, 30 µL s^{-1}, and 50 µL s^{-1}, respectively. Therefore, the mean size of the CS-Cu(II) catalyst particles is not significantly different with increasing flow rate. On the other hand, the size distribution increases with the flow rate. Therefore, the flow rate for the reagent solution of 10 µL s^{-1} was selected.

The parameters of the flow system studied above are summarized in Table 2. The selected conditions are given in the last column of Table 2.

Table 2. The concentrations of reagents and flow rates for preparation of CS-Cu (II) catalyst particles.

Parameter	Studied Range	Selected Value
1. Concentration of chitosan (% w/v)	0.5–1.5	1.5
2. Concentration of copper acetate (M)	1–3	2
3. Concentration of NaOH (M)	1–3	2
4. Flow rate (µL s^{-1})	10–50	10

2.3. Comparison of Methods for Synthesizing CS-Cu(II) Catalyst Particles: Flow Methods vs. Manual Addition

Three methods for the preparation of the CS-metal catalyst particles were investigated, i.e., flow method with a vibrating rod (see Figure 1A), flow without the vibrating rod, and the manual dropping method (see Figure 1B). The resulting products were collected, dried, and sampled, and the size distribution measured as described in Section 3.3. Figure 4 shows the distributions of the particle size of the CS-Cu(II) catalyst and images of the particles as prepared by the three methods. A much smaller particle size (150–210 µm) was obtained by the flow method using the vibrating rod. There was greater uniformity of particle shape and narrower size distribution. The vibrating rod assisted in producing smaller droplets with a 4-fold reduction in the particle size than that obtained without vibration (780 ± 20 µm vs. 180 ± 30 µm), see Figure 4B,C.

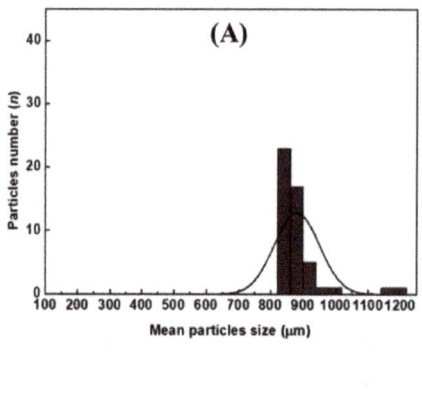
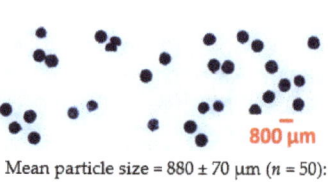

Mean particle size = 880 ± 70 µm (n = 50):

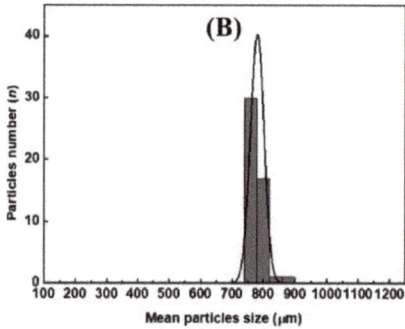
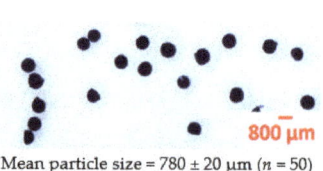

Mean particle size = 780 ± 20 µm (n = 50)

Figure 4. *Cont.*

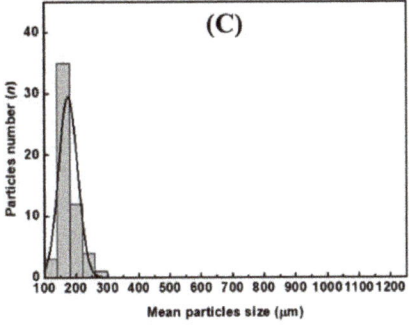
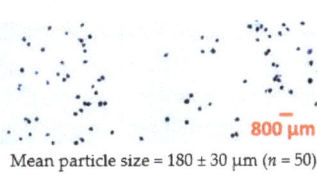

Mean particle size = 180 ± 30 μm (n = 50)

Figure 4. Particle size distributions with images of representative particles obtained using different flow methods. (**A**) Manual dropping method, (**B**) flow method without vibration and (**C**) flow method with the vibrating rod. The scale bar for 800 μm length is shown.

The manual dropping method is the conventional method for synthesis of the particles. The mean particle size was 880 ± 70 μm (see Figure 4A). Drawbacks of this method are the difficulty in controlling the resulting size and shape of particles and the low production rate. The size and number of drops depend on the manual pressure applied to the rubber bulb. The throughput of this method is about 40 drops min^{-1}.

The flow-based method provides advantages such as precision of flow rate and automation of operation. Therefore, the flow method with the vibrating rod was employed for producing micron-size uniformly shaped particles of CS-Cu(II) catalyst particles.

Comparison of the performance between manual dropping method and flow methods, with and without vibrating rod, is shown in Table 3.

Table 3. Comparison of method performance between manual dropping method, flow method without vibration, and flow method with vibrating rod.

Parameter	Manual Method	Flow Method Without Vibration	Flow Method with Vibrating Rod
1. Operation mode	Manual	Computer control	Computer control
2. Mean particles size (n = 50)	880 ± 70 μm	780 ± 20 μm	180 ± 30 μm
3. Size distribution	810–950 μm	760–800 μm	150–210 μm
4. Shape	Quasi-spherical	Spherical	Spherical
5. Throughput	40 drops min^{-1}	60 drops min^{-1}	100 drops min^{-1}

2.4. Characterization of CS-Cu(II) Catalyst Particles

A SEM picture of the micron-size CS-Cu(II) catalyst is shown in Figure S1a, Supplementary Material A [28]. The surface morphology of the particle shows unique streaking patterns with uniform surface structure. Figure S1b and S1c show characteristic FTIR spectrum and XRD pattern of the catalyst, respectively. The FTIR spectrum has bands at 3430, 2868, 2364, 1622, 1379, and 1097, 1013, and 599 cm^{-1}. These are the "fingerprint" of chitosan and the strong broad peak at 3430 cm^{-1} is the hydrogen-bonded O-H stretching band. The XRD pattern reveals low-intensity peaks at $2\theta = 35.4°$ and 38.8°, designated to (002) and (111) reflections of CuO, respectively. Nitrogen adsorption/desorption isotherms, as shown in Figure S1d, indicate non-porous structure with low surface area.

2.5. Assessment of the Catalytic Activity of the Chitosan-Cu(II) Particles in the Reduction Reaction of p-Nitrophenol with Excess Borohydride

The reduction of p-nitrophenol (p-NP) with excess borohydride is a common chemical reaction employed for testing the catalytic activity of heterogeneous metal catalysts [29]. The net chemical reaction is shown in Figure 5A. When a *p*-nitrophenol solution is mixed with a solution of NaBH$_4$, *p*-nitrophenol is first rapidly deprotonated to give the *p*-nitrophenolate anion as an intermediate, which is then reduced to *p*-aminophenol. The UV–visible spectra of *p*-nitrophenolate anion and *p*-aminophenol are shown in Figure 5A. The reaction can be monitored by measurement of absorbance at 400 nm. The rate of the reaction without catalyst particles is slow. Metal catalysts, such as Au [30], Ag [31], and Cu [9], have thus been used to accelerate the reaction.

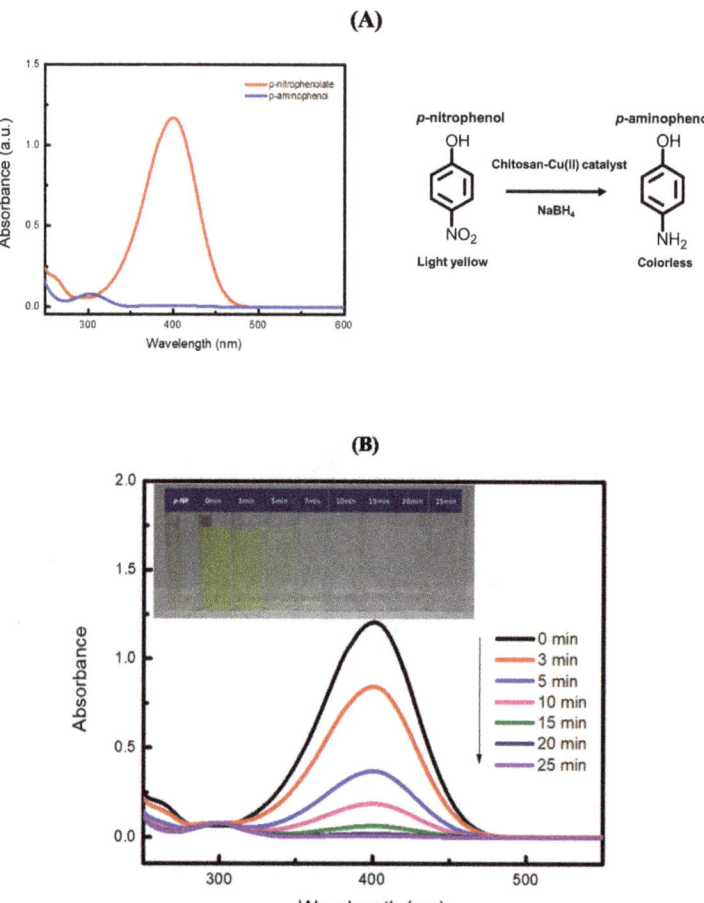

Figure 5. (**A**) Spectra of the reduction reaction of *p*-nitrophenol with 100-fold excess sodium borohydride and 0.2 mg/mL CS-Cu(II) catalyst particles (180 µm mean size). The red spectrum is the initial spectrum of the *p*-nitrophenolate anion (83 µM) produced rapidly on addition of borohydride. The blue spectrum is the spectrum at the end of the reaction. (**B**) UV–Visible absorption spectra and photographs (inset) of the filtered and diluted (5-fold) reaction solution sampled at various times. The initial concentrations are 0.5 mM *p*-NP with 0.05 M NaBH$_4$ solutions containing 0.2 mg mL^{-1} CS-Cu(II) catalyst particles (180 µm mean size).

In this work, CS-Cu(II) particles were assessed as catalyst for the above reduction reaction. Spectra of a solution of *p*-NP containing the catalyst were measured at various time intervals, as shown in Figure 5B. The *p*-nitrophenol solution is pale yellow. On addition of sodium borohydride (NaBH$_4$), the color of the solution is changed to bright yellow due to the rapid formation of the *p*-nitrophenolate anion (see the first and second cuvettes from the left in the inset of Figure 5B. There is a red shift of the wavelength at the maximum absorbance from 317 nm to 400 nm (data not shown). The absorbance of p-nitrophenolate at 400 nm decreases with time with a small concomitant increase of the absorbance of the p-aminophenol product at 300 nm. Percent conversion of p-nitrophenol was calculated from $\frac{(A_0-A_t)}{A_0} \times 100$, where A_0 is the initial absorbance of p-nitrophenolate anion, and A_t is the absorbance of p-nitrophenolate at different time intervals. Figure 5B shows the photographs (inset) of the solutions of 1.0 mM p-nitrophenol with CS-Cu(II) catalyst before addition of borohydride and after addition at 0, 3, 5, 10, 15, 20, and 25 min, respectively. The measured UV–visible spectra at 0–25 min are also shown. At 25 min, there was 98% conversion ($t_{1/2}$ ca. 4 min) as compared with less than 14% conversion ($t_{1/2}$ ca. 115 min) without the catalyst (data not shown). Our conversion efficiency (>90%) was comparable with previous reports using the same reaction of p-nitrophenol catalyzed by other synthesized catalysts, such as activated carbon-supported gold nanocatalysts (Au/AC) [32] and magnetic chitosan-supported silver nanoparticles [33]. Further detailed kinetics study, together with employment of other compounds, are part of future studies.

3. Materials and Methods

3.1. Chemicals and Preparation of Chitosan and Cu(II) Solutions

All chemicals used in the experiments were of analytical grade (AR). Chitosan powder, copper(II) acetate monohydrate (98% assay), and sodium hydroxide were from Sigma-Aldrich (St. Louis, MO, USA). Dimethyl sulfoxide (99.9% assay) and hydrochloric acid (37% assay) were obtained from RCI Labscan (Bangkok, Thailand). p-Nitrophenol (99.5% assay) was purchased from Merck (Darmstadt, Germany). Sodium borohydride (98% assay) was from Alfa Aesar (Tewksbury, MA, USA). Ultrapure water was from a Milli-Q® Advantage A10 water purification system (Darmstadt, Germany).

The chitosan and copper acetate solutions were prepared freshly. Chitosan solution (1.5% *w/v*) was prepared by dissolving chitosan powder in 10.00 mL of 0.2 M HCl with continuous stirring for 30 min. Then, 4.0 mL of the Cu(II) solution (2.0 M) was added dropwise to the chitosan solution while stirring continuously for 45 min. Then, 0.5 mL DMSO was added into the CS-Cu(II) solution, which was then stirred for 15 min and sonicated for an hour.

3.2. Preparation of CS-Cu(II) Catalyst Particles

3.2.1. Configuration and Operation of the Flow System

Figure 1A shows the manifold of the flow system with the vibrating tool for the preparation of micron-size CS-Cu(II) catalyst particles. The flow system comprises a syringe pump with a 5.0 mL syringe barrel (Norgren Kloehn Inc., Las Vegas, NV, USA). A 3-way switching valve is used to draw the sample into the holding coil (HC) (FEP tubing, 2.0 mm i.d., 159 cm long). The flow line connecting the SV and the HC is a PTFE tubing (0.5 mm i.d., 26 cm long). The 3-way switching valve is also connected to a PEEK (polyether ether ketone) tubing (0.75 mm. i.d., 48 cm long), which passes through a hole at the paddle end of the vibrating rod. The length of the PEEK tubing extending from the vibrating rod is set at 5.0 cm. The tip of the PEEK tubing is placed above the container of the NaOH solution (see Figure 1A).

The vibrator is a commercial electric toothbrush (Systema Sonic, Lion, Japan). The step-by-step operation is set through the software controller and is listed in Table 1. A 4000 µL solution of CS-Cu(II) reagent is pumped slowly (10 µL s^{-1}) to form a small drop at the tip of the PEEK tubing. The drop then breaks off into the NaOH solution, leading to the formation of hydrogel droplets. The resulting hydrogels were washed and dried to obtain CS-Cu(II) catalyst particles. The dry product was analyzed for particle size distribution as described in Section 3.3 (see Supplementary Material B (Figure S2) and Table S1 for detail of ImageJ software).

3.2.2. Manual Dropping Method

Schematic of the manual dropping method for preparation of the particles is shown in Figure 1B. The CS-Cu(II) solution was added to the NaOH solution dropwise with a Pasteur pipette using a rubber teat. The hydrogel particles were then filtered and washed with deionized water prior to analysis for shape and size distribution (see Section 3.3).

3.3. Sampling and Measurement of Particle Size of Synthesized CS-Cu(II) Catalyst

The freshly synthesized CS-Cu(II) particles were added to deionized water to give a concentration of 10.8% (w/v) and vortexed. Then, three 0.3 mL aliquots of the suspension were pipetted and transferred into 3 microtubes (1.5 mL) and 0.70 mL ultrapure water added to each tube. After vortexing, 0.3 mL of the suspension from each tube was pipetted onto 3 separate filter paper (Whatman no. 2) to absorb the water and then dried in an oven at 60 °C for 2 h (see Figure 2A). The dry particles were divided into 4 sections, and particles in each section collected in individual vials. Images of particles from each vial were recorded on a stereomicroscope at 2× magnification (OLYMPUS SZ51, Tokyo, Japan). The digital image was analyzed by ImageJ software to obtain the mean size of the particles. Threshold number of the ImageJ software was adjusted ± 30 from the default threshold number (see details in Supplementary Material B).

3.4. Method of Characterization of CS-Cu(II) Catalyst Particles

Characterization of CS-Cu(II) catalyst was carried out using various techniques, e.g., scanning electron microscopy (SEM) using a Hitachi S-2500 instrument, Fourier-transform spectroscopy (FTIR) employing KBr disc with a PerkinElmer spectrometer, x-ray diffraction (XRD) using a Bruker D8 Advance diffractometer, and nitrogen adsorption/desorption isotherms and pore size distribution using Quantachrome AUTOSORB-1 system. The results are shown in Supplementary Material A [28].

3.5. Procedure for Measuring Activity of CS-Cu(II) Particles Using Reduction Reaction of p-Nitrophenol

A solution of 1.0 mM p-nitrophenol was prepared by dissolving 0.0696 g of p-nitrophenol powder in ultrapure water and then adjusted to volume in a 500.0 mL volumetric flask. Sodium borohydride (0.1 M $NaBH_4$) was used as the reducing agent by dissolving 0.1892 g $NaBH_4$ in 50.0 mL of DI water. A 25.0 mL volume of 1 mM p-nitrophenol solution was mixed with 25.0 mL of 0.1 M $NaBH_4$ solution and continuously stirred for 30 s. Then, 10.0 mg of the dry 180 μm-size CS-Cu(II) particles were added to the reaction solution with continuous stirring and the timing of the reaction started. At various time intervals (0, 3, 5, 10, 15, 20, and 25 min), 1.5 mL aliquot of the reaction solution was removed and filtered with a syringe filter. Then, 1.00-mL of the filtrate was mixed with 5.00 mL of ultrapure water and the absorbance measured on a UV–visible spectrophotometer at 400 nm. The reaction solution was observed to change from bright yellow to colorless over time (see inset Figure 5B). The UV–visible spectra of the solutions at various time intervals are shown in Figure 5B.

4. Conclusions

This work presents the development of a simple flow method for the synthesis of uniform micron-size CS-Cu(II) catalyst particles, together with testing of their catalytic activity for the reduction reaction of p-nitrophenol with excess borohydride. The flow system comprises a reversible-flow syringe pump together with a 3-port switching valve and a holding coil. Various flow parameters, viz. chitosan concentration, NaOH concentration, and flow rate, were investigated to produce the smallest micron-size with narrow size distribution of the dry synthesized material. The selected conditions were: 1.5% (w/v) chitosan, 2.0 M copper acetate, and 2 M NaOH. The most suitable flow rate for adding the reagent solution into the sodium hydroxide was 10 μL s^{-1}. The performance of the flow method employing the vibrating rod was compared to that without the vibrator and also to a manual dropping method.

A sampling method of the synthesized particles was employed to obtain representative samples for size measurement. Images of the samples were recorded on an optical microscope and the size and shape of the particles analyzed using ImageJ software. The mean sizes of the catalyst particles were 180 ± 30 µm, 780 ± 20 µm, and 880 ± 70 µm, produced using the flow system with and without the vibrating tool and the manual dropping method, respectively. The flow method with the sonic tool provided narrower size distribution of the particles (150–210 µm) and uniform quasi-spherical particles. The flow methods have higher throughput for preparation of the catalyst particles, with rates of 100 drops min^{-1} using the sonic tool and 60 drops min^{-1} without, respectively. The 180 µm-size CS-Cu(II) particles, at concentration of 0.2 mg mL^{-1}, were found to catalyze the reduction of p-nitrophenol (p-NP) with excess borohydride by more than 30-fold compared to reduction without the catalyst.

Supplementary Materials: The following are available online at http://www.mdpi.com/1420-3049/25/8/1798/s1, Figure S1: Characterization of chitosan-Cu(II) catalyst particles: (a) SEM image; (b) FTIR spectrum of chitosan-Cu(II) catalyst showing bands at 3430, 2868, 2364, 1622, 1379, and 1097, 1013 and 599 cm^{-1}, respectively; (c) XRD pattern; (d) Nitrogen adsorption/desorption isotherms and (e) Pore size distribution. Figure S2: Overview of steps including: (1) Preparation of CS-Cu(II) catalyst particles, (2) recording of particles image by an optical microscope, and (3) measurement of particles for size by Image J software. Table S1: Step-by-step procedure using Image J software for particle size analysis.

Author Contributions: R.C. and P.I. conceived and designed the experiments. A.I. performed the experiments. J.T. and T.C. provided materials and information of the catalyst including catalyst characterization. A.I. and R.C. analyzed the data, A.I. wrote the original draft, and R.C. revised the manuscript. P.W. advised and edited the manuscript. All authors have read and agreed to the published version of the manuscript.

Funding: The authors would like to thank the Faculty of Science, Mahidol University for grants and financial support. Support of equipment from Faculty of Science, Mahidol University and the Center of Excellence for Innovation in Chemistry (PERCH-CIC), Commission on Higher Education, Ministry of Education, and the Office of Higher Education Commission and Mahidol University is gratefully acknowledged. J.T. and T.C. thank the Royal Golden Jubilee Ph.D. Scholarship (PHD/0171/2556). We also acknowledge P. Phoopraintra and A. Obma for their kind help in preparation of some figures.

Conflicts of Interest: The authors declare no conflict of interest.

References

1. Parshetti, G.K.; Suryadharma, M.S.; Pham, T.P.T.; Mahmood, R.; Balasubramanian, R. Heterogeneous catalyst-assisted thermochemical conversion of food waste biomass into 5-hydroxymethylfurfural. *Bioresour. Technol.* **2015**, *178*, 19–27. [CrossRef] [PubMed]
2. Raja, R.; Thomas, J.M.; Greenhill-Hooper, M.; Ley, S.V.; Almeida Paz, F.A. Facile, One-step production of niacin (Vitamin b3) and other nitrogen-containing pharmaceutical chemicals with a single-site heterogeneous catalyst. *Chem. Eur.* **2008**, *14*, 2340–2348. [CrossRef] [PubMed]
3. Lee, M.; Chen, B.Y.; Den, W. Chitosan as a natural polymer for heterogeneous catalysts support: A short review on its applications. *Appl. Sci.* **2015**, *5*, 1272–1283. [CrossRef]
4. Guibal, E. Heterogeneous catalysis on chitosan-based materials: A review. *Prog. Polym. Sci.* **2005**, *30*, 71–109. [CrossRef]
5. Ravi Kumar, M.N.V. A review of chitin and chitosan applications. *React. Funct. Polym.* **2000**, *46*, 1–27. [CrossRef]
6. Ma, J.; Sahai, Y. Chitosan biopolymer for fuel cell applications. *Carbohydr. Polym.* **2013**, *92*, 955–975. [CrossRef]
7. Pillai, C.K.S.; Paul, W.; Sharma, C.P. Chitin and chitosan polymers: Chemistry, solubility and fiber formation. *Prog. Polym. Sci.* **2009**, *34*, 641–678. [CrossRef]
8. Baig, R.B.N.; Nadagouda, M.N.; Varma, R.S. Ruthenium on chitosan: A recyclable heterogeneous catalyst for aqueous hydration of nitriles to amides. *Green Chem.* **2014**, *16*, 2122–2127. [CrossRef]
9. Li, M.; Su, Y.J.; Hu, J.; Geng, H.J.; Wei, H.; Yang, Z.; Zhang, Y.F. Hydrothermal synthesis of porous copper microspheres towards efficient 4-nitrophenol reduction. *Mater. Res. Bull.* **2016**, *83*, 329–335. [CrossRef]
10. Fisch, A.G.; dos Santos, J.H.Z.; Secchi, A.R.; Cardozo, N.S.M. Heterogeneous catalysts for olefin polymerization: Mathematical model for catalyst particle fragmentation. *Ind. Eng. Chem. Res.* **2015**, *54*, 11997–12010. [CrossRef]

11. Sheldon, R.A.; Downing, R.S. Heterogeneous catalytic transformations for environmentally friendly production. *Appl. Catal. A* **1999**, *189*, 163–183. [CrossRef]
12. Ding, Y.B.; Zhu, L.H.; Huang, A.Z.; Zhao, X.R.; Zhang, X.Y.; Tang, H.Q. A heterogeneous Co3O4-Bi2O3 composite catalyst for oxidative degradation of organic pollutants in the presence of peroxymonosulfate. *Catal. Sci. Technol.* **2012**, *2*, 1977–1984. [CrossRef]
13. Fedorczyk, A.; Ratajczak, J.; Kuzmych, O.; Skompska, M. Kinetic studies of catalytic reduction of 4-nitrophenol with NaBH4 by means of Au nanoparticles dispersed in a conducting polymer matrix. *J. Solid State Electrochem.* **2015**, *19*, 2849–2858. [CrossRef]
14. Prusse, U.; Dalluhn, J.; Breford, J.; Vorlop, K.D. Production of spherical beads by JetCutting. *Chem. Eng. Technol.* **2000**, *23*, 1105–1110. [CrossRef]
15. Desai, K.G.H.; Park, H.J. Preparation of cross-linked chitosan microspheres by spray drying: Effect of cross-linking agent on the properties of spray dried microspheres. *J. Microencapsul.* **2005**, *22*, 377–395. [CrossRef] [PubMed]
16. Preibisch, I.; Niemeyer, P.; Yusufoglu, Y.; Gurikov, P.; Milow, B.; Smirnova, I. Polysaccharide-Based Aerogel Bead Production via Jet Cutting Method. *Materials* **2018**, *11*, 1287. [CrossRef]
17. Barreiro-Iglesias, R.; Coronilla, R.; Concheiro, A.; Alvarez-Lorenzo, C. Preparation of chitosan beads by simultaneous cross-linking/insolubilisation in basic pH: Rheological optimisation and drug loading/release behaviour. *Eur. J. Pharm. Sci.* **2005**, *24*, 77–84. [CrossRef]
18. Kramareva, N.V.; Stakheev, A.Y.; Tkachenko, O.P.; Klementiev, K.V.; Grunert, W.; Finashina, E.D.; Kustov, L.M. Heterogenized palladium chitosan complexes as potential catalysts in oxidation reactions: Study of the structure. *J. Mol. Catal. A. Chem.* **2004**, *209*, 97–106. [CrossRef]
19. Makhmalzadeh, B.; Moshtaghi, F.; Rahim, F.; Aakhgari, A. Preparation and evaluation of sodium diclofenac loadedchitosan controlled release microparticles using factorial design. *Int. J. Drug Dev. Res.* **2010**, *2*, 468–475.
20. Xu, S.; Wang, Z.; Gao, Y.; Zhang, S.; Wu, K. Adsorption of Rare Earths (III) Using an efficient sodium alginate hydrogel cross-linked with poly-γ-glutamate. *PLoS ONE* **2015**, *10*, e0124826. [CrossRef]
21. Yang, L.; Jiang, L.; Hu, D.; Yan, Q.; Wang, Z.; Li, S.; Chen, C.; Xue, Q. Swelling induced regeneration of TiO2-impregnated chitosan adsorbents under visible light. *Carbohydr. Polym.* **2016**, *140*, 433–441. [CrossRef] [PubMed]
22. Quadrado, R.F.N.; Fajardo, A.R. Microparticles based on carboxymethyl starch/chitosan polyelectrolyte complex as vehicles for drug delivery systems. *Arab. J. Chem.* **2020**, *13*, 2183–2194. [CrossRef]
23. He, P.; Davis, S.S.; Illum, L. Chitosan microspheres prepared by spray drying. *Int. J. Pharm.* **1999**, *187*, 53–65. [CrossRef]
24. Trojanowicz, M. Flow chemistry in contemporary chemical sciences: A real variety of its applications. *Molecules* **2020**, *25*, 1434. [CrossRef] [PubMed]
25. Ye, R.-P.; Lin, L.; Li, Q.; Zhou, Z.; Wang, T.; Russell, C.K.; Adidharma, H.; Xu, Z.; Yao, Y.-G.; Fan, M. Recent progress in improving the stability of copper-based catalysts for hydrogenation of carbon–oxygen bonds. *Catal. Sci. Technol.* **2018**, *8*, 3428–3449. [CrossRef]
26. Punniyamurthy, T.; Rout, L. Recent advances in copper-catalyzed oxidation of organic compounds. *Coord. Chem. Rev.* **2008**, *252*, 134–154. [CrossRef]
27. Kong, X.; Zhu, H.; Chen, C.; Huang, G.; Chen, Q. Insights into the reduction of 4-nitrophenol to 4-aminophenol on catalysts. *Chem. Phys. Lett.* **2017**, *684*, 148–152. [CrossRef]
28. Chutimasakul, T.; Na Nakhonpanom, P.; Tirdtrakool, W.; Intanin, A.; Bunchuay, T.; Chantiwas, R.; Tantirungrotechai, J. Uniform Cu/chitosan beads as green and reusable catalyst for facile synthesis of imines via oxidative coupling reaction. Unpublished work. 2020.
29. Nasrollahzadeh, M.; Sajadi, S.M.; Rostami-Vartooni, A.; Bagherzadeh, M.; Safari, R. Immobilization of copper nanoparticles on perlite: Green synthesis, characterization and catalytic activity on aqueous reduction of 4-nitrophenol. *J. Mol. Catal. A: Chem.* **2015**, *400*, 22–30. [CrossRef]
30. Chang, Y.-C.; Chen, D.-H. Catalytic reduction of 4-nitrophenol by magnetically recoverable Au nanocatalyst. *J. Hazard. Mater.* **2009**, *165*, 664–669. [CrossRef]
31. Rostami-Vartooni, A.; Nasrollahzadeh, M.; Alizadeh, M. Green synthesis of perlite supported silver nanoparticles using Hamamelis virginiana leaf extract and investigation of its catalytic activity for the reduction of 4-nitrophenol and Congo red. *J. Alloy. Compd.* **2016**, *680*, 309–314.4. [CrossRef]

32. Kumar, A.; Belwal, M.; Maurya, R.R.; Mohan, V.; Vishwanathan, V. Heterogeneous catalytic reduction of anthropogenic pollutant, 4-nitrophenol by Au/AC nanocatalysts. *Mat. Sci. Eng.Technol.* **2019**, *2*, 526–531. [CrossRef]
33. Hasan, K.; Shehadi, I.A.; Al-Bab, N.D.; Elgamouz, A. Magnetic chitosan-supported silver nanoparticles: A heterogeneous catalyst for the reduction of 4-nitrophenol. *Catalysts* **2019**, *9*, 839. [CrossRef]

Sample Availability: Samples of the compounds are not available from the authors.

© 2020 by the authors. Licensee MDPI, Basel, Switzerland. This article is an open access article distributed under the terms and conditions of the Creative Commons Attribution (CC BY) license (http://creativecommons.org/licenses/by/4.0/).

Article

Flow-Based Dynamic Approach to Assess Bioaccessible Zinc in Dry Dog Food Samples

Bruno J. R. Gregório [1], Ana Margarida Pereira [2], Sara R. Fernandes [1], Elisabete Matos [3], Francisco Castanheira [4], Agostinho A. Almeida [1], António J. M. Fonseca [2], Ana Rita J. Cabrita [2] and Marcela A. Segundo [1,*]

1. LAQV, REQUIMTE, Departamento de Ciências Químicas, Faculdade de Farmácia, Universidade do Porto, Rua de Jorge Viterbo Ferreira n 228, 4050-313 Porto, Portugal; bruno.jr.gregorio@gmail.com (B.J.R.G.); saraferns@sapo.pt (S.R.F.); aalmeida@ff.up.pt (A.A.A.)
2. LAQV, REQUIMTE, Instituto de Ciências Biomédicas de Abel Salazar (ICBAS), Universidade do Porto, Rua de Jorge Viterbo Ferreira n 228, 4050-313 Porto, Portugal; amargaridabp@gmail.com (A.M.P.); ajfonseca@icbas.up.pt (A.J.M.F.); arcabrita@icbas.up.pt (A.R.J.C.)
3. SORGAL, Sociedade de Óleos e Rações S.A., Estrada Nacional 109, Lugar da Pardala, 3880-728 S. João Ovar, Portugal; Elisabete.matos@sojadeportugal.pt
4. Alltechaditivos—Alimentação Animal Lda., Parque de Monserrate, Av. Dr. Luis Sá n 9 - Arm. A, 2710-089 Abrunheira, Portugal; fcastanheira@Alltech.com
* Correspondence: msegundo@ff.up.pt; Tel.: +351-22042-8676

Academic Editor: Pawel Koscielniak
Received: 1 March 2020; Accepted: 11 March 2020; Published: 15 March 2020

Abstract: This work proposes a simple and easy-to-use flow-through system for the implementation of dynamic extractions, aiming at the evaluation of bioaccessible zinc and the characterization of leaching kinetics in dry dog food samples. The kinetic profile of Zn extraction was determined by flame atomic absorption spectroscopy and the results were fitted in an exponential function ($R^2 > 0.960$) compatible with a two first-order reactions model. Values of fast leachable Zn ranged from 83 ± 1 to 313 ± 5 mg of Zn per kg of sample, with associated rate constants ranging from 0.162 ± 0.004 to 0.290 ± 0.014 min^{-1}. Similar results were observed compared to the static batch extraction. The percentage of bioaccessible Zn ranged from 49.0 to 70.0%, with an average value of 58.2% in relation to total Zn content. Principal component analysis regarding the variables fast leachable Zn, associated rate constant, total Zn, and market segment, has shown that 84.6% of variance is explained by two components, where the second component (24.0%) presented loadings only for the fast leachable Zn and associated rate constant. The proposed method is suitable for the fast evaluation (<1 h) of leaching kinetics and bioaccessibility in dry dog food.

Keywords: bioaccessibility; dog food; dog nutrition; dynamic extraction; flow analysis; kinetic profile; zinc

---

1. Introduction

Zinc, an essential trace element for dogs, is a component of several metalloenzymes that influence the metabolism of carbohydrates, lipids, proteins and nucleic acids [1]. It is also important for cellular immunity, reproductive and skin function, and wound healing [2,3]. Nowadays, companion animals' tutors look for high quality pet food, that ensure the required energy and nutrients for the healthy growth and life of pets [4]. For zinc in particular, the minimum recommended level in complete dog food is 7.20 mg per 100 g of dry matter for dogs which are three to seven years of age and 8.34 mg per 100 g of dry matter for dogs over seven years of age, with a higher level for puppies (10.00 mg per 100 g of dry matter) [5]. There is also a maximum legal limit established in the EU, corresponding to 22.7 mg per 100 g of dry matter [6].

In this context, during the development of a new compound feed for animals, it is not only important to consider the nutritional guidelines and the legal limits, but also to assess the bioaccessibility and bioavailability of elements like Zn, because their source and the composition of the matrix influences both factors [7–12]. Zinc can be present in dog food as free or inorganic zinc, and also complexed with amino acids (e.g., histidine, methionine, glutamate, and glycine) or with other low-molecular-weight organic molecules (e.g., citrate, ascorbate, picolinate, and propionate) [2]. Bioaccessibility refers to the concentration of the nutrient that is released from the food matrix to the gastrointestinal (GI) tract and is available for absorption [13], being considered the first step towards bioavailability (the fraction that reaches systemic circulation from the GI tract), also representing the maximum value that can be achieved [9,14]. This parameter can be evaluated by using in vivo or in vitro methods [15]. However, in vitro studies present several benefits, such as being faster, less laborious and inexpensive [16,17], without concerns on animal welfare or ethical issues [18]. Consequently, static [15,19–23] and dynamic [9,24–26] in vitro models have been developed to evaluate the bioaccessibility of metals in different types of solid samples, such as food, soil and incineration ashes.

Dynamic methods, where online continuous leaching occurs, have numerous advantages like (i) a minimal sample manipulation, which reduces contamination, (ii) less time required for each extraction [27], (iii) possibility of automation of the extraction procedure [28], (iv) mimicking of naturally-occurring processes that are dynamic and not static [29], (v) reduction of the readsorption effects of the element to the surface of the matrix [30], and (vi) study of the leaching kinetics [24], allowing the discrimination of fast and slow leachable fractions [25,31,32]. Moreover, the constant pumping of extraction solution through the sample drives the dissolution equilibrium to the right, thereby providing information about the maximum amount of bioaccessible analyte [33]. The information obtained is important to help manufacturers choose the best ingredients for their products, taking into account the current legislation and aiming at complying with these limits without compromising the nutritional requirements. This means the formulation of dog food with a high amount of bioaccessible Zn within the legal limits established for total Zn.

The present work reports the development of a new, simple and easy-to-use flow-through system for on-line continuous leaching experiments and its application to the evaluation of bioaccessible Zn in complete dry dog food samples. The results obtained were compared with the static batch procedure. The characterization of the leaching kinetics of this metal was also targeted.

2. Results and Discussion

2.1. Extraction Chamber Configuration

The initial configuration of the extraction chamber (EC) was based on the scheme proposed by Maia et al. [25], using two equal filters to build it. The results obtained showed poor repeatability, with variable extraction profiles for the same sample and an RSD > 40% for the total amount of zinc extracted ($n = 4$). Compared to the previous work, a smaller amount of sample (ca. 35 mg) was used here and the size of the sample particles was larger (0.5 vs 0.25–0.35 mm). Moreover, complete dog foods are composed of several ingredients (cereals, meat, fish, etc.), creating a heterogeneous product (Supplementary Information, Table S1) [34]. The association of these factors may explain the low repeatability obtained.

Since it was not possible to increase the mass of sample in the initial EC, the configuration was changed (Figure 1 and Supplementary Information, Figure S1a). In the new configuration, the sample was placed inside a larger polypropylene disk holder, through the wider opening (Supplementary Information, Figure S1b), and kept in place by the Millex® syringe filter (Figure 1). Initially, 35 mg of sample were placed inside the holder, but non-reproducible results were still obtained, showing variable extraction profiles and RSD > 25% for the total amount of Zn extracted. Then, the amount of sample was increased to 70 mg, so that the lower part of the chamber was filled with it. The results obtained showed good repeatability regarding both the extraction profile and extracted zinc values (RSD < 2%, $n = 2$), due to the decrease of the void volume of the EC. This situation has been observed previously using other extraction chambers when the void volume was filled with a solid diluent (e.g., cellulose) [25]. This new configuration is simple, allowing the use of larger amounts of sample, and the disk holder can be reused in the following experiments, just having to be cleaned with the replacement of the filter membrane.

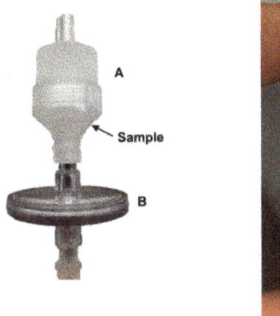

Figure 1. Extraction chamber (EC), composed by a polypropylene disk holder (**A**) with 25 mM of internal diameter containing a Fluoropore™ membrane filter (polytetrafluoroethylene), with a 1.0 µm pore, and by a Millex® syringe filter (polyvinyl chloride housing and polyvinylidene fluoride membrane), with a 5.0 µm pore (**B**). Dry dog food sample is placed inside the EC, as indicated by the arrow, through the wide opening of the disk holder (Supplementary Information, Figure S1).

2.2. Study of the Extraction Procedure

The initial approach for the static batch protocol simulated the digestion in the stomach, followed by the digestion in the intestine. Using this approach, it was not possible to detect any amount of Zn after the gastric phase, since it precipitates in a pH > 6.0, as zinc oxide and/or zinc hydroxides, as reported elsewhere [35], and it was removed by the centrifugation and filtration steps before FAAS analysis. Hence, the following experiment mimicked only the gastric extraction phase, where it was possible to quantify Zn in the solution (Table 1). To demonstrate this rationale, one of the samples was subjected to both gastric and intestinal extractions and, before the centrifugation step, the extract was acidified to pH 2.0 with concentrated HCl. This way, the precipitated Zn was solubilized again, and its quantification was feasible, attaining similar results. Thus, in the following dynamic extraction experiments, only the gastric extraction procedure was performed.

Table 1. Values of bioaccessible Zn [1] using the flow-based dynamic extraction and the static batch extraction.

Sample [2]	Dynamic Extraction	Batch Extraction	
		Gastric	Gastric + Intestinal
Sample #A	107 ± 5	102 ± 7	ND
Sample #B	224 ± 4	194 ± 3	ND
Sample #C	222 ± 4	210 ± 9	ND

[1] Values expressed as mg of Zn per kg of sample, $n = 2$; [2] Economic dry dog food (Sample #A), medium type dry dog food (Sample #B), and premium dry dog food (Sample #C); ND, Not detected.

In order to evaluate the bioaccessibility of Zn, the first step was to study the flow rate of the extraction solution. Flow rates of 0.5, 0.75 and 1.0 mL min^{-1} were tested. The experiment with 1.0 mL min^{-1} led to leakage of the extraction fluid, due to excessive backpressure in the system.

The flow rates of 0.5 and 0.75 mL min^{-1} showed a similar extraction profile (Supplementary Information, Figure S2). Consequently, the flow rate of 0.5 mL min^{-1} was chosen to proceed with sample analysis.

The static batch protocol included the use of the gastric enzyme pepsin [36], so its influence in the dynamic extraction procedure was also assessed (Supplementary Information, Figure S3). The results obtained showed that the difference in the amount of Zn extracted in the presence or absence of pepsin is not statistically significant (paired t-test, $p > 0.05$, $n = 4$). The extraction protocol using pepsin was more complex, longer, and entailed the use of more reagents, leading to fluid leakages and increased system backpressure. So, as the results were similar, this enzyme was not included in the dynamic extraction protocol.

2.3. Zinc Bioaccessibility Assessment

With the purpose of evaluating the applicability of the proposed flow-based dynamic extraction scheme towards the assessment of bioaccessible Zn from dog food and the characterization of the extraction kinetics, three dry dog food samples of different market segments were analyzed via both dynamic and static methods. The results obtained are shown in Table 1. The relative difference values of Zn per kg found for the tested samples after 30 min of extraction ranged from 6% to 16% when compared to gastric batch extraction. Hence, the results have shown that the proposed procedure provided an alternative evaluation of the bioaccessible Zn when compared to the longer and more complex static batch extraction. The dynamic approach also brings the opportunity to perform the extraction in non-exhaustive conditions, as the extraction fluid is constantly renovated.

Then, a total of 14 dry dog food samples were tested by the dynamic extraction procedures, with the results obtained for 4 different samples shown in Figure 2. The extraction profiles of all samples, divided by market segment, are shown in Figure S4 of the Supplementary Information.

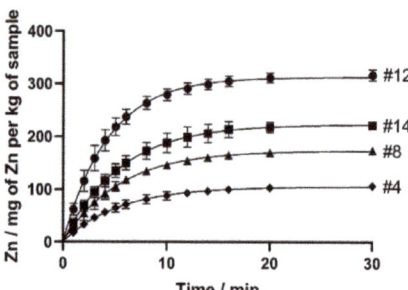

Figure 2. Kinetic profiles of bioaccessible Zn obtained for samples #4, #8, #12, and #14. Sample number is adjacent to the respective curve.

The kinetic extraction provides two types of information: (i) the bioaccessibility of the metal and (ii) the kinetics of the metal leaching [32]. The cumulative amount of bioaccessible Zn extracted at time t (C(t), mg of Zn per kg of sample) is showed to fit an exponential function $C(t) = A \times (1 - e^{-Bt})$ (Table 2, $R^2 > 0.960$ for all experiments). This model is in good agreement with the two first-order reactions model reported by other authors [31,32], for readily extractable compounds. In this mathematical model, A represents the fast-leachable amount of Zn in the sample, and B is the associated rate constant. The values obtained for all tested samples are shown in Table 2.

Values of fast leachable Zn ranged from 83 ± 1 to 313 ± 5 mg per kg of sample, with associated rate constants ranging from 0.162 ± 0.004 to 0.290 ± 0.014 min^{-1}. Globally, 77.1% to 91.5% of leachable Zn was released in the first 10 min of the experiment, showing a fast bioaccessibility in gastric acidic media. Good precision was attained, with RSD values ranging from 0.6% to 2.8% for the values of fast leachable Zn and from 1.7% to 5.9% for the associated rate constants.

water (1:10 dilution). For the sample analysis, 400 µL of each fraction was mixed with 6 µL of nitric acid and diluted in water to a final volume of 3.0 mL (final concentration of nitric acid of 0.2% (v/v)). Blank samples were prepared alongside the test samples by replacing the extracted fraction by gastric extraction juice, and they were used for the correction of the analytical signals.

Absorbance values were measured at a wavelength of 213.86 nm. The values obtained for samples were directly interpolated in the calibration curve of matrix matched Zn standards (mg L^{-1}). The content of Zn in mg per kg was calculated considering the volume of the fraction collected, V (µL) (Supplementary Information, Table S4), and the mass of sample weighed into the disk holder, m(mg), as follows: [Zn] = [(Abs − Intercept) × Dilution factor × V (µL)] / [Slope × M (mg)]. To define the kinetic profile of the extraction process, Zn content was plotted as the cumulative leached value (mg of Zn per kg of sample) as a function of time. Each dry dog food sample was analyzed in two independent experiments, comprising the collection of 13 fractions in each experiment (total number of sampling points was 26). For each sampling point, FAAS analysis was performed in triplicate, totalizing a number of 78 readings for each curve fitting. Precision was estimated as the relative standard deviation, calculated from the standard error and the mean value obtained for parameters A and B after curve fitting to the exponential function $C(t) = A \times (1 - e^{-Bt})$.

3.8. Statistical Analysis

The comparison of the mean value of Zn extracted in the flow-based dynamic extraction using (or not) pepsin was performed by applying a paired t-test. The comparison of the mean value of bioaccessible and total amount of Zn between samples of different market segments was done by applying a one-way ANOVA. The characterization of the leaching kinetics was done by fitting the data via nonlinear regression using a first order mathematical model. These operations were completed using GraphPad Prism 7 software (GraphPad Software, San Diego, CA, USA). Principal component analysis (PCA) was performed using IBM SPSS Statistics 26 software (IBM, Armonk, NY, USA), with a maximum of 25 iterations for convergence for matrix extraction and no rotation.

4. Conclusions

The dynamic extraction protocol proposed here provides a simple, fast and accurate assessment of bioaccessible Zn from dry dog food samples and their leaching kinetics, including both free and molecule-associated forms. This procedure has several advantages when comparing to traditional static batch methods, with fewer steps and more flexibility, as the combination of number of fractions and collection time can be changed, in order to perform the intended characterization of the extraction profile. Additionally, dynamic methods work on non-exhaustive conditions, since the extraction solutions are continuously propelled through the sample, driving the dissolution equilibrium to the right, providing information about the worst-case scenario, and not under equilibrium conditions that do not mimic naturally occurring processes.

The total amount of Zn in samples ranged from 169 ± 9 to 526 ± 14 mg per kg, while the percentage of bioaccessible Zn ranged from 49.0 to 70.0%, with an average of 58.2%. Similar results for bioaccessible Zn were obtained for samples tested using the proposed method and the time-consuming batch method. Moreover, most of the samples presented total Zn levels higher than the EU legal limit, but this value was not surpassed if the bioaccessible Zn content is considered (except for one sample). Despite the limited number of analyzed samples, the average results suggest a higher amount of total Zn in the premium segment, which will provide an absolute higher amount of bioaccessible Zn, because of the similar percentage of leachable Zn in the samples of the different market segments.

Finally, the reusable disk holder is an important improvement compared to previous flow systems proposed for bioaccessibility assessment, since it contributes to a more environmentally sustainable and cost-effective analysis. The extraction chamber proposed could also be utilized to perform extractions of other components in different products, such as other solid food materials, and environmental and pharmaceutical solid samples.

Supplementary Materials: The following are available online. Table S1. Main ingredients present on the tested dry dog food samples according to label information, Table S2. Total variance explained obtained by principal component analysis (PCA), Table S3. Component matrix obtained after the extraction method of the PCA, Table S4. Fraction collection time, fraction volume and total extraction volume for the extraction procedure, Figure S1. (a) Extraction chamber. A and D, polypropylene disk holder; B, O-ring; and C, Fluoropore™ membrane filter (polytetrafluoroethylene) with a 1.0 μm pore. (b) After the assembly of all parts, sample is placed inside the A moiety, through its wider opening, Figure S2. Kinetic profiles of bioaccessible Zn obtained for the dynamic extraction using flow rates of 0.5 mL min^{-1} and 0.75 mL min^{-1}, Figure S3. Comparison of the extraction profiles with and without pepsin, $n = 2$, Figure S4. Kinetic profiles of bioaccessible Zn obtained for all 14 samples, representing different types of market segment: (a) economic dry dog food, (b) medium type dry dog food, and (c) premium dry dog food, $n = 2$, Figure S5. Scree plot obtained by PCA.

Author Contributions: All authors have read and agree to the published version of the manuscript. Conceptualization, E.M., F.C., A.J.M.F., A.R.J.C. and M.A.S.; formal analysis, B.J.R.G., A.M.P. and S.R.F.; data curation, B.J.R.G., A.M.P., S.R.F., A.A.A. and M.A.S.; writing—original draft preparation, B.J.R.G. and M.A.S.; writing—review and editing, E.M., F.C., A.A.A., A.J.M.F., A.R.J.C. and M.A.S.; supervision, A.R.J.C. and M.A.S.; project administration, E.M.; funding acquisition, E.M., F.C., A.J.M.F., A.R.J.C. and M.A.S.

Funding: This work was financed by Project MinDog, funded by Portugal 2020, financed by the European Regional Development Fund (FEDER) through the Operational Competitiveness Program (COMPETE)—reference number 017616. Financial support FCT/MCTES through national funds (UIDB/50006/2020) is also acknowledged. BJR Gregório and SR Fernandes thank FCT and POCH (Programa Operacional Capital Humano) for their PhD grants, SFRH/BD/137224/2018 and SFRH/BD/130948/2017, respectively. AM Pereira thanks FCT, SANFEED Doctoral Programme, Soja de Portugal and Alltech for her PhD grant PD/BDE/114427/2016.

Conflicts of Interest: The authors declare no conflict of interest.

References

1. Case, L.P.; Carey, D.P.; Hirakawa, D.A.; Daristotle, L. *Canine and Feline Nutrition: A Resource for Companion Animal Professionals*; Mosby Inc.: St Louis, MO, USA, 2000; pp. 41–42.
2. Cummings, J.E.; Kovacic, J.P. The ubiquitous role of zinc in health and disease. *J. Vet. Emerg. Crit. Care* **2009**, *19*, 215–240. [CrossRef] [PubMed]
3. Pereira, A.M.; Pinto, E.; Matos, E.; Castanheira, F.; Almeida, A.A.; Baptista, C.S.; Segundo, M.A.; Fonseca, A.J.M.; Cabrita, A.R.J. Mineral composition of dry dog foods: Impact on nutrition and potential toxicity. *J. Agric. Food Chem.* **2018**, *66*, 7822–7830. [CrossRef] [PubMed]
4. Perring, L.; Nicolas, M.; Andrey, D.; Rime, C.F.; Richoz-Payot, J.; Dubascoux, S.; Poitevin, E. Development and validation of an ED-XRF method for the fast quantification of mineral elements in dry pet food samples. *Food Anal. Meth.* **2017**, *10*, 1469–1478. [CrossRef]
5. FEDIAF. *The European Pet Food Industry Federation, Nutritional Guidelines For Complete and Complementary Pet Food for Cats and Dogs*; FEDIAF: Bruxelles, Belgium, 2019.
6. European Comission. Commission Implementing Regulation (EU) 2016/1095 of 6 July 2016 concerning the authorisation of Zinc acetate dihydrate, Zinc chloride anhydrous, Zinc oxide, Zinc sulphate heptahydrate, Zinc sulphate monohydrate, Zinc chelate of amino acids hydrate, Zinc chelate of protein hydrolysates, Zinc chelate of glycine hydrate (solid) and Zinc chelate of glycine hydrate (liquid) as feed additives for all animal species and amending Regulations (EC) No 1334/2003, (EC) No 479/2006, (EU) No 335/2010 and Implementing Regulations (EU) No 991/2012 and (EU) No 636/2013. Available online: https://eur-lex.europa.eu/eli/reg_impl/2016/1095/oj (accessed on 1 March 2020).
7. Gabaza, M.; Shumoy, H.; Muchuweti, M.; Vandamme, P.; Raes, K. Baobab fruit pulp and mopane worm as potential functional ingredients to improve the iron and zinc content and bioaccessibility of fermented cereals. *Innov. Food Sci. Emerg. Technol.* **2018**, *47*, 390–398. [CrossRef]
8. Ramírez-Ojeda, A.M.; Moreno-Rojas, R.; Sevillano-Morales, J.; Cámara-Martos, F. Influence of dietary components on minerals and trace elements bioaccessible fraction in organic weaning food: A probabilistic assessment. *Eur. Food Res. Technol.* **2016**, *243*, 639–650. [CrossRef]
9. Leufroy, A.; Noel, L.; Beauchemin, D.; Guerin, T. Bioaccessibility of total arsenic and arsenic species in seafood as determined by a continuous online leaching method. *Anal. Bioanal. Chem.* **2012**, *402*, 2849–2859. [CrossRef]
10. Intawongse, M.; Dean, J.R. In-vitro testing for assessing oral bioaccessibility of trace metals in soil and food samples. *Trac-Trends Anal. Chem.* **2006**, *25*, 876–886. [CrossRef]

11. Gabaza, M.; Shumoy, H.; Louwagie, L.; Muchuweti, M.; Vandamme, P.; Du Laing, G.; Raes, K. Traditional fermentation and cooking of finger millet: Implications on mineral binders and subsequent bioaccessibility. *J. Food Compos. Anal.* **2018**, *68*, 87–94. [CrossRef]
12. Iturbide-Casas, M.A.; Molina-Recio, G.; Camara-Martos, F. Manganese preconcentration and speciation in bioaccessible fraction of enteral nutrition formulas by cloud point extraction (CPE) and atomic absorption spectroscopy. *Food Anal. Meth.* **2018**, *11*, 2758–2766. [CrossRef]
13. Cardoso, C.; Afonso, C.; Lourenço, H.; Costa, S.; Nunes, M.L. Bioaccessibility assessment methodologies and their consequences for the risk–benefit evaluation of food. *Trends Food Sci. Technol.* **2015**, *41*, 5–23. [CrossRef]
14. Moreda-Piñeiro, J.; Moreda-Piñeiro, A.; Romarís-Hortas, V.; Moscoso-Pérez, C.; López-Mahía, P.; Muniategui-Lorenzo, S.; Bermejo-Barrera, P.; Prada-Rodríguez, D. In-vivo and in-vitro testing to assess the bioaccessibility and the bioavailability of arsenic, selenium and mercury species in food samples. *Trac-Trends Anal. Chem.* **2011**, *30*, 324–345. [CrossRef]
15. Minekus, M.; Alminger, M.; Alvito, P.; Ballance, S.; Bohn, T.; Bourlieu, C.; Carriere, F.; Boutrou, R.; Corredig, M.; Dupont, D.; et al. A standardised static in vitro digestion method suitable for food - an international consensus. *Food Funct.* **2014**, *5*, 1113–1124. [CrossRef] [PubMed]
16. Etcheverry, P.; Grusak, M.A.; Fleige, L.E. Application of in vitro bioaccessibility and bioavailability methods for calcium, carotenoids, folate, iron, magnesium, polyphenols, zinc, and vitamins B(6), B(12), D, and E. *Front. Physiol.* **2012**, *3*, 317. [CrossRef] [PubMed]
17. Lucas-González, R.; Viuda-Martos, M.; Pérez-Alvarez, J.A.; Fernández-López, J. In vitro digestion models suitable for foods: Opportunities for new fields of application and challenges. *Food Res. Int.* **2018**, *107*, 423–436. [CrossRef] [PubMed]
18. Bohn, T.; Carriere, F.; Day, L.; Deglaire, A.; Egger, L.; Freitas, D.; Golding, M.; Le Feunteun, S.; Macierzanka, A.; Menard, O.; et al. Correlation between in vitro and in vivo data on food digestion. What can we predict with static in vitro digestion models? *Crit. Rev. Food Sci. Nutr.* **2018**, *58*, 2239–2261. [CrossRef] [PubMed]
19. van Zelst, M.; Hesta, M.; Alexander, L.G.; Gray, K.; Bosch, G.; Hendriks, W.H.; Du Laing, G.; De Meulenaer, B.; Goethals, K.; Janssens, G.P.J. In vitro selenium accessibility in pet foods is affected by diet composition and type. *Br. J. Nutr.* **2015**, *113*, 1888–1894. [CrossRef]
20. Santos, W.P.C.; Ribeiro, N.M.; Santos, D.; Korn, M.G.A.; Lopes, M.V. Bioaccessibility assessment of toxic and essential elements in produced pulses, Bahia, Brazil. *Food Chem.* **2018**, *240*, 112–122. [CrossRef]
21. Devaraju, S.K.; Thatte, P.; Prakash, J.; Lakshmi, J.A. Bioaccessible iron and zinc in native and fortified enzyme hydrolyzed casein and soya protein matrices. *Food Biotechnol.* **2016**, *30*, 233–248. [CrossRef]
22. Theodoropoulos, V.C.T.; Turatti, M.A.; Greiner, R.; Macedo, G.A.; Pallone, J.A.L. Effect of enzymatic treatment on phytate content and mineral bioacessability in soy drink. *Food Res. Int.* **2018**, *108*, 68–73. [CrossRef]
23. Chi, H.; Zhang, Y.; Williams, P.N.; Lin, S.; Hou, Y.; Cai, C. In vitro model to assess arsenic bioaccessibility and speciation in cooked shrimp. *J. Agric. Food Chem.* **2018**, *66*, 4710–4715. [CrossRef]
24. Lamsal, R.P.; Beauchemin, D. Estimation of the bio-accessible fraction of Cr, As, Cd and Pb in locally available bread using on-line continuous leaching method coupled to inductively coupled plasma mass spectrometry. *Anal. Chim. Acta* **2015**, *867*, 9–17. [CrossRef] [PubMed]
25. Maia, M.A.; Soares, T.R.P.; Mota, A.I.P.; Rosende, M.; Magalhaes, L.M.; Miro, M.; Segundo, M.A. Dynamic flow-through approach to evaluate readily bioaccessible antioxidants in solid food samples. *Talanta* **2017**, *166*, 162–168. [CrossRef] [PubMed]
26. Rosende, M.; Miró, M.; Cerdà, V. Fluidized-bed column method for automatic dynamic extraction and determination of trace element bioaccessibility in highly heterogeneous solid wastes. *Anal. Chim. Acta* **2010**, *658*, 41–48. [CrossRef] [PubMed]
27. Gomes, A.; Furtado, G.d.F.; Cunha, R.L. Bioaccessibility of lipophilic compounds vehiculated in emulsions: Choice of lipids and emulsifiers. *J. Agric. Food Chem.* **2019**, *67*, 13–18. [CrossRef]
28. Cave, M.R.; Rosende, M.; Mounteney, I.; Gardner, A.; Miró, M. New insights into the reliability of automatic dynamic methods for oral bioaccessibility testing: A case study for BGS102 soil. *Environ. Sci. Technol.* **2016**, *50*, 9479–9486. [CrossRef]
29. Rosende, M.; Miró, M. Recent trends in automatic dynamic leaching tests for assessing bioaccessible forms of trace elements in solid substrates. *Trac-Trends Anal. Chem.* **2013**, *45*, 67–78. [CrossRef]
30. Chomchoei, R.; Miró, M.; Hansen, E.H.; Shiowatana, J. Sequential injection system incorporating a micro-extraction column for automatic fractionation of metal ions in solid samples: Comparison of the

extraction profiles when employing uni-, bi-, and multi-bi-directional flow plus stopped-flow sequential extraction modes. *Anal. Chim. Acta* **2005**, *536*, 183–190.
31. Fangueiro, D.; Bermond, A.; Santos, E.; Carapuça, H.; Duarte, A. Kinetic approach to heavy metal mobilization assessment in sediments: Choose of kinetic equations and models to achieve maximum information. *Talanta* **2005**, *66*, 844–857. [CrossRef]
32. Labanowski, J.; Monna, F.; Bermond, A.; Cambier, P.; Fernandez, C.; Lamy, I.; van Oort, F. Kinetic extractions to assess mobilization of Zn, Pb, Cu, and Cd in a metal-contaminated soil: EDTA vs. citrate. *Environ. Pollut.* **2008**, *152*, 693–701. [CrossRef]
33. Rosende, M.; Magalhaes, L.M.; Segundo, M.A.; Miro, M. Assessing oral bioaccessibility of trace elements in soils under worst-case scenarios by automated in-line dynamic extraction as a front end to inductively coupled plasma atomic emission spectrometry. *Anal. Chim. Acta* **2014**, *842*, 1–10. [CrossRef]
34. Kersey, J.H.; Carter, R.A.; Buff, P.R.; Bauer, J.E. Natural pet food: A review of natural diets and their impact on canine and feline physiology. *J. Anim. Sci.* **2014**, *92*, 3781–3791.
35. Top, A.; Cetinkaya, H. Zinc oxide and zinc hydroxide formation via aqueous precipitation: Effect of the preparation route and lysozyme addition. *Mater. Chem. Phys.* **2015**, *167*, 77–87. [CrossRef]
36. Hervera, M.; Baucells, M.D.; González, G.; Pérez, E.; Castrillo, C. Prediction of digestible protein content of dry extruded dog foods: Comparison of methods. *J. Anim. Physiol. Anim. Nutr.* **2009**, *93*, 366–372. [CrossRef] [PubMed]
37. Davies, M.; Alborough, R.; Jones, L.; Davis, C.; Williams, C.; Gardner, D.S. Mineral analysis of complete dog and cat foods in the UK and compliance with European guidelines. *Sci. Rep.* **2017**, *7*, 17107. [CrossRef] [PubMed]

Sample Availability: Samples of the compounds used in this work are available from the authors.

© 2020 by the authors. Licensee MDPI, Basel, Switzerland. This article is an open access article distributed under the terms and conditions of the Creative Commons Attribution (CC BY) license (http://creativecommons.org/licenses/by/4.0/).

Review

Dynamic Flow Approaches for Automated Radiochemical Analysis in Environmental, Nuclear and Medical Applications

Jixin Qiao

Department of Environmental Engineering, Technical University of Denmark, DTU Risø Campus, 4000 Roskilde, Denmark; jiqi@env.dtu.dk; Tel.: +45-4677-5367

Received: 27 February 2020; Accepted: 22 March 2020; Published: 24 March 2020

Abstract: Automated sample processing techniques are desirable in radiochemical analysis for environmental radioactivity monitoring, nuclear emergency preparedness, nuclear waste characterization and management during operation and decommissioning of nuclear facilities, as well as medical isotope production, to achieve fast and cost-effective analysis. Dynamic flow based approaches including flow injection (FI), sequential injection (SI), multi-commuted flow injection (MCFI), multi-syringe flow injection (MSFI), multi-pumping flow system (MPFS), lab-on-valve (LOV) and lab-in-syringe (LIS) techniques have been developed and applied to meet the analytical criteria under different situations. Herein an overall review and discussion on these techniques and methodologies developed for radiochemical separation and measurement of various radionuclides is presented. Different designs of flow systems with combinations of radiochemical separation techniques, such as liquid–liquid extraction (LLE), liquid–liquid microextraction (LLME), solid phase extraction chromatography (SPEC), ion exchange chromatography (IEC), electrochemically modulated separations (EMS), capillary electrophoresis (CE), molecularly imprinted polymer (MIP) separation and online sensing and detection systems, are summarized and reviewed systematically.

Keywords: flow techniques; radionuclides; automation; radiochemical separation; environmental monitoring; nuclear emergency preparedness; radioactive waste characterization; medical isotope production

1. Introduction

Radiochemical analysis of natural and anthropogenic radionuclides plays an important role in (1) radioactivity monitoring in the environment and surroundings of nuclear installations; (2) nuclear emergency preparedness to identify the composition of a radioactive source and evaluate the impact of a nuclear accident/incident; (3) characterization of wastes from operations and decommissioning of nuclear facilities to ensure safe and cost-effective waste management; and (4) medical isotope production to achieve required purity and quality assurance. In all cases, rapid and effective radioanalytical approaches are desirable to cope with the growing demands of improving analytical speed and sample throughput and reducing labor intensity and cost.

Flow analysis is considered as an efficient and universal chemical analysis method, which provides, usually, low sample consumption and possibilities of online sample processing in the flow system by effortless extension of the construction with additional units. Another essential feature of the flow analysis is its automation ability with full control over the fluid flow, volumes, flow rates, timing and detection conditions. This improves the analytical efficiency, provides satisfactory reproducibility and also minimizes human errors [1].

Many radiochemical analyses consist of a series of identical chemical separation steps with little or no variation from sample to sample, which makes them feasible for automation via the implementation of versatile flow techniques. A number of review papers have been published, focusing either on flow techniques for automation of certain radiochemical separation processes [2,3], development of radionuclide sensors [4] or methods for selected radionuclides/sample types [1,5–7]. This work presents the development and application of flow techniques for radiochemical analysis in different situations with focuses on technical design, assembly and performance of the flow systems. The application status, advantages, limitations and future perspectives for exploiting diverse flow systems in radiochemical analysis are critically reviewed. More than 100 publications were extracted mainly from data base websites such as the Web of Science, Google Scholar and Scopus from the 1950s until present, with keywords including "radionuclide," "radiochemical analysis," "automated," "flow technique," etc. Patents are not included for the review.

2. Basic Concept in Flow Analysis

The development of flow analysis laboratory methods began with the research of so-called segmented flow analysis (SFA) conducted in the 1950s by Skeggs [8], followed by significant technical progress in flow injection analysis (FIA) pioneered by Ruzicka and Hansen [9]. A basic FIA fluidic system is typically equipped with one peristaltic pump as the fluid driver, a tubing manifold, an injection valve with an injection loop to load a sample into the system and a detector. A schematic illustration of an FIA system is shown in Figure 1a. In more complex FIA setups there are also different modules for online sample processing incorporated into the flow systems. A sequential injection analysis (SIA) is considered to be a new generation of the FIA method, which, compared to FIA, can be considered more flexible because it introduces bidirectional flow and scales well for handling milliliter size to microliter size samples with precise control of volumes, flow rates and timing. An SIA system, as shown in Figure 1b, typically consists of a syringe pump, a multi-position rotary valve, a tubing manifold with a holding and a reactor coil and a detector.

Over the past 50 years, flow techniques have been prompted in chemical analysis with the development of a number of highly specialized concepts, including multi-commuted flow injection analysis (MCFIA) [10,11], multi-syringe flow injection analysis (MSFIA) [12,13], multi-pumping flow system (MPFS) [14,15] and the recent lab-on-valve (LOV) and lab-in-syringe (LIS) systems [16–18].

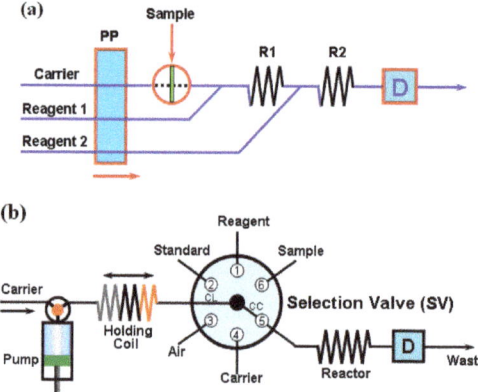

Figure 1. Diagram of a flow injection analysis (FIA) (**a**) and a sequential injection analysis (SIA) system (**b**) [19] (PP—peristaltic pump; R1, R2—reaction coils 1 and 2; D—detector).

3. Application of Flow Techniques in Radiochemical Analysis

The implementation of flow techniques for the determination of radionuclides is a relatively new and not very common field of application. The very first attempt at developing flow systems involving radiometric detection was proposed in the late 1960s for the determination of mercury in biological samples using neutron activation analysis (NAA) [20]. The term radiometric flow injection analysis (RFIA), relating to FIA systems combined with radiometric detectors, was suggested by Myintu et al. [21], and later extended to flow injection radiorelease analysis (FIRRA) and flow injection activation analysis (FIAA) [22]. The first study on RFIA constructed four types of radiometric cells using Geiger–Muller (GM) counters (end-window and liquid-type) and scintillation (NaI (Tl)) counters (cylindrical and well-type) for analysis of ^{131}I and ^{32}P [23]. The well-type scintillation (NaI(Tl)] cell was thereafter applied as a successful FIRRA for vanadate (V) determination by counting radioactive ^{110m}Ag released through redox reaction between VO_3^- and Ag (s) in a micro-column containing ^{110m}Ag labelled silver [24].

Recent testing showed that flow techniques can be used for radiochemical analysis in many situations, as summarized in Table 1, including monitoring of environmental radioactivity for radiological risk assessment and remediation, nuclear emergency preparedness, characterization of radioactive materials in nuclear decommissioning and waste management, and production of radioactive isotopes for medical applications.

Environmental radioactivity monitoring covers both the general environment and the surrounding environment of nuclear installations; e.g., nuclear power plants, nuclear waste storage facilities and disposal sites. In this case, sample types include environmental samples, such as air, precipitation, water, soil, sediment and biota, and effluents (e.g., waste discharges) from nuclear facilities. Environmental radioactivity monitoring focuses on both natural (e.g., ^{210}Po, ^{210}Pb, ^{222}Rn, ^{226}Ra and ^{228}Ra) and artificial (e.g., ^{3}H, ^{14}C, ^{89}Sr, ^{90}Sr and actinides) radionuclides [25]. Typical analytical challenges involved in environmental radioactivity monitoring are trace or ultra-trace levels of radioactivity, large sample volume and large number of samples.

For nuclear emergency preparedness, biological and environmental samples, including milk, urine, air, drinking water and soil are mostly analyzed. Radionuclides often required to be measured in emergency situations include ^{89}Sr, ^{90}Sr, ^{137}Cs, $^{239, 240, 241}$Pu and ^{241}Am [26,27]. The requirement of a rapid response and the unknown composition of radionuclides (interferences) are major challenges in such situations.

For radioactive material characterization in nuclear decommissioning and waste management, constructional and operational materials (e.g., concrete, graphite, steel, ion exchange resin and coolant from nuclear reactors) are typically required to be analyzed for a number of radionuclides (e.g., ^{3}H, ^{14}C, ^{36}Cl, ^{41}Ca, ^{55}Fe, ^{63}Ni, ^{90}Sr, ^{99}Tc, Pu isotopes, ^{241}Am and ^{244}Cm) [28,29]. The large variations in radioactivity levels and sample matrix compositions occur often as challenges in the relevant radiochemical analyses.

In medical isotope production, short-lived radioisotopes (e.g., ^{18}F, ^{64}Cu, ^{99m}Tc, ^{131}I, ^{85}Sr, ^{89}Zr, ^{90}Y, ^{68}Ga, ^{188}Re, ^{213}Bi) are produced in a cyclotron or nuclear reactor for diagnosis and treatment. Thorough radiochemical separation/analysis of the produced radioisotopes from the target materials (e.g., organic solvent or metal foil) is required to ensure their purity [30–34] and to monitor their entry into the environment [35,36].

Table 1. Overview of flow approaches developed for radionuclide determination.

Purpose	Radionuclides	Sample Type	Flow System Design	Sample Processing Mode	Chemical Separation	Measurement Technique	Performance	Ref
Environmental radioactivity monitoring	^{90}Sr	Groundwater (0.35 L)	FI	Single sample	SuperLig 620 column	LSC (Cherenkov counting)	Chemical yields: 99.9 ± 2.8% LOD: 0.057 Bq/L Turnover time: 41.5 h (27 h for ^{90}Y ingrowth and 13.5 h for counting)	[37]
	^{90}Sr	Water, powdered milk, soil (2 mL of sample solution)	SI flow-reversal wetting-film extraction	Single sample	Wetting-film of BCHC in 1-octanol	LBPC *	Applied to measure ^{90}Sr ranging in 0.07–0.30 Bq Chemical yield: up to 80% Precision: <3% RSD (n = 10)	[38]
	^{89}Sr, ^{90}Sr, ^{226}Ra	Milk (1000 mL), water (800 mL)	Semi-automated FI combined with HPLC	Multi-sample (8 samples)	Cation exchange chromatography (16 mL of Dowex 50W-X8) + HPIC (PRP-X400 poly (styrene-divinylbenzene)-sulfonate cation-exchange)	LSC *	Chemical yield: >95% for Sr, ca.100% for Ra MDC: 30 mBq/L of ^{89}Sr, 20 mBq/L of ^{90}Sr, 2 mBq/L of ^{226}Ra Turnover time: 4–5 h	[39]
	^{99}Tc	Groundwater (150 mL)	SI-minicolumn sensor	Single sample	Anion exchange chromatography (AG 4 × 4)	Flow-through scintillation counter	-	[40]
	^{99}Tc	Seawater (50–200 L)	SI	Single sample	Tandem extraction chromatography (two 1.5-mL TEVA columns)	ICP-MS *	Chemical yield: 60–75% LOD (200 L seawater): 7.5 µBq/L of ^{99}Tc Turnover time: 24 h (for a batch sample (n > 4))	[41]
	^{99}Tc	Soil (0.5 g), water (0.1–100 mL)	LOV-SI	Single sample Renewable column	Extraction chromatography (32 mg TEVA resin)	ICP-MS *	Chemical yield: 94–98% LOD: 5 pg of ^{99}Tc Precision: 3.8% (n = 5) Repeatability: 2% (n = 10) Turnover time: 2–5 h	[42]
	^{226}Ra	Leachate from phosphogypsum	LOV-MSFIA	Single sample Renewable column	MnO$_2$ coated on macroporous bead cellulose (0.3 g)	LBPC *	Chemical yield of ^{226}Ra: > 90%	[43]
	226 Ra	Drinking, natural water	LOV-MSFIA	Single sample Renewable column	MnO$_2$ coated on macroporous bead cellulose	LSC * LBPC *	Chemical yield: > 90% MDA: 4 mBq/L (LSC), 20 mBq/L (LBPC) Precision: 1.7% RSD Turnover time: 20 min	[44]

Table 1. Cont.

Purpose	Radionuclides	Sample Type	Flow System Design	Sample Processing Mode	Chemical Separation	Measurement Technique	Performance	Ref
Environmental radioactivity monitoring	^{232}Th, ^{238}U	Sediment, water (sample solution up to 30 mL for U, up to 8 mL for Th)	LOV-MSFIA	Single sample Renewable column	Extraction chromatography (0.03 g UTEVA)	Spectrophotometry with arsenazo-III	LOD: 5.9 ng/L of U, 60 ng/L of Th. Repeatability: 1.6% ($n = 10$) Turnover time: 11–50 min for U, 10–20 min for Th	[45]
	^{238}U	Seawater (10 mL)	FI	Single sample	Styrene-divinylbenzene copolymer resin, Bio-Beads SM-2	Spectrophotometry with Chlorophosphonazo III	Chemical yields: 95–99% LOD: 130 ng/L Turnover time: 2.6 min	[46]
	^{238}U	Soil, sediment, water, phosphogypsum	LIS-MSA-MSFIA	Single sample	LIS-LLME	LWCC spectrophotometry	Chemical yield: close to 100% LOD: 3.2 mg/L. Precision: 3.3% RSD	[47]
	^{238}U	Phosphogypsum, sediment, water	LOV-MSFIA	Single sample Renewable column	Extraction chromatography (0.03 g UTEVA)	Spectrophotometry with arsenazo-III	Chemical yield: > 90% LOD: 10.3 ng/L of U. Repeatability: 1.6% ($n = 10$) Turnover time: 11–50 min	[48]
	^{239}Pu, ^{240}Pu	Soil and sediment (0.5–1 g)	FI	Single sample	Tandem chromatography (0.5 mL Sr resin and 0.17 mL TEVA resin)	ICP-MS	Chemical yield: > 70% LOD: 9.2 mBq of ^{239}Pu, 25 mBq of ^{240}Pu and 0.87 mBq of ^{242}Pu Turnover time: 5 h	[49]
	$^{239+240}$Pu, ^{210}Po, ^{210}Pb	Soil (10 g), phosphogypsum (0.5 g)	FI	Multi-sample (2 samples)	Anion exchange and extraction chromatography (Dowex 1 × 8 resin, 100–200 mesh and Sr resin)	Alpha spectrometry * LSC *	Chemical yield: 87 ± 8% for Pu, 86 ± 6% for ^{210}Pb, 82 ± 6% for ^{210}Po Turnover time (online separation): 4.8 h for ^{210}Po and ^{210}Pb, 5.0 h for Pu	[50]
	^{239}Pu, ^{240}Pu	Seawater (1 L)	FI	Single sample	Co-precipitation and ion exchange *	ICP-MS	LOD: 5 mBq/L Precision: 12% RSD	[51]
	^{239}Pu, ^{240}Pu	Seawater (3–10 L)	FI	Single sample	Tandem chromatography (Sr resin and TEVA resin)	ICP-MS	LOD: 1.5 mBq/L of ^{239}Pu, 1.6 mBq/L of ^{240}Pu Precision: <3.4% RSD ($n = 7$) for ^{239}Pu and <5% RSD ($n = 7$) for ^{240}Pu Turnover time: 4 h	[52]

Table 1. Cont.

Purpose	Radionuclides	Sample Type	Flow System Design	Sample Processing Mode	Chemical Separation	Measurement Technique	Performance	Ref
Environmental radioactivity monitoring	$239+240$Pu, 241Am	Soil, vegetable ashes leachate, urine, blood	MSFIA-MPFS	Single sample	Extraction chromatography (0.08 g TRU)	Low-background proportional counter	Chemical yield: <90% for both Pu and Am LOD: 4 Bq/L Precision: 3% Turnover time (online separation): 40 min.	[53]
	90Sr, 238Pu	Seawater (1 or 10 L)	FI	Single sample	Tandem chromatography (4 or 35 mL Sr resin and 4 or 6 mL TEVA resin)	LSC* Alpha spectrometry*	Chemical yield: 87.8 ± 6.5% for Sr, 62.5 ± 10.4% for Pu Turnover time (online separation): 3.2 h for 1 L seawater, 9.4 h for 10 L seawater	[54]
	237Np	Soil/sediment (1–10 g) and seaweed (20 g)	SI	Nice samples in sequential mode	Anion exchange chromatography (2 mL AG 1 × 4 resin)	ICP-MS*	Chemical yield: 60–70% for Np Turnover time (in-line anion exchange chromatography): <2.5 h	[55]
Environmental radioactivity monitoring, nuclear emergency preparedness	237Np, 239Pu, 240Pu	Soil (10 g) and seaweed (20 g)	SI	Single sample	Extraction chromatography (2 mL TEVA resin)	ICP-MS*	Chemical yield: 80–105% LOD (for 10 g soil): 1.5 mBq/kg of 239Pu, 5.3 mBq/kg of 240Pu, 16 mBq/kg of 237Np Turnover time (in-line extraction chromatography): <1.5 h	[56]
	237Np, 239Pu, 240Pu	Soil/sediment (0.5–100 g) and seaweed (20 g)	SI	Nice samples in sequential mode	Anion exchange chromatography (2 mL AG MP-1M resin)	ICP-MS*	Chemical yield (100 g soil): 85 ± 10% for Pu, 79 ± 10% for Np Turnover time (in-line anion exchange chromatography): <3.5 h	[57]
	239Pu, 240Pu	Soil/sediment (10–200 g), seaweed (20 g), seawater (200 L)	SI	Single sample	Extraction chromatography (2 mL TEVA resin)	ICP-MS*	Chemical yield: 80–105%DFs for U, Th, Hg and Pb: > 10^4. Duration for in-line extraction chromatography: <1.5 h	[58]
	239Pu, 240Pu	Soil/sediment (5–100 g), seaweed (20 g)	SI	Nice samples in sequential mode	Anion exchange chromatography (2 mL AG 1 × 4 resin)	ICP-MS*	Chemical yield: up to 90% Turnover time (in-line anion exchange chromatography): <2.5 h	[59]

Table 1. Cont.

Purpose	Radionuclides	Sample Type	Flow System Design	Sample Processing Mode	Chemical Separation	Measurement Technique	Performance	Ref
Environmental radioactivity monitoring, nuclear safeguards	^{238}U, ^{242}Pu	Urine (1 mL) and tap water (10 L)	FI	Single sample	Co-precipitation and extraction chromatography (TEVA) for water sample *	ICP-MS	LOD: 0.09 fg of ^{238}U and 0.015 fg of ^{242}Pu	[60]
Environmental radioactivity monitoring, nuclear waste management	^{90}Sr	Rain water and reactor coolant	LOV-MSFIA	Single sample Renewable column	Extraction chromatography (0.35 mL Sr resin)	ICP-MS	Chemical yield: 53–100% Turnover time: 16–24 min for 5 mL sample, 60 min for 100 mL sample, 6 h for 1 L sample	[61]
	^{99}Tc	Ground water (250 mL)	FI	Multi-sample (4 samples)	Extraction chromatography (1.4 g TEVA resin)	ICP-MS *	Chemical yield: 96 ± 2% LOD: 0.2 ng/L ^{99}Tc Turnover time: 81 min	[62]
Environmental radioactivity monitoring, nuclear safeguards, radioecology and tracer studies	^{236}U, ^{237}Np, ^{239}Pu, ^{240}Pu	Seawater (10 L)	SI	Single sample	Tandem chromatography (2 mL TEVA resin and 1 UTEVA resin)	ICP-MS * AMS *	Chemical yields: 70–100% Turnover time: 8 h	[63]
Environmental radioactivity monitoring, emergency preparedness, radioecology and tracer studies	^{99}Tc, ^{237}Np, ^{239}Pu, ^{240}Pu, ^{238}U	Seawater (200 L)	FI	Multi-sample (4 samples)	Extraction and anion exchange chromatography (TEVA, AG MP-1M, UTEVA resin)	ICP-MS * AMS *	Chemical yield: 50–70% LOD: 8 µBq/L of ^{99}Tc, 0.26 nBq/L of ^{237}Np, 23 nBq/L of ^{239}Pu, 84 nBq/L of ^{240}Pu and 0.6 µBq/L of ^{238}U Turnover time: 3–4 day	[64]
Medical isotope production	^{89}Zr **	Cyclotron bombarded Y foil	SI	Single sample	Tandem chromatography (AG MP-1 M and hydroxamate resin)	Gamma spectrometry	Chemical yield: 95.1 ± 1.3%	[31]
	^{90}Y	Water, urine and blood	MSFIA coupling online column-based LLE	Single sample	LLME in a column (0.32 mL) containing HDEHP absorbed on C18 (0.11 g)	LBPC *	Chemical yield: 100 ± 2.3% (n = 10). LLD: 5 mBq of ^{90}Y	[35]

Table 1. Cont.

Purpose	Radionuclides	Sample Type	Flow System Design	Sample Processing Mode	Chemical Separation	Measurement Technique	Performance	Ref
Medical isotope production	^{99}Tc	Urine, saliva and hospital residues	LIS-DLLME	Single sample	LIS-DLLME with 22.5% of Aliquat®336 in acetone	LSC *	MDA: 75 mBq Turnover time (extraction): 7.5 min	[36]
	^{99m}Tc	Cyclotron bombarded Mo target	Vacuum pumping flow system	Single sample	Triple tandem chromatography (ABEC-2000, SCX and Al resin)	Gamma spectrometry	Chemical yield: close to 90% Turnover time: 27 ± 2 min	[30]
	^{68}Ga, ^{99m}Tc, ^{188}Re, ^{213}Bi **	Parent radionuclides ^{68}Ge for ^{68}Ga, ^{99}Mo for ^{99m}Tc, ^{188}W for ^{188}Re, ^{225}Ac for ^{213}Bi	SI	Single sample	Tandem chromatography ^{68}Ge/^{68}Ga: 50W × 8 +UTEVA ^{99}Mo/^{99m}Tc: ABEC – 2000 + 50W × 8/Diphonix ^{188}W/^{188}Re: ABEC – 2000 + 50W × 8/Diphonix ^{225}Ac/^{213}Bi: UTEVA + 50W × 8/pre-filter	Gamma spectrometry * LSC *	Chemical yield: 87 ± 3% for ^{213}Bi, 95 ± 1% for ^{68}Ga, 88 ± 2% for ^{99m}Tc and 93 ± 3% for ^{188}Re Turnover time: 19–58 min.	[34]
	^{213}Bi **	Parent radionuclide ^{225}Ac	SI	Single sample	Anion exchange chromatography	–	Chemical yield: 85–93% Turnover time: 6 min.	[33]
Nuclear emergency preparedness	^{89}Sr, ^{90}Sr	Milk	FI	Multi-sample (4 samples)	Cation exchange chromatography (Dowex 50W × 8 – 100) * Extraction chromatography (5 mL Sr resin)	LSC *	Chemical yield: 80% MDA: 0.7 Bq/L of ^{89}Sr, 0.3 Bq/L of ^{90}Sr Precision: 5% RSD Turnover time: <1 day	[26]
	^{237}Np, ^{239}Pu	Urine (0.2–1 L)	LOV-SI	Single sample Renewable column	Extraction chromatography (ca. 300 mg TEVA resin, 100–150 µm)	ICP-MS *	Chemical yield: 88.7 ± 11.6% for Pu, 94.2 ± 2.0% for Np LOD: 1.0–1.5 pg/L for both ^{237}Np and ^{239}Pu Turnover time: 6 h	[65]

92

Table 1. Cont.

Purpose	Radionuclides	Sample Type	Flow System Design	Sample Processing Mode	Chemical Separation	Measurement Technique	Performance	Ref
Nuclear emergency preparedness	^{239}Pu	Urine (1 L)	LOV-SI	Single sample Renewable column	Extraction chromatography (ca. 300 mg TEVA resin, 100–150 μm)	ICP-MS *	Chemical yield: > 90% LOD: 1.0–1.5 pg/L of ^{239}Pu Turnover time: 6 h	[27]
Nuclear safeguards	^{239}Pu, ^{240}Pu, ^{241}Pu, ^{242}Pu, ^{244}Pu isotope ratios	Spiked working solution	FI	Single sample	Electrochemically modulated separation	ICP-MS	LOD: 0.055 fg of ^{239}Pu Precision: 31.1% RSD for ^{239}Pu/^{244}Pu, 14.5% RSD for ^{240}Pu/^{244}Pu, 83.8% RSD for ^{241}Pu/^{244}Pu, 11.2% RSD for ^{242}Pu/^{244}Pu	[66]
	^{90}Sr	Aged nuclear waste samples from the Hanford site	SI	Single sample	Extraction chromatography (0.35 mL Sr resin)	Flow-through LSC	Chemical yield: 94 ± 5%. LOD: 2.62 Bq of ^{90}Sr Turnover time: <40 min.	[67]
	^{90}Sr, ^{241}Am, ^{99}Tc	Aged nuclear wastes	SI	Single sample Renewable column	Extraction chromatography (50 μL Sr resin, TRU resin and TEVA resin)	Flow-through LSC	Chemical yield: 92 ± 2% for ^{90}Sr, 99 ± 5% for ^{99}Tc	[68]
	^{99}Tc	Nuclear waste samples from the Hanford site	SI	Single sample	Extraction chromatography (0.83 mL TEVA, 20–50 μm)	Flow-through LSC	LOD: 2 ng of ^{99}Tc Turnover time: 20–40 min.	[69]
Nuclear waste management	^{99}Tc	Nuclear waste simulant solutions and aged nuclear waste	SI coupling online microwave-assisted sample treatment	Single sample	Anion exchange chromatography (0.83 mL AG MP-1M, 38–75 μm)	Flow-through solid scintillator detector	-	[70]
	^{99}Tc	Nuclear waste simulant solutions and Hanford tank waste sample	SI coupling online microwave-assisted sample treatment	Single sample	Anion exchange column (AG MP-1M)	Flow-through solid scintillator detector	LOD: 23.5 kBq/L of ^{99}Tc Precision: <10% RSD Turnover time: 12.5 min	[71]
	^{99}Tc	Aged nuclear wastes	SI	Single sample Renewable column	Extraction chromatography (212 μL TEVA resin)	Flow-through LSC	LOD: 6 Bq/L Turnover time: 30 min	[72]

Table 1. Cont.

Purpose	Radionuclides	Sample Type	Flow System Design	Sample Processing Mode	Chemical Separation	Measurement Technique	Performance	Ref
Nuclear waste management	^{99}Tc, ^{230}Th, ^{234}Th	Soil (0.25–5 g)	FI	Single sample	Extraction chromatography (ca. 30 mg TEVA resin and ca. 30 mg TRU resin)	ICP-MS	LOD: 11 Bq/kg of ^{99}Tc, 3.7 Bq/kg of ^{230}Th, 0.74 Bq/kg ^{234}Th	[73]
	^{230}Th, ^{233}U, ^{239}Pu, ^{241}Am	Spiked sample solution in 2 M HNO$_3$	FI	Single sample	Extraction chromatography (0.63 mL TRU resin, 20–50 µm)	Flow-through LSC * LSC * Alpha spectrometry *	Chemical yield: up to 102 ± 4% for ^{241}Am up to 101 ± 3% for ^{239}Pu up to 93 ± 4% for ^{233}U up to 88 ± 3% for ^{230}Th	[74]
	^{237}Np, ^{242}Pu	Ground water at Gorleben site	FI	Single sample	Capillary electrophoresis	ICP-MS	LOD: 50 µg/L Turnover time: <15 min	[75]
	^{237}Np, ^{238}Pu, $^{239+240}$Pu, ^{241}Am	Dissolved vitrified nuclear waste	SI	Single sample	Extraction chromatography (0.63 mL TRU resin, 20–50 µm)	ICP-MS	U decontamination factor (for Pu determination): 3.0×10^5	[76]
	^{238}Pu, $^{239+240}$Pu, ^{241}Am, $^{243+244}$Cm, ^{242}Cm	Vitrified glass waste, aged irradiated nuclear fuel and waste from Hanford site	SI	Single sample	Extraction chromatography (0.63 mL TRU resin, 20–50 µm)	Flow-through LSC * LSC * Alpha spectrometry *	Chemical yield: 85% for Pu, 86% for Am	[77]
Radioecology and tracer studies	^{99}Tc	Soil (1–10 g)	FI	Single sample	Tandem chromatography (0.75 mL TEVA resin and 0.17 mL TEVA resin)	ICP-MS	Chemical yield: 63–73% LOD: 50 mBq/L Precision: <4% RSD Turnover time: 3–5 h	[78]
	^{236}U	Seawater (10 L)	FI	Multi-sample (4 samples)	Extraction chromatography (2 mL UTEVA resin, 100–150 µm)	ICP-MS * AMS *	Chemical yield: 80–100% LOD: 6.6×10^{-11} of ^{236}U/^{238}U atomic ratio Turnover time: 4 h	[79]

* Offline separation or measurement. ** The flow system is used for the radionuclide purification. Abbreviations: AMS: accelerator mass spectrometry; DF: decontamination factor; FI: flow injection; ICP-MS: inductively coupled plasma mass spectrometry; HPIC: high performance liquid chromatography; LBPC: low background proportional counter; LIS: lab-in-syringe; LLD: lower limit of detection; LLME: liquid–liquid microextraction; LOD: limit of detection; LOV: lab-on-valve; LSC: liquid scintillation counting; LWCC: long waveguide capillary cell; MSFIA: multi-syringe flow injection analysis; MSA: magnetic-stirring-assisted; MDA: minimum detectable activity; SI: sequential injection.

4. Implementation of Flow Approaches in Radiochemical Analysis

The overall procedure for radionuclide determination is presented schematically in Figure 2. Most gamma emitters can be directly measured by gamma spectrometers after suitable sample preparation (e.g., homogenization and packing). For alpha and beta emitters, so-called difficult-to-measure radionuclides, the analytical procedure can be divided into four steps: initial sample pretreatment, chemical separation/purification, source preparation and detection. Different approaches utilized in each step and their connections with flow approaches are discussed in the context with relevant examples taken from published articles.

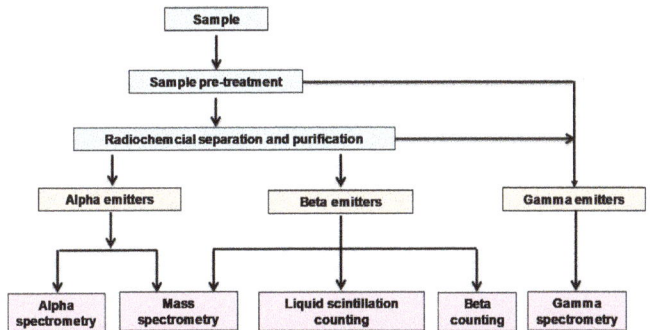

Figure 2. Schematic illustration of the overall procedure for radionuclide determination.

4.1. Sample Pretreatment

Sample pretreatment is necessary to ensure sample homogeneity and appropriate conditions for quantification. Drying, grinding, sieving and ashing are often sequentially performed for solid samples when analyzing non-volatile radionuclides. For volatile radionuclides (e.g., ^{3}H, ^{14}C), fresh samples should be processed without drying and ashing, and for semi-volatile radionuclides (e.g., ^{210}Po) ashing under high temperature should be avoided. For the extraction of most non-volatile radionuclides from solid samples—acid digestion using a mixture of mineral acids in open systems, with a pressure vessel or microwave assistance—is commonly applied [80]. Nevertheless, to ensure a complete release of radionuclides into the aqueous phase, alkaline fusion is often required to totally decompose the sample matrix. For liquid samples, preconcentration is performed either in-situ or in the laboratory. Typically, evaporation, co-precipitation or chelation can be used to remove most sample matrix elements. The evaporation involves reduction of sample volume by careful heating, whereas the selection of co-precipitation or chelation approach depends on the chemical property of the target radionuclide [81].

To study the dynamic release of ^{226}Ra from phosphogypsum (PG), a lab-on-valve multi-syringe flow injection analysis (LOV-MSFIA) system was developed for the fully automated ^{226}Ra lixiviation from PG [43]. The system coupled a homemade cell for online leaching of ^{226}Ra, followed by preconcentration/purification of ^{226}Ra using a renewable sorbent (MnO_2) and its posterior co-precipitation with $BaSO_4$. The $BaSO_4$ co-precipitation was formed by dispensing Na_2SO_4 and acetate buffer/Ba^{+2} into the ^{226}Ra fraction collector.

Online microwave assisted sample pretreatment incorporated in a sequential injection (SI) system was reported for ^{99}Tc determination in nuclear waste [70]. The sample digestion was automatically performed using an open-vessel microwave digestion system. The flow reaction cell in the microwave system was constructed using concave-bottom digestion vessel. The automated fluid-handling system was configured using two syringe pumps equipped with the multi-position distribution valves. A two-way six-port injection valve was used to introduce the sample and two three-position selection valves were used upstream and downstream from the digestion cell to facilitate the delivery of sample/reagents and agitation gas to the reaction cell and uptake of digested sample to further sample purification on an anion exchange column.

Despite numerous advantages offered by flow analysis, it is still rarely implemented in online sample pretreatment for radiochemical analysis. This might be related to the complicated sample pretreatment processes which are difficult to be fulfilled in a fully automated manner in flow systems. As a consequence, there is a lack of commercialized equipment with detailed procedures of such applications provided by manufacturers. This is also a bottle-neck in developing integrated and fully automated flow systems for practical implementation to process samples from their original phases.

4.2. Chemical Separation and Purification

Chemical separation and purification is often necessary for unambiguous and reliable quantification of individual radionuclides. In addition, concentrating analyte and removing matrix/interferences will typically improve sensitivity and detection limits. Individual, group or radionuclide/matrix separations represent an important part of the overall radionuclide determination scheme (Figure 2). Numerous operations in chemical separation and purification can be introduced into flow systems; e.g., liquid–liquid extraction (LLE), liquid–liquid microextraction (LLME), solid phase extraction chromatography (SPEC), ion exchange chromatography (IEC), electrochemically modulated separations (EMS), capillary electrophoresis (CE) and molecularly imprinted polymer (MIP) separation [37,38,41,42,48,52,58,64,69,79,82–89].

4.2.1. Liquid–Liquid Extraction/Microextraction

Liquid–liquid extraction (LLE) is among the oldest of the preconcentration and matrix isolation techniques in analytical chemistry. LLE in a flow-based system can be carried out in a pipette tip, in-syringe, by a pseudo stationary phase or on a coating film consisting of the extractant adhered on an inert support. A flow-reversal wetting-film extraction approach towards the radionuclide separation in an SI system was reported for ^{90}Sr determination in environmental samples [38]. The film coated on the walls of a tubular open reactor for selectively retained strontium ions was composed of 4,4′-(5′)-bis (tetra-butylcyclohexane)-18-crown-6 (BCHC) in 1-octanol. The noteworthy aspects of using a wetting-film phase instead of a solid-phase material are the reduction of crown ether consumption and the simplification of the operational sequence to avoid analyte carryover and reduce the resin capacity factor caused by irreversible interferences. A online LLE process for ^{90}Y determination in environmental and biological samples has been carried out using a column containing di-2-ethylhexylphosphoric acid (HDEHP) adsorbed on a C18 support integrated in an MSFIA system [35]. In this way, the extraction process is carried out in a pseudostationary phase or a coating film which is generated by passing HDHEP solution through the column, and removed by washing the column with 96% ethanol.

To improve the efficiency and cost-effectiveness of conventional LLE, liquid–liquid microextraction (LLME) and dispersive liquid–liquid microextraction (DLLME), among others, have been developed and applied in flow systems for radiochemical analysis. LLME is based on the usage of small volumes of organic solvents as extractants, which leads to high enrichment factors, even with limited sample volumes. A fully automated lab-in-syringe (LIS) LLME method (Figure 3) with magnetic stirring assistance (MSA) and spectrophotometric detection was developed and applied to U determination in environmental samples (soil, sediment, water and phosphogypsum) [47]. Uranium was extracted online and back-extracted with cyanex-272 in dodecane and hydrochloric acid, respectively, prior to reaction with arsenazo-III for the detection. A multisyringe burette coupled to a selection valve was used to implement the whole method, facilitating the U determination in a single instrumental assembly. The LIS technique permitted the simple automation of LLME methods with enhanced reproducibility and the capability of handling small volumes with satisfactory accuracy and precision.

DLLME is a fast microextraction technique based on the use of a ternary mixture, composed by an aqueous phase, an organic phase (extractant) and an additional organic solvent denoted as a disperser solvent, which is miscible in both phases. Extractant and disperser solvent are usually mixed and injected rapidly into the sample, producing a turbulent mixture due to the formation of small droplets of the extractant throughout the aqueous sample, thereby enhancing the effective surface area of extraction. This technique has attracted much attention due to its simplicity and the improved enrichment factors achieved. Furthermore, extraction times are usually short in DLLME since the extraction equilibrium is quickly reached due to the enhanced transfer area for the extraction. An approach exploiting LIS-DLLME for ^{99}Tc extraction and preconcentration from biological samples (urine, saliva and liquid residues from treated patients) has been developed [36]. This system is very simple, comprising an eight-port multiposition selection valve connected to a multisyringe burette equipped with a 5 mL glass syringe. There are many other formats of LLE, yet they not often used in radiochemical analysis; e.g., direct-immersion single-drop microextraction (DI-SDME) and in-drop stirring SDME reported for the determination of nanomolar concentrations of lead using the automated LIS technique [90].

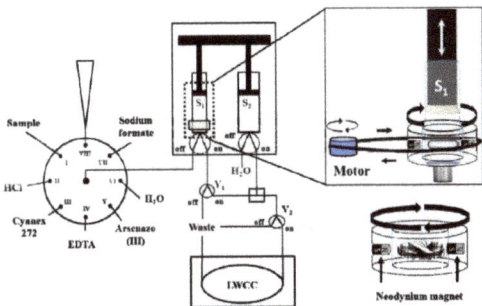

Figure 3. Schematic depiction of a flow system incorporating lab-in-syringe liquid–liquid microextraction (LIS-LLME) with magnetic stirring assistance (MSA) for radionuclide determination [47]. LWCC: liquid waveguide capillary cell, V: solenoid valve, S: syringe.

LLE and LLME offers the advantages of simplicity, flexibility and cost-effectiveness in flow systems for radiochemical analysis; however, they are deemed less selective and often require consecutive extraction, and thereby the analytical process is prolonged. Besides, hazardous organic liquid waste is generated during the analysis. Compared to the rapid development of chromatographic techniques, LLE and LLME are less popular in flow-based radiochemical analysis.

4.2.2. Chromatographic Separation

Single-Column Chromatographic Separation

One of the first attempts to use flow injection based solid phase extraction chromatography (SPEC) for radiochemical separation was by Grate and co-workers [67]. The authors developed an SI system incorporating a Sr resin for determination of ^{90}Sr in nuclear waste samples [67]. Later on they applied the SI system for actinides separated by TRU resin [74,76,77,91] and ^{99}Tc by TEVA resin (Egorov et al., 1998). In the work for ^{99}Tc determination [70,71], ion exchange chromatography (IEC) was also applied using macroporous anion exchange resin AGMP-1M through an implementation of a reversing elution, which ensured an effective separation process in a short time.

For the determination of ^{239}Pu, ^{240}Pu and ^{237}Np in environmental samples, SPEC using TEVA resin has been applied in SI systems to obtain high decontamination factors for interfering radionuclides, especially ^{238}U [56,58]. Simultaneous determination of ^{241}Am and $^{239+240}$Pu was reported by coupling SPEC using TRU resin in an MPFS (Figure 4), which is constituted of a multi-syringe buret equipped with four syringes as flow drivers [53]. Each syringe has a three-way solenoid valve at the head, which facilitates the application of multi-commutation schemes. The developed system was successfully used in analysis of real environmental and biological samples.

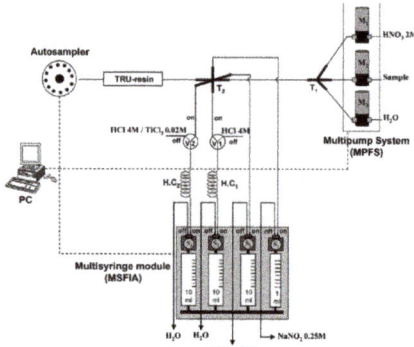

Figure 4. Schematic depiction of a multi-syringe flow injection analysis-multi-pump flow system (MSFIA–MPFS) [53].

Tandem-Column Chromatographic Separation

In some cases, one chromatographic separation is not sufficient to purify the target radionuclides. Therefore, assembly of tandem-column chromatographic separation manifolds is necessary. For example, to improve the purification of ^{99}Tc from large volume seawater samples, an SI method based on the use of two TEVA columns was developed [41]. The system consisted of one syringe pump as a flow driver and five selection valves for flexible connections between the two columns and for the delivery of samples/reagents (see Figure 5). Between the two column separations, a pH adjustment was performed via collecting Tc eluate (in 8 M HNO$_3$) from the first TEVA column into a vial containing NaOH solution, in order to obtain a final solution of 0.1 M HNO$_3$ for loading on the second TEVA column. An FI system for ^{239}Pu and ^{240}Pu determination in environmental samples was developed via tandem SPEC (Sr and TEVA resin) and online inductively coupled plasma mass spectrometry (ICP-MS) detection [49]. Sr resin in the first column was used to remove many interferences, including ^{238}U, from the environmental sample, while the TEVA column was used to further remove ^{238}U from Pu isotopes to eliminate its interference in the ICP-MS measurement.

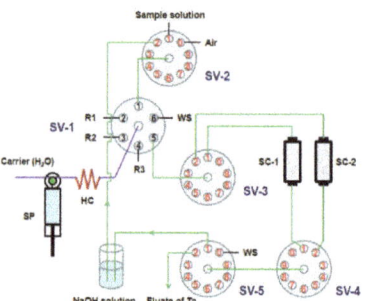

Figure 5. Schematic diagram of the sequential injection (SI) system coupling tandem chromatographic columns [41]. HC: holding coil, R: reagents, SC: separation column, SP: syringe pump, SV: selection valve, WS: waste.

A tandem column purification method was also reported in medical isotope production, such as for the preparation of high-purity ^{89}Zr (IV) oxalate [31] and purification of cyclotron-produced ^{99m}TcO$_4^-$ [30]. In the ^{89}Zr preparation system, the primary column was a microporous, strongly basic anion exchange resin onstyrene divinylbenzene co-polymer, while the secondary column was packed with hydroxamate resin. The ability to transfer ^{89}Zr from one column to the next allows two sequential column clean-up to be performed prior to the final elution of ^{89}Zr (IV) oxalate. In the ^{99m}Tc purification system, triple tandem columns (SPEC packed with ABEC-2000, strong cation exchange (SCX) and aluminum (Al) columns) were applied to ensure a complete separation of $^{99m}TcO_4^-$ from MoO_4^-, wherein a mini-vacuum pump was used as the fluid driver [30].

In many other cases, tandem-column systems provide advantages of sequential separation of multi-radionuclide from the same sample. For example, an SI system coupling a tandem TEVA and UTEVA column was reported for sequential separation of $^{239,240}Pu/^{237}Np$ and ^{236}U in seawater [63]. After loading the sample onto the tandem TEVA/UTEVA column, the two columns were disconnected for further purification of Pu/Np on TEVA and U on UTEVA, respectively. The flexible connection of the two columns was realized via the use of a 10-port two-position injection valve (see Figure 6). An FI system was developed for the separation of ^{238}Pu and ^{90}Sr in seawater with the use of TEVA and Sr resin [54]. The sample was firstly loaded on the tandem TEVA/Sr resin; thereafter the Sr and TEVA column was manually switched in the system for further purification of ^{238}Pu and ^{90}Sr.

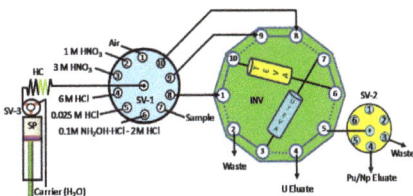

Figure 6. Schematic depiction of a sequential system (SI) incorporating a tandem-column for multi-radionuclide (Pu, Np and U) determination [63]. HC: holding coil, INV: injection vale, SV: selection valve, SP: syringe pump.

In both single and tandem column chromatographic separations, essential problems related to the stability of the separation resins in multiple retention/elution cycles in flow systems were encountered in many works [35,57,58,67,69,77,78,87]. Even though the resin bead's surface can be renewed chemically by washing with, e.g., complexing reagents or weak acids, the limited lifetime of each resin constrains its infinite reuse in the flow systems. Physical or chemical deformations of the resin during the regeneration process will deteriorate its separation performance (capacity, selectivity, etc.), leading to a carry-over effect, and influencing the flow dynamics in the flow system. For example, it was reported that TEVA resin could be reused up to 40 times for analyzing Pu isotopes in environmental soil (10 g), while after 20 times reuse, the flow system was forced to stop due to high backpressure caused by the compression of the TEVA column [58]. Therefore, in all cases of chromatographic separation, repacking columns with fresh separation material is necessary in order to ensure stable analytical performance of the flow systems.

Renewable-Column Chromatographic Separation

Several renewable separation column (RSC) flow systems were developed with the aim of improving the analytical throughput. An RCS-SI system was reported for ^{99}Tc water analysis, wherein the column was packed with a selective scintillating microsphere for absorbing and reacting with ^{99}Tc for online detection [72]. The use of a dual-functional microsphere combined selective sportive and scintillating properties within a single bead. The microsphere in the column was renewed by fluidic replacement of the beads. A multipurpose SI system equipped with an RSC was developed for determination of different radionuclides in nuclear wastes [68]. Depending on the particular target analyte, the RSC was automatically packed with Sr resin for selective separation of ^{90}Sr, TEVA resin for ^{99}Tc or TRU-resin for ^{241}Am. The RSC setup was controlled within a two-position valve, modified with a frit restriction, directly connected to the bottom of the column body (Figure 7).

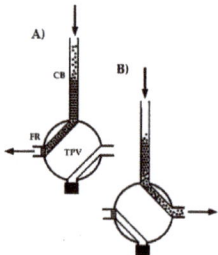

Figure 7. Schematic deposition of a renewable separation column (RSC) using a two valve [68] (**A**) column packing operation. (**B**) Disposal of separation material, CB: column body, FR: frit restriction, TPV: two-position valve.

The LOV concept, introduced in 2000, allied to SIA, has emerged as an appealing downscaled analytical tool and provided more possibilities to renew the separation column in a flow system [17,18]. A number of LOV bead injection (BI) approaches have been applied for determination of actinides [27, 45,48,65], ^{99}Tc [42], ^{226}Ra [44] and ^{90}Sr [88]. The design of the LOV platform is normally based on a multi-port selection valve, where one upper port is connected to the reservoir of the separation material, and one lower port after certain modifications is used directly as the separation column or connected to an extended separation column (Figure 8).

With the use of RSC, the time needed to change the resin and instrument conditions is saved. It provides the possibility for RSC flow systems to perform multi-sample or multi-radionuclide analysis in a consecutive manner with minimized carryover effect. The RSC format is typically miniaturized, favorable for cost-effective and efficient sample processing.

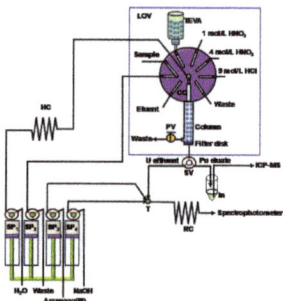

Figure 8. Schematic deposition of lab-on-valve (LOV) sequential injecting system renewable chromatographic separation [27]. HC: holding coil, PV: pinch valve, RC: reaction coil, SP: syringe pump, SV: solenoid valve, T: confluence point.

Multi-Sample Chromatographic Separation

SI approaches coupling SPEC or IEC for processing nine samples in a sequential mode showed high sample throughput for ^{239}Pu, ^{240}Pu and ^{237}Np environmental and biological assays [55,57,59,92]. The system (Figure 9) consists of one syringe pump as the fluid driver and five 10-port selection valves to integrate nine chromatographic columns (TEVA or AG 1 resin). A multi-sample processing FI system was developed for separation of $^{239+240}$Pu, ^{210}Po and ^{210}Pb in environmental samples [50]. The separation was conducted in two parallel lines for two samples, which was respectively applied to Pu with Dowex 1× 8 anion exchange resin, and ^{210}Po and ^{210}Pb with Sr resin in an independent sequence.

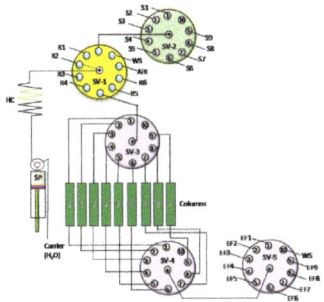

Figure 9. Schematic deposition of a flow system for four sample simultaneous processes [92]. SV: selection valve.

A modular automated radionuclide separator (MARS) has been manufactured and applied to determine ^{99}Tc in groundwater and ^{89}Sr/^{90}Sr in milk samples [26,62]. The separator is capable of processing four samples in parallel with four integrated SPEC columns (TEVA or Sr resin). The separator consists of a four-channel peristaltic pump as the fluid driver, a 6-port selective valve for selecting different reagents, a 5-way flow distribution connector to distribute reagents into the four separation lines, four 2-port selection valves to select sample or reagent delivered to the columns and four 3-way distribution valves to select eluate/waste after the column separation. A multi-sample processing flow system simultaneously handling four samples has also been used for the determination of ^{239}Pu, ^{240}Pu, ^{237}Np, ^{236}U, ^{238}U and ^{99}Tc in seawater [64,79] (Figure 10). The system is more compact via the use of two 12-port injection valves, in the front and bottom end of the chromatographic columns to facilitate the respective selection of sample/reagent and eluate/waste. A semi-automated method handling eight samples in parallel has been reported for monitoring ^{89}Sr, ^{90}Sr and ^{226}Ra in milk and drinking water samples [39]. The method used a 2-stage purification process during which the first purification step using strong cation exchange (SCX) chromatography was performed within an FI system, followed by the second purification using high-performance ion chromatography (HPIC).

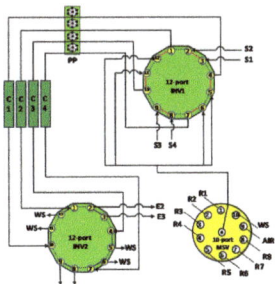

Figure 10. Schematic deposition of a flow system for four-sample simultaneous processing [64,79]. SV: selection valve.

The flow-based multi-sample processing methods alleviate the analytical workloads compared to error-prone and batch-wise manual methods. The application of automation is important for obtaining good analytical repeatability and constant sample throughput. One drawback related to the use of a peristaltic pump in the FI systems could be the aging/deformation of peristaltic pump tubing, resulting in changes of flow rate during operation. Therefore, precalibration of flow rate is necessary each time prior to the analysis, which could be avoided by replacing peristaltic pumps with multichannel syringe pumps in the flow systems.

4.2.3. Other Separation Methods

An FI system employing online electrochemically modulated separations (EMS) was developed for determination of Pu isotope ratios [66]. The flow-through voltammetric cell was used to accumulate Pu by anodic oxidation of Pu(III) to Pu(IV and VI), and then to release them at a controlled potential. Due to more negative potentials being required for U(IV), the separation of Pu from the U interference was possible. Capillary electrophoresis (CE) in combination with ICP-MS has been used for the separation of Pu ions in the oxidation states III-VI and Np ions in the oxidation states IV and V. The method was applied to study the redox behavior of Pu in a natural groundwater rich in humic substances under anaerobic conditions, providing advantages of short separation time and a high separating efficiency [75].

Molecularly imprinted polymers (MIPs) are widely regarded as ideal recognition elements for sensor applications because of their stability, selectivity and affinity [93]. Metal ion imprinting (IIP), based on molecular imprinting technology, is used for preparing materials that can recognize metal ions. Proof-of-concept applications of IIP materials for radionuclide separation have been reported; e.g., the selective removal of ^{60}Co from wastewater [94] and selective extraction of ^{90}Y and ^{152}Eu for medical applications and nuclear power plant monitoring [95]. Yet, MIP as a solid phase extraction (SPE) reactor in a flow setup has not been applied to the radioanalytical field [96,97].

4.3. Detection of Radionuclides

The detection of radionuclides is normally based on quantifying their characteristic radiations, i.e., radiometric methods, or directly courting their atoms, i.e., mass spectrometric methods. In some cases, radionuclides can also be determined spectrophotometrically based on their reactions with complexing agents.

4.3.1. Radiometric Detection

Radiometric detection techniques which have been applied in flow systems include proportional counter, ionization chamber and liquid scintillation counter (LSC) for alpha emitters; Geiger–Müller counters; LSC and Cerenkov cells for beta emitters; and gamma spectrometry for gamma emitters [1]. The first online detection for ^{90}Sr was based on the use of flow-through LSC in an SI system, wherein the purified ^{90}Sr after chromatographic separation was mixed with scintillation liquid and transported to the LSC [67]. Stopped-flow mode LSC detection was reported for online measuring of ^{99}Tc in an SI system [69].

A sensor device integrating LSC has been developed for analysis of ^{99}Tc in groundwater [72]. Dual function sensor beads or the mixture of sorbent (TEVA resin) and scintillator beads were arranged in a mini-column located between the two photo-multiplier (PM) tubes of the scintillation detection system. Upon retention of pertechnetate ions on the resin, the scintillation pulses produced by the radioactive decay of ^{99}Tc are counted. The detection absolute efficiency was 56%, which is sufficiently high for a practical analytical application. This composite bed approach also allows the use of SPE sorbents that can not be readily converted to scintillators by impregnation techniques. A mini-column sensor with the use of a packed bed containing a mixture of anion-exchange resin and scintillating plastic beads was also applied for ^{99}Tc online detection in water [40].

An automated fluid handling system coupled to a Cherenkov radiation detector for measuring ^{90}Sr via the high-energy decay of its daughter, ^{90}Y, has been assembled and applied to Hanford groundwater analysis [37]. A SuperLig 620 column in the system enables preconcentration and separation of ^{90}Sr in the sample, and creates a pure ^{90}Sr source from which subsequent ^{90}Y ingrowth can be measured. ^{90}Y is fluidically transferred from the column to the Cherenkov detection flow cell configured between dual PM tubes for quantification and calculation of the original ^{90}Sr concentration (Figure 11).

A prototype apparatus for at-line/online monitoring of ^{99}Tc in nuclear wastes demonstrated an analytical turnover time of less than 15 min. [71]. The apparatus integrates microwave-assisted sample preparation, anion exchange column separation and detection with a flow scintillation detector in one fully automated sequence. The authors used standard addition method to ensure a matrix-matched measurement to calibrate the process. The ^{99}Tc standard was delivered by a syringe pump to the digestion vessel immediately following sample delivery to the microwave digestion chamber. Standard addition was automatically performed after every sample during research and after every fourth sample during extended monitor operation.

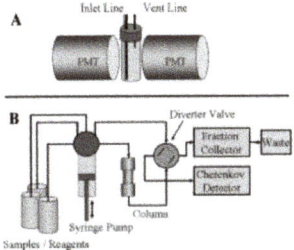

Figure 11. Schematic deposition of a flow-system-coupling Cherenkov detection flow cell configured between dual photo-multiplier tubes (PMTs) [37].

Even though radiometric detection techniques are being willingly introduced to flow analysis, they are mostly applied to online monitoring of high radioactive samples for nuclear waste management. For low-level environmental samples, the radiometric measurements are used mostly offline due to the requirement of relatively long counting time.

4.3.2. Mass Spectrometry

With the rapid development of mass spectrometry, especially ICP-MS, its application for the online detection of radionuclides is being prompted. The unique advantages of using ICP-MS include short analytical time (within several minutes), multi-radionuclides measurement capability and simple source preparation which facilitates complete automation of the determination processes in versatile flow systems.

A number of examples in the literature demonstrate ICP-MS as a widespread detection technique applied in flow analysis [51,52,60,76,78,98,99]. The introduction of FI-ICP-MS in 1986 opened the way to automate classical ICP-MS methods using FI techniques [100]. The first effective attempt to design a flow system coupling ICP-MS for radionuclide analysis was intended for the determination of ^{99}Tc, ^{230}Th and ^{234}Th in soil [73]. The radionuclides were separated from matrix components with TEVA and TRU resins in the FI system and then transported directly to an ICP-MS detector.

The use of a highly specialized ICP-MS instrument enhanced the sensitivity significantly, especially for long-lived radionuclides, such as ^{99}Tc and actinides, allowing for precise trace and ultra-trace level radiochemical analyses. An FI system with online preconcentration/separation with TEVA resin and ICP-MS detection enabled the completion of a Pu urine assay in 15 min. with a detection limit of 0.2 mBq/L for ^{239}Pu and ^{240}Pu [85]. An FI-ICP-MS system developed for ^{99}Tc determination in soil achieved a detection limit of 50 mBq/L ^{99}Tc and an analytical turnover time of 3–5 h [78].

Despite the numerous advantages of ICP-MS mentioned above, some limitations of this method should be noticed, such as sensitivity to spectral and non-spectral interference; e.g., ^{238}U to ^{238}Pu, ^{238}U^{1}H to ^{239}Pu, ^{99}Ru and ^{98}Mo^{1}H to ^{99}Tc. Currently, the majority application of FI-ICP-MS is still limited to long-lived radionuclides; e.g., ^{99}Tc and actinides. For detection of short-lived radionuclides, e.g., ^{90}Sr, a higher detection limit inherent to ICP-MS is a disadvantage compared to radiometric methods. The method detection limit (MDL) for ^{90}Sr has been reported to be 14.5 Bq/L (which was sufficient for ^{90}Sr determination in real nuclear reactor coolant) with online ICP-MS detection connected in a LOV-SI system [61].

4.3.3. Spectrophotometric Measurement

The basis of spectrophotometric measurements is usually the reaction of a determined element with a complexing agent. This leads to formation of a color product, detectable by a conventional spectrophotometer. The most reported flow systems coupling spectrophotometric detection were applied to determination of U and Th. For example, an FI method with online spectrophotometric detection was successfully applied to U determination in seawater [46]. Prior to spectrophotometric detection, the seawater sample injected into the FI system was concentrated with a column packed with styrene-divinylbenzene copolymer resin (Bio-Beads SM-2) modified with dodecylamidoxime. The system provided very high sample throughput (23 per hour) and chemical yield (95% to 99%).

A hyphenated LOV-MSFIA system, coupled to a long path length liquid waveguide capillary cell (LWCC), performed spectrophotometric determination of U and Th in different types of environmental samples [45]. Online separation of U and Th is carried out in a UTEVA column, which is automatically regenerated within the LOV platform. Following the separation, U and Th are spectrophotometrically detected after reaction with arsenazo-III.

4.4. Automation of Flow Systems

During the development of a prototype device or a virtual instrument, the indispensable requirement for the automation of any flow systems relies on expertise in mechanical design and assembly, electronics and software development. The automation of most flow systems presented in this work generally is fulfilled via two approaches: (1) assembly of flow systems based on commercially available integrated flow setups or individual components (e.g., valves and pumps), including electronic control modules, in combination with commercial software for instrument control and data acquisition. (2) The design and assembly of flow systems based on commercial components, in combination with in-house developed electronic boards, instrument controls and data acquisition software.

At present there are several commercialized software for flow systems, including Perkin Elmer AA Winlab [89,101,102], Atlantis [67], FIAlab [56,58,68,69,72,77,87,90], etc. One of the main constraints in most commercial software packages is the high specificity for a certain configuration. To avoid the inconvenience, some in-house developed software based on LabVIEW [26,62,82], LabWindows [37,70,103,104] and Delphi plus Visual C++ (Autoanalysis software) [12,27,36,42–45,47,53,61,105,106] respectively, has been widely applied in the radioanalytical field. In most cases of online radionuclide detection, signals from the detectors (e.g., LSC and ICP-MS) are directly collected and processed by standalone software associated with the detection instruments. It is still not common that radionuclide signals are directly digitized by the computer-based digital oscilloscope card and processed by the same software for controlling the flow system. Besides, most software is prioritized toward mechanical control of the flow systems, whereas an integrated software package controlling the whole radioanalytical procedure covering sample pretreatment, chemical separation, detection and data processing is limited.

4.5. Perspectives on the Future Development of Flow Approaches for Radiochemical Analysis

Table 1 presents an overview of the applications of flow approaches in radiochemical analyses in different fields as well as their analytical performances in terms of chemical yield, limit of detection, repeatability, precision and turnover time. Even though it is not very straightforward to set benchmarks for different flow techniques, as they are applied in different situations with different analytical criteria, most flow approaches demonstrate satisfactory sensitivity, cost-effectiveness, robustness and efficiency under the circumstances presented in the original research. For medical applications, most applications of flow approaches serve as platforms for automated separation and purification of the produced radioisotopes, rather than radiochemical analysis.

It is noted that most applications are focused on "classical" radionuclides with the majority consisting of actinides and fission products (e.g., ^{99}Tc, ^{90}Sr) appearing in environmental monitoring, emergency preparedness and nuclear industry. In recent years, more and more countries launched comprehensive plans for decommissioning of nuclear installations, including nuclear power plants, research reactors and nuclear fuel reprocessing facilities [107–112]. It is apparent that the heavy demands in nuclear decommissioning and waste management require solid technical support for sample characterization to facilitate the categorization and management of radioactive waste. Demands in determining hard-to-measure radionuclides, especially for several newly-appearing long-lived radionuclides (e.g., ^{93}Zr, ^{93}Mo, ^{79}Se, ^{126}Sn, ^{135}Cs) in nuclear decommissioning are becoming more and more notable [113]. To achieve economical and efficient waste characterization, the development of effective flow based radioanalytical methods for wider ranges of radionuclides and sample types (e.g., construction materials of nuclear reactors) is necessary.

In many applications, it is desirable to perform measurements on-site or in-situ using analyzer or sensor instruments. This sets a high requirement for radionuclide analyzers: to perform all the functions carried out in the laboratory rapidly and efficiently in an automated device. Current flow systems developed for radiochemical analysis are primarily laboratory-based, which are not well suited to real-time monitoring. Challenges still exist for creating a promising prospect for applying flow techniques to on-site or in-situ monitoring of radioactive content, such as (1) having an automated sample pretreatment process; (2) creating a highly selective purification approach; and (3) ensuring quick online detection with sufficiently low detection limits via portable detectors. Besides, the compactness, robustness and flexibility of the system are also important to ensure the mobility and applicability in real circumstances. To tackle current limitations in the automation of radiochemical analysis, development of thoughtful flow approaches in combination with state-of-the-art technologies, such as microfluidics, artificial intelligence, big data and neural networks should be considered.

5. Conclusions

Flow analysis is a useful tool in the hands of analysts and it improves the determination of radionuclides by constructing various configurations of flow systems with satisfactory effectiveness. Versatile flow approaches have been utilized in different steps for radiochemical analysis, including sample pretreatment, chemical separation/purification, source preparation and detection. The automation of the analysis leads to improvement of the functional parameters by increasing the reliability of the results and reducing the duration of measurements, and makes the analysis less laborious and safer because of less exposure to radioactivity.

Nevertheless, continuous development of more advanced flow approaches is necessary to cope with the growing demands for radiochemical analysis in different fields, especially nuclear decommissioning, considering not only "classical" but also "emerging" radionuclides and sample types. It is also desirable to develop smart and cost-effective real-time monitoring mobile devices for online chemical processes and detection, with the possibility to transmit results using wireless communication to a central server where the data could be stored and analyzed.

Funding: This research received no external funding.

Acknowledgments: J. Qiao is grateful for the support from all colleagues in the Radioecology and Tracer Studies Section, Department of Environmental Engineering, Technical University of Denmark.

Conflicts of Interest: The authors declare no conflict of interest.

References

1. Kołacińska, K.; Trojanowicz, M. Application of flow analysis in determination of selected radionuclides. *Talanta* **2014**, *125*, 131–145. [CrossRef]
2. Grate, J.W.; Egorov, O.B. Automating analytical separations in radiochemistry. *Anal. Chem.* **1998**, *70*, 779A–788A. [CrossRef]
3. Rodríguez, R.; Avivar, J.; Leal, L.O.; Cerdà, V.; Ferrer, L. Strategies for automating solid-phase extraction and liquid-liquid extraction in radiochemical analysis. *TrAC Trends Anal. Chem.* **2016**, *76*, 145–152. [CrossRef]

4. Grate, J.W.; Egorov, O.B.; O'Hara, M.J.; DeVol, T.A. Radionuclide sensors for environmental monitoring: From flow injection solid-phase absorptiometry to equilibrium-based preconcentrating minicolumn sensors with radiometric detection. *Chem. Rev.* **2008**, *108*, 543–562. [CrossRef]
5. Cerdà, V. Automation of radiochemical analysis by flow techniques -. A review. *TrAC Trends Anal. Chem.* **2019**, 352–367.
6. Trojanowicz, M.; Kołacińska, K.; Grate, J.W. A review of flow analysis methods for determination of radionuclides in nuclear wastes and nuclear reactor coolants. *Talanta* **2018**, *183*, 70–82. [CrossRef]
7. Fajardo, Y.; Avivar, J.; Ferrer, L.; Gómez, E.; Casas, M.; Cerdá, V.; Cerdà, V.; Casas, M. Automation of radiochemical analysis by applying flow techniques to environmental samles. *Trends Anal. Chem.* **2010**, *29*, 1399–1408. [CrossRef]
8. Skeggs, L. An automatic method for colorimetric analysis. *Am. J. Clin. Pathplogy* **1957**, 311–322. [CrossRef]
9. Ruzicka, J.; Hansen, E.H. Flow Injeciton analysis part 1. a new concept of fast continuous flow analysis. *Anal.Chim. Acta* **1975**, *78*, 145–157.
10. Rocha, F.R.P.; Reis, B.F.; Zagatto, E.A.G.; Lima, J.L.F.C.; Lapa, R.A.S.; Santos, J.L.M. Multicommutation in flow analysis: Concepts, applications and trends. *Anal. Chim. Acta* **2002**, *468*, 119–131. [CrossRef]
11. Catalá Icardo, M.; García Mateo, J.V.; Martínez Calatayud, J. Multicommutation as a powerful new analytical tool. *TrAC Trends Anal. Chem.* **2002**, *21*, 366–378. [CrossRef]
12. Cerdà, V.; Estela, J.M.; Forteza, R.; Cladera, A.; Becerra, E.; Altimira, P.; Sitjar, P. Flow techniques in water analysis. *Talanta* **1999**, *50*, 695–705. [CrossRef]
13. Miró, M.; Cerdà, V.; Estela, J.M. Multisyringe flow injection analysis: Characterization and applications. *TrAC Trends Anal. Chem.* **2002**, *21*, 199–210. [CrossRef]
14. Lima, J.L.F.C.; Santos, J.L.M.; Dias, A.C.B.; Ribeiro, M.F.T.; Zagatto, E.A.G. Multi-pumping flow systems. An automation tool. *Talanta* **2004**, *64*, 1091–1098. [CrossRef]
15. Santos, J.L.M.; Ribeiro, M.F.T.; Dias, A.C.B.; Lima, J.L.F.C.; Zagatto, E.E.A. Multi-pumping flow systems: The potential of simplicity. *Anal. Chim. Acta* **2007**, *600*, 21–28. [CrossRef]
16. Ruzicka, J. Lab-on-valve: Universal microflow analyzer based on sequential and bead injection. *Analyst* **2000**, *125*, 1053–1060. [CrossRef]
17. Miró, M.; Hansen, E.H. Recent advances and future prospects of mesofluidic Lab-on-a-Valve platforms in analytical sciences . A critical review. *Anal. Chim. Acta* **2012**, *750*, 3–15. [CrossRef]
18. Yu, Y.L.; Jiang, Y.; Chen, M.L.; Wang, J.H. Lab-on-valve in the miniaturization of analytical systems and sample processing for metal analysis. *TrAC Trends Anal. Chem.* **2011**, *30*, 1649–1658. [CrossRef]
19. Hansen, E.H.; Wang, J. Implementation of suitable flow injection/sequential injection-sample separation/preconcentration schemes for determination of trace metal concentrations using detection by electrothermal atomic absorption spectrometry and inductively coupled plasma mass. *Anal. Chim. Acta* **2002**, *467*, 3–12. [CrossRef]
20. Růžička, J.; Lamm, C.G. Automated determination of traces of mercury in biological materials by substoichiometric radioisotope dilution. *Talanta* **1969**, *16*, 157–168. [CrossRef]
21. Myint, U.; Han, B.; Myoe, K.M.; Kywe, A.; Thida; Tölgyessy, J. Radiometric detection in flow-injection analysis (Radiometric flow-injection analysis). *J. Radioanal. Nucl. Chem.* **1994**, *187*, 117–122. [CrossRef]
22. Myint, U.; Tölgyessy, J. Radiometric flow injection analysis. *J. Radioanal. Nucl. Chem.* **1995**, *191*, 413–426. [CrossRef]
23. Grudpan, K.; Nacapricha, D.; Wattanakanjana, Y. Radiometric detectors for flow-injection analysis. *Anal. Chim. Acta* **1991**, *246*, 325–328. [CrossRef]
24. Grudpan, K.; Nacapricha, D. Flow-injection radiorelease analysis for vanadium. *Anal. Chim. Acta* **1991**, *246*, 329–331. [CrossRef]
25. Qiao, J.; Nielsen, S. Radionuclide Monitoring. In *Encyclopedia of Analytical Chemistry*; Elsevier: Amsterdam, The Netherlands, 2019; Volume 9, pp. 31–39. ISBN 9780124095472.
26. Chung, K.H.; Kim, H.; Lim, J.M.; Ji, Y.Y.; Choi, G.S.; Kang, M.J. Rapid determination of radiostrontium in milk using automated radionuclides separator and liquid scintillation counter. *J. Radioanal. Nucl. Chem.* **2015**, *304*, 293–300. [CrossRef]
27. Qiao, J.; Hou, X.; Roos, P.; Miró, M. Bead injection extraction chromatography using high-capacity lab-on-valve as a front end to inductively coupled plasma mass spectrometry for urine radiobioassay. *Anal. Chem.* **2013**, *85*, 2853–2859. [CrossRef]

28. Hou, X. Radioanalysis of ultra-low level radionuclides for environmental tracer studies and decommissioning of nuclear facilities. *J. Radioanal. Nucl. Chem.* **2019**, 1217–1245. [CrossRef]
29. Croudace, I.W.; Russell, B.C.; Warwick, P.W. Plasma source mass spectrometry for radioactive waste characterisation in support of nuclear decommissioning. A review. *J. Anal. At. Spectrom.* **2017**, *32*. [CrossRef]
30. Morley, T.J.; Dodd, M.; Gagnon, K.; Hanemaayer, V.; Wilson, J.; McQuarrie, S.A.; English, W.; Ruth, T.J.; Bénard, F.; Schaffer, P. An automated module for the separation and purification of cyclotron-produced 99mTcO 4-. *Nucl. Med. Biol.* **2012**, *39*, 551–559. [CrossRef]
31. O'Hara, M.J.; Murray, N.J.; Carter, J.C.; Kellogg, C.M.; Link, J.M. Tandem column isolation of zirconium-89 from cyclotron bombarded yttrium targets using an automated fluidic platform: Anion exchange to hydroxamate resin columns. *J. Chromatogr. A* **2018**, *1567*, 37–46. [CrossRef]
32. Bond, A.H.; Horwitz, E.P.; Hines, J.J.; Young, J.E. A compact automated radionuclide separation system for nuclear medical applications. *Czechoslov. J. Phys.* **2003**, *53*, A717–A723. [CrossRef]
33. Bray, L.A.; Tingey, J.M.; DesChane, J.R.; Egorov, O.B.; Tenforde, T.S. Development of a unique bismuth (Bi-213) automated generator for use in cancer therapy. *Ind. Eng. Chem. Res.* **2000**, *39*, 3189–3194. [CrossRef]
34. McAlister, D.R.; Philip Horwitz, E. Automated two column generator systems for medical radionuclides. *Appl. Radiat. Isot.* **2009**, *67*, 1985–1991. [CrossRef]
35. Fajardo, Y.; Gomez, E.; Garcias, F.; Cerda, V.; Casas, M. Multisyringe flow injection analysis of stable and radioactive yttrium in water and biological samples. *Anal. Chim. Acta* **2005**, *539*, 189–194. [CrossRef]
36. Villar, M.; Avivar, J.; Ferrer, L.; Borràs, A.; Vega, F.; Cerdà, V. Automatic in-syringe dispersive liquid-liquid microextraction of ^{99}Tc from biological samples and hospital residues prior to liquid scintillation counting. *Anal. Bioanal. Chem.* **2015**, *407*, 5571–5578. [CrossRef]
37. O'Hara, M.J.; Burge, S.R.; Grate, J.W. Automated Radioanalytical System for the Determination of Sr-90 in Environmental Water Samples by Y-90 Cherenkov Radiation Counting. *Anal. Chem.* **2009**, *81*, 1228–1237. [CrossRef]
38. Miro, M.; Gomez, E.; Estela, J.M.; Casas, M.; Cerda, V. Sequential injection Sr-90 determination in environmental samples using a wetting-film extraction method. *Anal. Chem.* **2002**, *74*, 826–833. [CrossRef]
39. St-Amant, N.; Whyte, J.C.; Rousseau, M.-E.; Lariviere, D.; Ungar, R.K.; Johnson, S. Radiostrontium and radium analysis in low-level environmental samples following a multi-stage semi-automated chromatographic sequential separation. *Appl. Radiat. Isot.* **2011**, *69*, 8–17. [CrossRef]
40. O'Hara, M.J.; Burge, S.R.; Grate, J.W. Quantification of technetium-99 in complex groundwater matrixes using a radiometric preconcentrating minicolumn sensor in an equilibration-based sensing approach. *Anal. Chem.* **2009**, *81*, 1068–1078. [CrossRef]
41. Shi, K.; Qiao, J.; Wu, W.; Roos, P.; Hou, X. Rapid determination of technetium-99 in large volume seawater samples using sequential injection extraction chromatographic separation and ICP-MS measurement. *Anal. Chem.* **2012**, *84*, 6783–6789. [CrossRef]
42. Rodríguez, R.; Leal, L.; Miranda, S.; Ferrer, L.; Avivar, J.; García, A.; Cerdà, V. Automation of 99Tc extraction by LOV prior ICP-MS detection: Application to environmental samples. *Talanta* **2015**, *133*, 88–93. [CrossRef] [PubMed]
43. Ceballos, M.R.; Borràs, A.; García-Tenorio, R.; Rodríguez, R.; Estela, J.M.; Cerdà, V.; Ferrer, L. 226Ra dynamic lixiviation from phosphogypsum samples by an automatic flow-through system with integrated renewable solid-phase extraction. *Talanta* **2017**, *167*, 398–403. [CrossRef] [PubMed]
44. Rodríguez, R.; Borràs, A.; Leal, L.; Cerdà, V.; Ferrer, L. MSFIA-LOV system for 226Ra isolation and pre-concentration from water samples previous radiometric detection. *Anal. Chim. Acta* **2016**, *911*, 75–81. [CrossRef] [PubMed]
45. Avivar, J.; Ferrer, L.; Casas, M.; Cerdà, V. Smart thorium and uranium determination exploiting renewable solid-phase extraction applied to environmental samples in a wide concentration range. *Anal. Bioanal. Chem.* **2011**, *400*, 3585–3594. [CrossRef] [PubMed]
46. Oguma, K.; Suzuki, T.; Saito, K. Determination of uranium in seawater by flow-injection preconcentration on dodecylamidoxime-impregnated resin and spectrophotometric detection. *Talanta* **2011**, *84*, 1209–1214. [CrossRef]
47. Rodríguez, R.; Avivar, J.; Ferrer, L.; Leal, L.O.; Cerdà, V. Uranium monitoring tool for rapid analysis of environmental samples based on automated liquid-liquid microextraction. *Talanta* **2015**, *134*, 674–680. [CrossRef]

48. Avivar, J.; Ferrer, L.; Casas, M.; Cerdà, V. Lab on valve-multisyringe flow injection system (LOV-MSFIA) for fully automated uranium determination in environmental samples. *Talanta* **2011**, *84*, 1221–1227. [CrossRef]
49. Kim, C.-S.S.K.C.-K.; Kim, C.-S.S.K.C.-K.; Lee, J.-I.I.; Lee, K.-J.J.; Kim CS Lee JI Kim C K, L.K.J. Rapid determination of Pu isotopes and atom ratios in small amounts of environmental samples by an on-line sample pre-treatment system and isotope dilution high resolution inductively coupled plasma mass spectrometry. *J. Anal. At. Spectrom.* **2000**, *15*, 247–255. [CrossRef]
50. Kim, C.-K.; Kim, C.-S.; Sansone, U.; Martin, P. Development and application of an on-line sequential injection system for the separation of Pu, 210Po and 210Pb from environmental samples. *Appl. Radiat. Isot.* **2008**, *66*, 223–230. [CrossRef]
51. Eroglu, A.E.; McLeod, C.W.; Leonard, K.S.; McCubbin, D. Determination of plutonium in seawater using co-precipitation and inductively coupled plasma mass spectrometry with ultrasonic nebulisation. *Spectrochim. Acta Part B At. Spectrosc.* **1998**, *53*, 1221–1233. [CrossRef]
52. Kim, C.; Kim, C.; Lee, K. Determination of pu isotopes in seawater by an on-line sequential injection technique with sector field inductively coupled plasma mass spectrometry. *Anal. Chem.* **2002**, *74*, 3824–3832. [CrossRef] [PubMed]
53. Fajardo, Y.; Ferrer, L.; Gómez, E.; Garcias, F.; Casas, M.; Cerdà, V. Development of an automatic method for americium and plutonium separation and preconcentration using an multisyringe flow injection analysis-multipumping flow system. *Anal. Chem.* **2008**, *80*, 195–202. [CrossRef] [PubMed]
54. Kim, H.; Chung, K.H.; Jung, Y.; Jang, M.; Kang, M.J.; Choi, G.S. A rapid and efficient automated method for the sequential separation of plutonium and radiostrontium in seawater. *J. Radioanal. Nucl. Chem.* **2015**, *304*, 321–327. [CrossRef]
55. Qiao, J.; Hou, X.; Roos, P.; Miró, M. Reliable determination of 237Np in environmental solid samples using 242Pu as a potential tracer. *Talanta* **2011**, *84*, 494–500. [CrossRef]
56. Qiao, J.; Hou, X.; Roos, P.; Miró, M. Rapid and simultaneous determination of neptunium and plutonium isotopes in environmental samples by extraction chromatography using sequential injection analysis and ICP-MS. *J. Anal. At. Spectrom.* **2010**, *25*, 1769. [CrossRef]
57. Qiao, J.; Hou, X.; Roos, P.; Miró, M. High-throughput sequential injection method for simultaneous determination of plutonium and neptunium in environmental solids using macroporous anion-exchange chromatography, followed by inductively coupled plasma mass spectrometric detection. *Anal. Chem.* **2011**, *83*, 374–381. [CrossRef]
58. Qiao, J.; Hou, X.; Roos, P.; Miro, M. Rapid Determination of Plutonium Isotopes in Environmental Samples Using Sequential Injection Extraction Chromatography and Detection by Inductively Coupled Plasma Mass Spectrometry. *Anal. Chem.* **2009**, *81*, 8185–8192. [CrossRef]
59. Qiao, J.; Hou, X.; Roos, P.; Miró, M. Rapid isolation of plutonium in environmental solid samples using sequential injection anion exchange chromatography followed by detection with inductively coupled plasma mass spectrometry. *Anal. Chim. Acta* **2011**, *685*, 111–119. [CrossRef]
60. Schaumloffel, D.; Giusti, P.; Zoriy, M.V.; Pickhardt, C.; Szpunar, J.; LobinSki, R.; Becker, J.S. Ultratrace determination of uranium and plutonium by nano-volume flow injection double-focusing sector field inductively coupled plasma mass spectrometry (nFI?ICP-SFMS). *J. Anal. At. Spectrom.* **2005**, *20*, 17. [CrossRef]
61. Kołacińska, K.; Chajduk, E.; Dudek, J.; Samczyński, Z.; Łokas, E.; Bojanowska-Czajka, A.; Trojanowicz, M. Automation of sample processing for ICP-MS determination of 90Sr radionuclide at ppq level for nuclear technology and environmental purposes. *Talanta* **2017**, *169*, 216–226. [CrossRef]
62. Chung, K.; Choi, S.; Choi, G.; Kang, M. Design and performance of an automated radionuclide separator: Its application on the determination of 99Tc in groundwater. *Appl. Radiat. Isot.* **2013**, *81*, 57–61. [CrossRef] [PubMed]
63. Qiao, J.; Hou, X.; Steier, P.; Golser, R. Sequential injection method for rapid and simultaneous determination of 236U, 237Np, and Pu isotopes in seawater. *Anal. Chem.* **2013**, *85*, 11026–11033. [CrossRef] [PubMed]
64. Qiao, J.; Shi, K.; Hou, X.; Nielsen, S.; Roos, P. Rapid multisample analysis for simultaneous determination of anthropogenic radionuclides in marine environment. *Environ. Sci. Technol.* **2014**, *48*, 3935–3942. [CrossRef] [PubMed]

65. Qiao, J.; Xu, Y.; Hou, X.; Miró, M. Comparison of sample preparation methods for reliable plutonium and neptunium urinalysis using automatic extraction chromatography. *Talanta* **2014**, *128*, 75–82. [CrossRef] [PubMed]
66. Liezers, M.; Lehn, S.A.; Olsen, K.B.; Farmer, O.T.; Duckworth, D.C.; Farmer, O.T., III; Duckworth, D.C. Determination of plutonium isotope ratios at very low levels by ICP-MS using on-line electrochemically modulated separations. *J. Radioanal. Nucl. Chem.* **2009**, *282*, 299–304. [CrossRef]
67. Grate, J.W.; Strebin, R.; Janata, J.; Egorov, O.; Ruzicka, J. Automated analysis of radionuclides in nuclear waste: Rapid determination of 90Sr by sequential injection analysis. *Anal. Chem.* **1996**, *68*, 333–340. [CrossRef]
68. Egorov, O.; O'Hara, M.J.; Grate, J.W.; Ruzicka, J. Sequential injection renewable separation column instrument for automated sorbent extraction separations of radionuclides. *Anal. Chem.* **1999**, *71*, 345–352. [CrossRef]
69. Egorov, O.; O'Hara, M.J.; Ruzicka, J.; Grate, J.W. Sequential injection separation system with stopped-flow radiometric detection for automated analysis of 99Tc in nuclear waste. *Anal. Chem.* **1998**, *70*, 977–984. [CrossRef]
70. Egorov, O.B.; O'Hara, M.J.; Grate, J.W. Microwave-assisted sample treatment in a fully automated flow-based instrument: Oxidation of reduced technetium species in the analysis of total technetium-99 in caustic aged nuclear waste samples. *Anal. Chem.* **2004**, *76*, 3869–3877. [CrossRef]
71. Egorov, O.; O'Hara, M.J.; Grate, J.W. Automated radioanalytical system incorporating microwave-assisted sample preparation, chemical separation, and online radiometric detection for the monitoring of total 99Tc in nuclear waste processing streams. *Anal. Chem.* **2012**, *84*, 3090–3098. [CrossRef]
72. Egorov, O.; Fiskum, S.; O'Hara, M.; Grate, J. Radionuclide Sensors Based on Chemically Selective Scintillating Microspheres: Renewable Column Sensor for Analysis of 99Tc in Water. *Anal. Chem.* **1999**, *71*, 5420–5429. [CrossRef] [PubMed]
73. Hollenbach, M.; Grohs, J.; Mamich, S.; Kroft, M.; Denoyer, E. Determination of technetium-99, thorium-230 and uranium-234 in soils by inductively coupled plasma mass spectrometry using flow injection preconcentration. *J. Anal. At. Spectrom.* **1994**, *9*, 927–933. [CrossRef]
74. Grate, J.W.; Egorov, O.B. Investigation and optimization of on-column redox reactions in the sorbent extraction separation of americium and plutonium using flow injection analysis. *Anal. Chem.* **1998**, *70*, 3920–3929. [CrossRef]
75. Kuczewski, B.; Marquardt, C.M.; Seibert, A.; Geckeis, H.; Kratz, J.V.; Trautmann, N. Separation of plutonium and neptunium species by capillary electrophoresis-inductively coupled plasma-mass spectrometry and application to natural groundwater samples. *Anal. Chem.* **2003**, *75*, 6769–6774. [CrossRef] [PubMed]
76. Egorov, O.; Grate, J.W.; O'Hara, M.J.; Farmer, O.T., III. Extraction chromatographic separations and analysis of actinides using sequential injection techniques with on-line inductively coupled plasma mass spectrometry (ICP MS) detection. *Analyst* **2001**, *126*, 1594–1601. [CrossRef]
77. Grate, J.W.; Egorov, O.B.; Fiskum, S.K. Automated extraction chromatographic separations of actinides using separation-optimized sequential injection techniques. *Analyst* **1999**, *124*, 1143–1150. [CrossRef]
78. Kim, C.; Kim, C.; Rho, B.; Lee, J. Rapid determination of 99Tc in environmental samples by high resolution ICP-MS coupled with on-line flow injection system. *J. Radioanal. Nucl. Chem.* **2002**, *252*, 421–427. [CrossRef]
79. Qiao, J.; Hou, X.; Steier, P.; Nielsen, S.; Golser, R. Method for 236U Determination in Seawater Using Flow Injection Extraction Chromatography and Accelerator Mass Spectrometry. *Anal. Chem.* **2015**, *87*, 7411–7417. [CrossRef]
80. Qiao, J.; Hou, X.; Miró, M.; Roos, P. Determination of plutonium isotopes in waters and environmental solids. A review. *Anal. Chim. Acta* **2009**, *652*, 66–84. [CrossRef]
81. Chen, Q.J.; Hou, X.L.; Yu, Y.X.; Dahlgaard, H.; Nielsen, S.P. Separation of Sr from Ca, Ba and Ra by means of Ca(OH)(2) and Ba(Ra)Cl-2 or Ba(Ra)SO4 for the determination of radiostrontium. *Anal. Chim. Acta* **2002**, *466*, 109–116. [CrossRef]
82. Chung, K.H.; Kang, D.; Rhee, D.S. Design and performance of an automated single column sequential extraction chromatographic system. *J. Radioanal. Nucl. Chem.* **2019**, *321*, 935–942. [CrossRef]
83. Andersson, K.G.; Qiao, J.; Hansen, V. On the requirements to optimise restoration of radioactively contaminated soil areas. In *Contaminated Soils: Environmental Impact, Disposal and Treatment*; Nova Science Publishers: Hauppauge, NY, USA, 2011; pp. 419–432. ISBN 9781607417910.
84. Qiao, J. *Rapid and automated determination of plutonium and neptunium in environmental samples*; Technical University of Denmark: Roskilde, Denmark, 2011; Volume 75.

85. Lariviere, D.; Cumming, T.A.; Kiser, S.; Li, C.; Cornett, R.J.; Larivière, D. Automated flow injection system using extraction chromatography for the determination of plutonium in urine by inductively coupled plasma mass spectrometry. *J. Anal. At. Spectrom.* **2008**, *23*, 352. [CrossRef]
86. Kim, C.S.; Kim, C.K.; Lee, K.J. Simultaneous analysis of 237Np and Pu isotopes in environmental samples by ICP-SF-MS coupled with automated sequential injection system. *J. Anal. At. Spectrom.* **2004**, *19*, 743–750. [CrossRef]
87. Grate, J.W.; Egorov, O.; Fadeff, S.K. Separation-optimized sequential injection method for rapid automated analytical separation of 90Sr in nuclear waste. *Analyst* **1999**, *124*, 203–210. [CrossRef]
88. Rodríguez, R.; Avivar, J.; Ferrer, L.; Leal, L.O.; Cerdà, V. Automated total and radioactive strontium separation and preconcentration in samples of environmental interest exploiting a lab-on-valve system. *Talanta* **2012**, *96*, 96–101. [CrossRef] [PubMed]
89. Benkhedda, K.; Larivière, D.; Scott, S.; Evans, D. Hyphenation of flow injection on-line preconcentration and ICP-MS for the rapid determination of 226Ra in natural waters. *J. Anal. At. Spectrom.* **2005**, *20*, 523.
90. Šrámková, I.H.; Horstkotte, B.; Fikarová, K.; Sklenářová, H.; Solich, P. Direct-immersion single-drop microextraction and in-drop stirring microextraction for the determination of nanomolar concentrations of lead using automated Lab-In-Syringe technique. *Talanta* **2018**, *184*, 162–172. [CrossRef]
91. Egorov, O.; Grate, J.W.; Ruzicka, J. Automation of radiochemical analysis by flow injection techniques. Am-Pu separation using TRU-resin sorbent extraction column. *J. Radioanal. Nucl. Chem.* **1998**, *234*, 231–235. [CrossRef]
92. Qiao, J.; Hou, X.; Roos, P.; Lachner, J.; Christl, M.; Xu, Y. Sequential injection approach for simultaneous determination of ultratrace plutonium and neptunium in urine with accelerator mass spectrometry. *Anal. Chem.* **2013**, *85*, 8826–8833. [CrossRef]
93. Erdem, Ö.; Saylan, Y.; Andaç, M.; Denizli, A. Molecularly Imprinted Polymers for Removal of Metal Ions. An Alternative Treatment Method. *Biomimetics* **2018**, *3*, 38. [CrossRef]
94. Yuan, G.; Tu, H.; Liu, J.; Zhao, C.; Liao, J.; Yang, Y.; Yang, J.; Liu, N. A novel ion-imprinted polymer induced by the glycylglycine modified metal-organic framework for the selective removal of Co(II) from aqueous solutions. *Chem. Eng. J.* **2018**, *333*, 280–288. [CrossRef]
95. Froidevaux, P.; Happel, S.; Chauvin, A.S. Ion-imprinted polymer concept for selective extraction of 90Y and 152Eu for medical applications and nuclear power plant monitoring. *Chimia (Aarau)* **2006**, *60*, 203–206. [CrossRef]
96. Boonjob, W.; Yu, Y.; Miró, M.; Segundo, M.A.; Wang, J.; Cerdà, V. Online Hyphenation of Multimodal Microsolid Phase Extraction Involving Renewable Molecularly Imprinted and Reversed-Phase Sorbents to Liquid Chromatography for Automatic Multiresidue Assays. *Anal. Chem.* **2010**, *82*, 3052–3060. [CrossRef] [PubMed]
97. Dias, A.C.B.; Figueiredo, E.C.; Grassi, V.; Zagatto, E.A.G.; Arruda, M.A.Z. Molecularly imprinted polymer as a solid phase extractor in flow analysis. *Talanta* **2008**, *76*, 988–996. [CrossRef] [PubMed]
98. Becker, J.S.; Dietze, H.-J.J. Application of double-focusing sector field ICP mass spectrometry with shielded torch using different nebulizers for ultratrace and precise isotope analysis of long-lived radionuclides - Invited lecture. *J. Anal. At. Spectrom.* **1999**, *14*, 1493–1500. [CrossRef]
99. Godoy, M.L.D.P.; Godoy, J.M.; Kowsmann, R.; dos Santos, G.M.; Petinatti da Cruz, R. 234U and 230Th determination by FIA-ICP-MS and application to uranium-series disequilibrium in marine samples. *J. Environ. Radioact.* **2006**, *88*, 109–117. [CrossRef]
100. Thompson, J.; Houk, R. Inductively coupled plasma mass spectrometric detection for multielement flow injection analysis and elemental speciation by reversed-phase liquid chromatography. *Anal. Chem.* **1986**, *58*, 2541–2548. [CrossRef]
101. Epov, V.N.; Douglas Evans, R.; Zheng, J.; Donard, O.F.X.; Yamada, M. Rapid fingerprinting of 239Pu and 240Pu in environmental samples with high U levels using on-line ion chromatography coupled with high-sensitivity quadrupole ICP-MS detection. *J. Anal. At. Spectrom.* **2007**, *22*, 1131–1137. [CrossRef]
102. Epov, V.N.; Benkhedda, K.; Cornett, R.J.; Evans, R.D. Rapid determination of plutonium in urine using flow injection on-line preconcentration and inductively coupled plasma mass spectrometry. *J. Anal. At. Spectrom.* **2005**, *20*, 424–430. [CrossRef]

103. Egorov, O.; O'Har, M.J.; Grate, J.W. Equilibration-Based Preconcentrating Minicolumn Sensors for Trace Level Monitoring of Radionuclides and Metal Ions in Water without Consumable Reagents. *Anal. Chem.* **2006**, *78*, 5480–5490. [CrossRef]
104. Egorov, O.B.; O'Hara, M.J.; Grate, J.W. Automated radiochemical analysis of total 99Tc in aged nuclear waste processing streams. *J. Radioanal. Nucl. Chem.* **2005**, *263*, 629–633. [CrossRef]
105. Becerra, E.; Cladera, A.; Cerdà, V. Design of a very versatile software program for automating analytical methods. *Lab. Robot. Autom.* **1999**, *11*, 131–140. [CrossRef]
106. Mar, J. Martinez Multisyringe flow injection spectrophotometric determination of uranium in water samples. *J. Radioanal. Nuclear Chem.* **2009**, *281*, 433–439.
107. Thierfeldt, S.; Schartmann, F. *Decomissioning of Nuclear Installations in Germany*, 3rd ed.; The Federal Ministry of Education and Research: Aachen, Germany, 2010.
108. European Commision. *COMMISSION RECOMMENDATION of 18 December 2003 on Standardised Information on Radioactive Airborne and Liquid Discharges into the Environment from Nuclear Power Reactors and Reprocessing Plants in Normal Operation*; European Commision: Brussels, Belgium, 2004.
109. Nellemann, T. Radiological Characterization and Decommissioning in Denmark. In Proceedings of the Workshop on Radiological Characterisation for Decommissioning, Nykoping, Sweden, 17–19 April 2012; Available online: https://www.oecd-nea.org/rwm/wpdd/rcd-workshop/A-3___OH_Radiological_Characterization_Denmark-Nellemann.pdf.pdf (accessed on 23 March 2020).
110. OECD Radioactive Waste Management Programmes in OECD/NEA Member Countries-Norway. Available online: https://www.oecd-nea.org/rwm/profiles/ (accessed on 23 March 2020).
111. IAEA Country Nuclear Power Profiles-Finland. Available online: https://cnpp.iaea.org/pages/index.htm (accessed on 23 March 2020).
112. Radiological Characterization and Decommissioning in Denmark. Available online: https://www.oecd-nea.org/rwm/wpdd/rcd-workshop/A-3___PAPER_Radiological_Characterization_Denmark-Nellemann.pdf.pdf (accessed on 23 March 2020).
113. Hou, X.; Olsson, M.; Togneri, L.; Englund, S.; Vaaramaa, K.; Askeljung, C.; Gottfridsson, O.; Hirvonen, H.; Ã–hlin, H.; ForsstrÃ¶m, M.; et al. Present status and perspective of radiochemical analysis of radionuclides in Nordic countries. *J. Radioanal. Nucl. Chem.* **2016**, *309*, 1283–1319. [CrossRef]

© 2020 by the author. Licensee MDPI, Basel, Switzerland. This article is an open access article distributed under the terms and conditions of the Creative Commons Attribution (CC BY) license (http://creativecommons.org/licenses/by/4.0/).

Article

An Automated SeaFAST ICP-DRC-MS Method for the Determination of ^{90}Sr in Spent Nuclear Fuel Leachates

Víctor Vicente Vilas [1], Sylvain Millet [1], Miguel Sandow [1], Luis Iglesias Pérez [2], Daniel Serrano-Purroy [1], Stefaan Van Winckel [1] and Laura Aldave de las Heras [1,*]

[1] European Commission, Joint Research Centre, Directorate for Nuclear Safety and Security, D-76125 Karlsruhe, Germany; Victor.VICENTE-VILAS@ec.europa.eu (V.V.V.); Sylvain.MILLET@ec.europa.eu (S.M.); Miguel.SANDOW@ec.europa.eu (M.S.); Daniel.SERRANO-PURROY@ec.europa.eu (D.S.-P.); Stefaan.VAN-WINCKEL@ec.europa.eu (S.V.W.)

[2] Karlsruhe Institute for Technology, Institute for Nuclear Waste Disposal, D-76021 Karlsruhe, Germany; luis.iglesias-perez@kit.edu

* Correspondence: Laura.ALDAVE-DE-LAS-HERAS@ec.europa.eu; Tel.: +49-7247-951-357

Academic Editor: Pawel Koscielniak
Received: 26 February 2020; Accepted: 18 March 2020; Published: 21 March 2020

Abstract: To reduce uncertainties in determining the source term and evolving condition of spent nuclear fuel is fundamental to the safety assessment. ß-emitting nuclides pose a challenging task for reliable, quantitative determination because both radiometric and mass spectrometric methodologies require prior chemical purification for the removal of interfering activity and isobars, respectively. A method for the determination of ^{90}Sr at trace levels in nuclear spent fuel leachate samples without sophisticated and time-consuming procedures has been established. The analytical approach uses a commercially available automated pre-concentration device (SeaFAST) coupled to an ICP-DRC-MS. The method shows good performances with regard to reproducibility, precision, and LOD reducing the total time of analysis for each sample to 12.5 min. The comparison between the developed method and the classical radiochemical method shows a good agreement when taking into account the associated uncertainties.

Keywords: nuclear waste; spent nuclear fuel; ß-emitting nuclides; ^{90}Sr; flow injection; ICP-DRC-MS

1. Introduction

To ensure long-term safety of spent nuclear fuel, deep geologic repository is the most accepted solution by the scientific community [1]. Within this concept, the spent fuel matrix is the first barrier in case of a canister breakage [2]. Once the groundwater enters into contact with the spent fuel, the radionuclides will be released into the geosphere [2]. The Instant Release Fraction (IRF) comprises the radionuclides segregated during the irradiation and with faster dissolution rates than the matrix and constitutes one of the main sources of radiological risks for the geological repository [3]. Some of the released nuclides in this IRF are both long-lived and geochemically mobile [4].

Gamma-emitters can be easily and selectively determined using gamma-spectroscopy, but there is a lack of analytical methods for the pure ß-emitting radionuclides. The standard radiochemical methods are time-consuming and there is an urgent need for faster methods. The use of ICP-MS in nuclear decommissioning has been widely discussed in the literature [5,6] and is the preferred technique for long-lived radionuclides due to its analytical characteristics [7]. However the fact that complex matrix may create non-spectral interferences in the plasma, may limit application and makes matrix removal steps necessary [8–11]. Therefore, an efficient sample preparation protocol is needed in order to remove other elements present in the sample attaining a final solution with low total dissolved solids. The improvement of analytical procedures, involving an enhancement of selectivity and sensitivity, and shortening the time of analysis are required for the determination of radionuclides usually present at ultra-trace levels and in complex matrices.

^{90}Sr is a hard to measure fission product and is of great interest due to its toxicity and high energy emission. ^{90}Sr is an IRF radionuclide as it is faster released than the uranium fuel matrix [12,13]. Depending on the matrix composition of the sample, isobaric interferences by molecular or atomic ions can be expected at m/z 90 affecting the detection limit, accuracy and precision of the determination of ^{90}Sr by ICP-MS. Complete removal of matrix-related interferences below background intensity is necessary to accurately quantify ^{90}Sr [14]. Low instrumental detection limits ($< $ pg·g^{-1}) are required because the high natural abundance of stable Sr, present in aqueous samples, limits pre-concentration [14].

Pre-concentration procedures including separation are efficient solutions to overcome matrix effects and, at the same time, improve sensitivity, selectivity, and precision of the measurement. The use of flow injection methods has been proposed to overcome the tedious, time-consuming and intensive radiochemical procedures [6,15–19]. Flow Injection automated pre-concentration systems such as SeaFAST (ESI, Omaha, NE, USA) coupled to an ICP-MS have been successfully used [20,21] for the determination of trace metals in seawater and have demonstrated their reliability in trace metal analysis [22].

The aim of this work is to develop a rapid, selective, and sensitive method for the separation and pre-concentration of ^{90}Sr in nuclear waste samples. The analytical approach uses a commercially available automated pre-concentration device (SeaFAST) coupled to an ICP-DRC-MS combining the use of a Sr-specific resin and the reaction with oxygen as reaction gas in a dynamic reaction cell (DRC) of the ICP-MS. The method has been applied for the determination of ^{90}Sr at trace levels in nuclear spent fuel leachate samples without sophisticated and time-consuming procedures.

2. Results

2.1. Analytical Parameters of the SeaFAST System with ICP-DRC-MS Method

Different working ranges were tested to find the best operational conditions for the separation and preconcentration of ^{90}Sr within the proposed SeaFAST system. The best results, in terms of linearity and target concentration expected in the SNF leachates, were obtained for a working range of 5 to 40 pg g^{-1}.

A series of standard solutions of 4 mol L^{-1} HNO$_3$ containing 5 to 40 pg g^{-1} of ^{90}Sr were analysed using the SeaFAST system with ICP-DRC-MS within the optimal conditions. The elution profiles obtained are reported in Figure 1. The result is a time resolved signal with 4 zones corresponding to the 4 position of the valves. The direct and pre-concentrated measurements correspond to B and C, respectively. The direct measurement (B) correlates the signal plateau height with concentration, whereas the pre-concentration measurement (C) is correlated with the signal's area. The direct mode was used to monitor the stability of the sample signals and the correct filling of the sample loop and syringes. The ^{90}Sr is eluted in a total time of 420 s and the analysis time for each sample (from injection to detection) is 12.5 min. This is significantly shorter than the conventional radiometric methods for measuring ^{90}Sr, which requires at least 14 days.

The peak of ^{90}Sr in pre-concentration mode shows a fronting profile likely due to the injected volume used in the proposed method. Indeed, the peak fronting is significantly less prominent when a sample coil of 300 μL is used. However, due to the dose constraints handling radioactive samples, which require important dilution factors allowing the transfer from the hot cell to a glove-box, a 2 mL sample coil is needed. The behavior of the fronting peak seems always the same and constant within the interval of concentration used and in all samples tested having no significant influence on the final result.

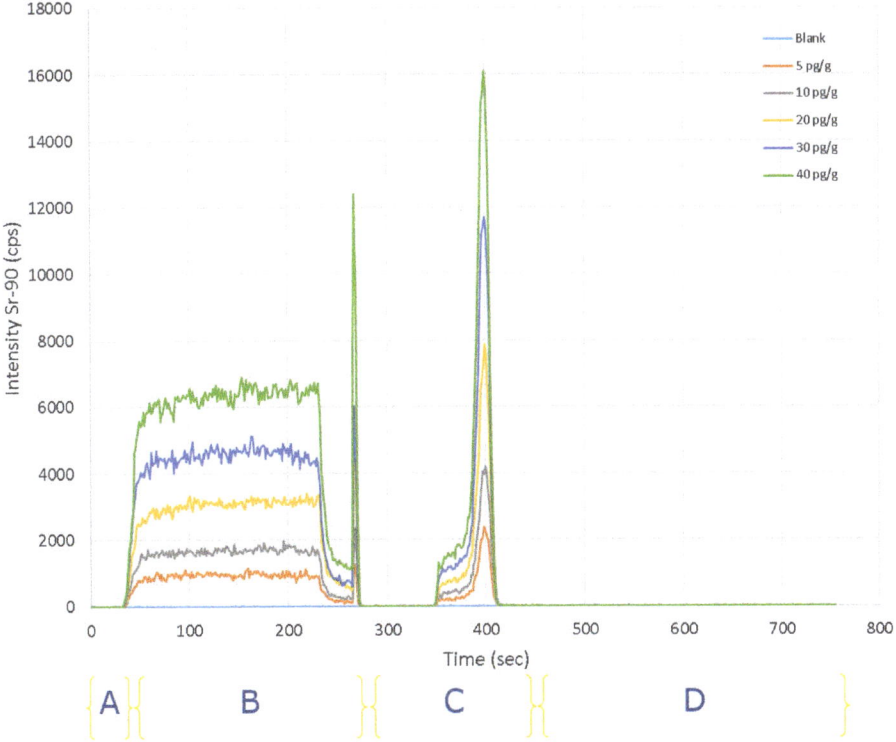

Figure 1. SeaFAST system with ICP-DRC-MS time resolved signal of the different ^{90}Sr standards solutions. (**A**) sample load to the loops; (**B**) direct mode; (**C**) pre-concentration mode; (**D**) cleaning and pre-conditioning.

A calibration curve was obtained by using the peak area of ^{90}Sr versus the total ^{90}Sr concentration. Linear regression was calculated using the Least squares linear regression method. The calibration curve is shown in Figure 2, the method shows good linearity in the concentration interval of 5 to 40 pg g^{-1} and is suitable for the quantitative determination of ^{90}Sr using the experimental conditions described in Section 4.5.

The main analytical parameters of the proposed method are summarized in Table 1.

Table 1. Main analytical parameters of the SeaFAST system with ICP-DRC-MS method.

Parameter	Value
Detection limit	0.02 pg
Quantification limit	0.05 pg
Regression coefficient	0.998
Repeatability ($n = 3$)	<1%
Intermediate precision	<2%
Linear working range	5 to 40 pg/g
Sensitivity	3500 cps/pg
Injection volume	2 mL
Recovery	>90%

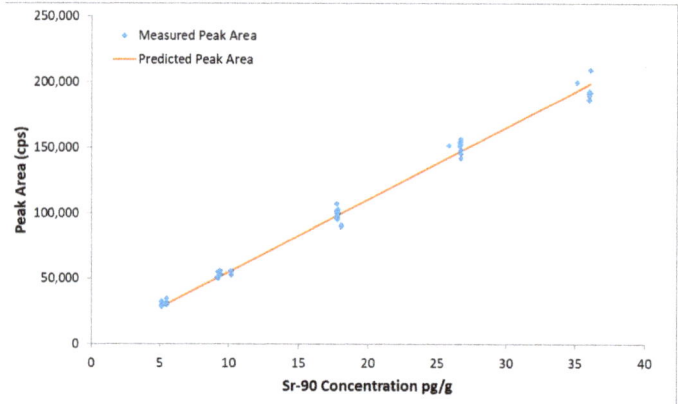

Figure 2. Liner regression for ^{90}Sr. The interval of concentration is from 5 to 40 pg g^{-1}. Blue diamonds are the data points included to calculate the linear regression. The line represents predicted values.

The detection limit of ^{90}Sr was calculated by means of repeated measurements of the blank and according to Currie [23]. The detection limit is 0.01 pg g^{-1} (that represents an absolute amount of 0.02 pg of ^{90}Sr. The repeatability of the method, based on the relative standard deviation of the peak area calculated on the basis of three repetitions is always less than 1% in this concentration interval.

The intermediate precision (within-lab reproducibility) of the method, determined from results obtained on different working days with different columns and standard solutions, is around 2% for the concentration interval studied.

Nuclear spent fuel leachate analogues with a uranium concentration from 1×10^{-7} to 10^{-5} mol L^{-1} and spiked with different ^{90}Sr concentrations were analyzed using the proposed method. Recoveries were satisfactory > 90% (±3%) in all cases. In previous experimental tests using the SeaFAST with a HR ICP-MS, the comparison of standard ^{90}Sr solutions in 4 mol L^{-1} HNO$_3$ and nuclear spent fuel leachate analogues with a uranium concentration from 1×10^{-7} to 10^{-5} mol L^{-1} spiked with ^{90}Sr at different concentrations showed similar results. In fact, the calibration slopes for the different matrix compositions where compared using a heuristic approach, and no influence of the matrix could be identified. Any observed difference between various matrices can be attributed to random experimental variations (data not shown).

The robustness of the proposed method was investigated by measuring ^{90}Sr reference solutions over a long period. Figure 3 shows the standard residual box-whisker plots for the ^{90}Sr references solution in function of time. The results cover ^{90}Sr concentration within the linear working range from 5 to 40 pg/g. All the results are within ±2%. The method can be maintained within statistical control and is thus suitable for the determination of ^{90}Sr in spent fuel leachates.

The functional lifetime of the column depends on the number of injections that can be done with the same resin present in the column without affecting its functionality. In the present case, more than 100 analyses were performed without detecting a loss of column functionality.

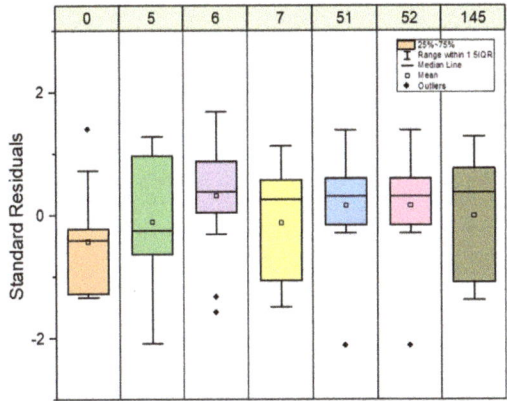

Figure 3. Standard residual box-whisker plots for Sr-90 reference solutions (5 to 40 pg/g) in function of time between 0 and 145 days.

2.2. Interferences in the Determination of ^{90}Sr in the SeaFAST System with ICP-DRC-MS Detection

An exhaustive list of the main isobaric interferences in the determination of ^{90}Sr by ICP-MS can be found elsewhere [15]. These interferences affect the detection limit, accuracy and precision of the determination of ^{90}Sr by ICP-MS. In spent nuclear fuel leachates, the determination of ^{90}Sr by ICP-MS is mainly affected by ^{90}Zr. Other potential isobaric interferences such as ^{58}Ni^{16}O$_2^+$, ^{74}Ge^{16}O$^+$, ^{52}Cr^{38}Ar$^+$, ^{50}V^{40}Ar$^+$, ^{54}Fe^{36}Ar$^+$, ^{50}Ti^{40}Ar$^+$, ^{180}W^{2+}, and ^{180}Hf^{2+} can affect the determination of ^{90}Sr as some transition metals are also present in the spent nuclear fuel leachates.

A screening analysis was performed with diluted spent nuclear fuel leachate samples using the NexION 300S ICP-MS. Each sample was measured in two different dilutions. Memory effect was avoided by acid washing in between the sample measurements. The screening analyses showed different levels of natural and fission Rb, Sr, and Zr in the leachates, as demonstrated in the spectra shown in Figure 4. The presence of fission Sr (consisting of ^{88}Sr + ^{90}Sr only) could qualitatively be confirmed. Semi-quantitative results can be obtained but major interference corrections are needed. These corrections can be derived from the isotopic profiles of the respective elements. However, whenever these corrections become too important, the uncertainties on the calculated end results are very high. The direct and precise determination of the ^{90}Sr concentration in the samples is not possible due to the high level of interference by natural zirconium coming mainly from the zircaloy. The elimination of the interference from Zr can be achieved by using the reaction of Zr with O_2 as gas in the reaction cell of the ICP-MS (Figure 4). The reaction of Sr, Zr, and Y with O_2 has already been reported [24] and the method applied to environmental samples [14,25].

Standard working solutions of strontium and zirconium were measured using the SeaFAST system with ICP-DRC-MS detection. Figure 5 shows the elution profile of Sr in the presence of 100 times more Zr. In the direct mode (Figure 5, A) the effect of the use of O_2 in the DRC can be observed but even if the efficiency of the reaction between Zr and O_2 reaches 99.9%, the isobar m/z 90 is still present. In pre-concentration mode, the comparison of the elution profiles reveals the different behavior between Sr and Zr in the Sr-resin column. Sr is retained while Zr is eliminated along with the other components of the matrix (Figure 5, B). Similar results have been obtained using nuclear spent fuel leachate analogues samples. Interference free determination of ^{90}Sr concentrations in the samples can be only achieved using the combination of matrix removal and DRC.

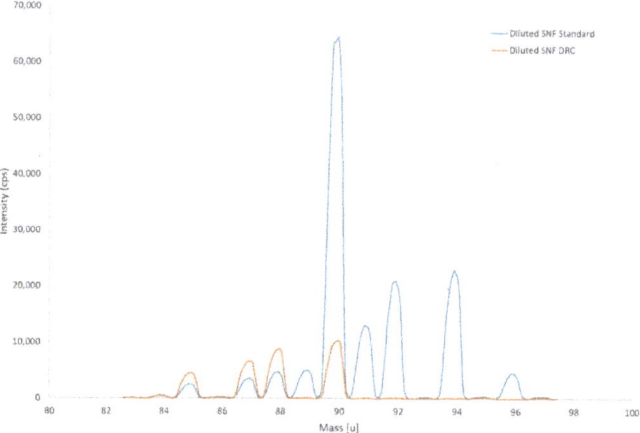

Figure 4. Mass spectra (from 84 till 100 u) of a diluted leachate sample; the blue spectrum shows predominantly natural Zr isotopes (90, 91, 92, 94, 96); and the presence of fission ^{88}Sr as $(^{88}Sr/^{86}Sr)_{meas}$ is much bigger than $(^{88}Sr/^{86}Sr)_{nat}$. The orange spectrum shows only the Sr isotopes and the near elimination of Zr isotopes when using the dynamic reaction cell (DRC).

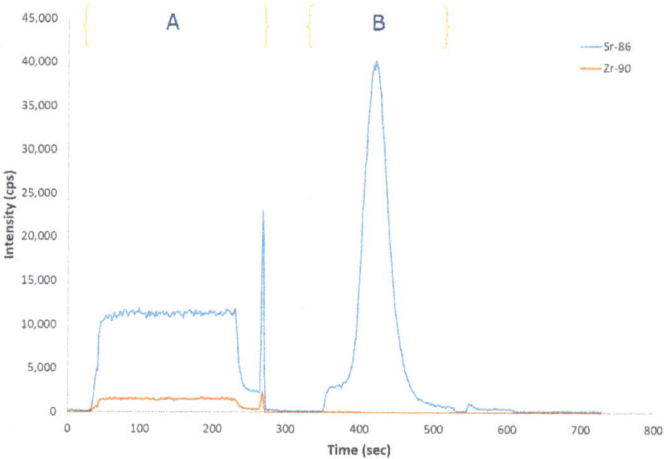

Figure 5. ^{86}Sr and ^{90}Zr profiles in direct (**A**) and pre-concentration mode (**B**). The standard working solution contains 80 pg g^{-1} from an enriched standard solution of ^{86}Sr and 8 µg L^{-1} of natural zirconium.

For the determination of ^{90}Sr in environmental samples, low instrumental detection limits (< ng L^{-1}) are required because of the high natural abundance of stable Sr. Indeed, high concentrations of stable Sr in the final sample solution require sufficient abundance sensitivity to resolve the peak tail of ^{88}Sr. In spent nuclear fuel leachates, this issue is less important as the fission ^{88}Sr/^{90}Sr ratio is 0.8, but an efficient sample preparation protocol is needed to avoid cross-contamination with natural Sr and remove other elements present in the sample to achieve a final solution with low total dissolved solids.

2.3. ^{90}Sr Determination in Spent Nuclear Fuel Leachates

The developed method was evaluated by analyzing different diluted spent nuclear fuel leachates and two spent nuclear fuel leachates analogues. Some samples were also analyzed by LSC. The main results of ^{90}Sr analysis for three replicates are shown in Table 2. The comparison between the developed method and the classical radiochemical method shows a good agreement when considering the associated uncertainties. Indeed, the results of the analysis of spent nuclear fuel leachates analogues present a satisfactory recovery being more than 90% in all the cases. The combined uncertainty was calculated using ISO/IEC Guide 98-3:2008 version of the Guide to the Expression of Uncertainty in Measurement (GUM). The overall uncertainty was obtained by identifying, quantifying and combining all individual contributions, mainly the ^{90}Sr standard reference solution (1.73%), sample and standard solutions weightings (0.054%), dilutions (0.072%), calibration (0.56%), and the sample repeatability (0.2% to 0.4%).

Table 2. Main results on the determination of ^{90}Sr in diluted spent fuel leachates using the SeaFAST system with ICP-DRC-MS detection and Liquid Scintillation Counting.

Sample	Spent Nuclear Leachates [1]									SNF Analogues [2]	
	54-1	54-2	54-3	54-4	54-5	54-6	54-7	54-8	54-9	SNF-18	SNF-35
SeaFAST Method											
Concentration pg g^{-1}	41.5	78.1	64.5	16.6	16.8	20.0	30.4	24.6	17.0	18.4	36.4
u%	2.0	2.0	3.2	1.9	3.1	2.6	1.9	3.6	0.9	2.2	4.4
Recovery%										92.5	93.6
Activity [3] Bq g^{-1}	216.0	406.8	335.8	86.3	87.4	104.0	158.5	128.0	88.6	95.7	189.8
u% [4]	2.0	2.0	3.2	1.9	3.1	2.6	1.9	3.6	0.9	2.2	2.4
LSC Method											
Activity Bq g^{-1}	216.0	396.5			86.0					88.6	177.7
u% [4]	3.7	4.0			4.7					2.3	4.4
Recovery%	92	92			92					92	92

[1] Diluted Spent nuclear leachates with dilution factors varying between 100 to 1000. [2] SNF analogues composition are described in Section 4.2. [3] ^{90}Sr specific activity is 5.21 × 10^{12} Bq g^{-1}. [4] Combined uncertainty.

The results on the spent nuclear fuel leachates are in good agreement with previously reported data in spent fuel leachates [3,12,13]. The fission ^{88}Sr/^{90}Sr ratio remained constant at 0.8 over the whole series of measurements showing good agreement with ORIGEN code calculations (ORIGEN-ARP, 2000) [26] and previous screening analysis [3,12,13].

3. Discussion

A method for the determination of ^{90}Sr based on the SeaFAST sample pre-concentration system with ICP-DRC-MS has been developed and tested in spent nuclear fuel leachates. The method is fully automated and provides an analysis time for each sample (from injection to detection) of 12.5 min. This is significantly shorter than the accepted radiometric method for measuring ^{90}Sr. In radiometric methods ^{90}Sr is normally measured through its daughter ^{90}Y, requiring 2–3 weeks in-growth time to reach secular equilibrium between ^{90}Sr and ^{90}Y. Indeed, the on-line separation and pre-concentration of ^{90}Sr from the matrix elements combined with the reaction with oxygen as reaction gas in a dynamic reaction cell (DRC) of ICP-MS removed the main interferences affecting the accurate determination of ^{90}Sr in the samples. The method showed good performances with regard to the main analytical figures of merit with a LOD of 0.01 pg g^{-1} (0.04 Bq g^{-1}). Although the ICP-DRC-MS is inferior to commonly used radiometric methods with respect to the minimum detectable activity (mBq level) it represents a time and cost-effective alternative technique for nuclear samples down to activities of about 1 Bq g^{-1}.

Furthermore, among other advantages the proposed method is very simple, minimizes the risk from contamination due to the limited sample handling, reduces the consumption of the chemical reagents and produces less chemical waste that it is an important aspect to consider when working with radioactive materials. The method is completely automated reducing the risk to staff when manipulating radioactive samples.

Other similar procedures have been described in the literature. An automated sequential injection separation with a flow liquid scintillation counter for on-line detection has been used for the determination of ^{90}Sr in nuclear waste [27]. The Sr-resin was used for the extraction chromatographic separation and the analytical time was reduced to 40 min, while the traditional method normally required 1–3 days. A method based on HPLC with ion chromatography coupled to LSC on-line detection was also used for the determination of ^{90}Sr in reactor water from the Gösgen power plant [28]. A SIA-LOV setup with an exchangeable SPE sorbent bed and an optimized procedure for sample processing for determination of ^{90}Sr using ICP-MS, has been used for monitoring samples of nuclear reactor coolant [15]. The limit of detection of this procedure depended on the configuration of the employed ICP-MS and on the available volume of the sample to be analyzed. For 1 L initial sample volume, the method detection limit (MDL) value was evaluated as 2.9 fg g^{-1} (14.5 Bq L^{-1}).

The source term for the mobilization of radionuclides from spent fuel typically used in the performance assessment of a geologic repository consists of two components: The instant release fraction (IRF) and the fraction released congruently with the matrix dissolution processes [29]. The IRF represents the fraction of the inventory of long-lived safety-relevant and highly mobile radionuclides, such as ^{36}Cl, ^{79}Se, ^{99}Tc, ^{126}Sn, ^{129}I, and ^{135}Cs that will be released upon first contact between fuel and groundwater, after breaching of all barriers. The need of reducing the uncertainties of difficult to measure radionuclides (DTM) is a continuous search for new analytical procedures, involving an enhancement of selectivity and sensitivity, and shortening the time of analysis.

The methodology proposed in this work is under development for the determination of other important difficult to measure radionuclides such as ^{135}Cs, ^{126}Sn, and ^{79}Se in nuclear waste characterization. Automated procedures based on the use of ICP-MS coupled to flow injection techniques for pre-concentration and/or separation of radionuclides present in spent nuclear fuel using specific stationary phases that selectively retains these key radionuclides are being tested with promising results.

Due to its long half-life and high fission yield, ^{135}Cs ($t_{1/2}$ = 2.3 Ma) is identified as a major contributor within ^{79}Se to the radiological risk for geological disposal/storage facilities [30]. Measurement of ^{135}Cs by mass spectrometry offers a considerable advantage because of the low specific radioactivity. However, a separation procedure is required to remove isobaric interferences from ^{135}Ba for the quantification of ^{135}Cs [31]. The automated SeaFAST coupled to the ICP-MS using the IonPac® CS12A (Thermo Fisher Scientific Inc., Waltham, MA, USA), a selective column for the analysis of alkali and alkaline earth metals, has been tested for the determination of ^{135}Cs (Figure 6A). The procedure allows the separation of Cs, Ba, and Sr in the same run showing its applicability for the determination of the three elements in spent nuclear fuel leachates and providing faster analytical results.

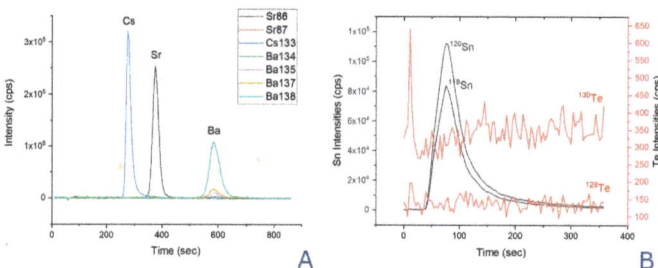

Figure 6. SeaFAST with ICP-DRC-MS time resolved signal of (**A**) a standard solution of Cs, Sr, and Ba (500 pg g^{-1}) on the Ion Pac CS12A column using methanesulfonic acid (MSA) 20 mmol L^{-1} as eluent (**B**) in HNO3 1 v/v% using a Nobias chelate-PA1 chelating column using as eluent Acetate 0.5 mol L^{-1} pH 6.

The major difficulty in ^{126}Sn measurement in samples on spent fuel is its low concentration due to its low fission yield of 0.065%. ^{126}Sn has a half-life of 230.000 years. There is a considerable lack of literature regarding the analysis of ^{126}Sn, and previous studies on ^{126}Sn measurements were mainly intended to confirm its half-life [32–34]. The content of ^{126}Sn in spent nuclear fuel samples has recently been reported [35]. The determination of ^{126}Sn by mass spectrometry suffers from isobaric interference of ^{126}Te which cannot be resolved even with modern high resolution mass spectrometers. Therefore, if Te is not fully chemically separated, it can yield incorrect ^{126}Sn concentrations in nuclear waste characterization. An analytical study carried out on TRU, TEVA, UTEVA, and DGA resins to identify a suitable solid phase extraction substrate to separate tin (Sn) from tellurium (Te) has recently been published [36]. Nobias chelate-PA1 chelating resin has been employed in an online pre-concentration system using the SeaFAST with ICP-MS detection for the analysis of Sn in spent nuclear fuel leachates. Sn and Te have a different behavior in the Nobias chelate-PA1 column (Figure 6B), Te is not retained on the column and ^{126}Sn can be determined successfully, indicating the potential for the determination of Sn in spent nuclear fuel leachates.

The accurate inventory estimation of long-lived fission products in spent nuclear fuel and high-level radioactive waste (HLW) is a major concern in the long-term safety assessment of a geological repository. Further improvement is still needed to prove the applicability of the proposed automated flow injection analysis coupled to ICP-MS for the on-line separation and/or pre-concentration on the radionuclides concerned to reduce matrix-related interferences and enhance sensitivity eliminating lengthy off-line sample preparation, providing faster analytical results and improving detection limits.

4. Materials and Methods

4.1. Reagents and Materials

High purity PFA columns with Sr-resin® 50–100 µm (Triskem International, Bruz, France) were used for the pre-concentration and analysis of ^{90}Sr.

A carrier-free ^{90}Sr standard reference solution containing 2046 Bq g^{-1} ± 3.2%, k = 2 from Eckert and Ziegler was used. Stock and working standards solutions were prepared gravimetrically in 4 mol L^{-1} HNO$_3$. Blanks with the same composition were also prepared.

For the preparation of all solutions, high-purity water (18.2 MΩ cm) from a Milli-Q Element system designed for ultratrace analysis (Millipore, Milford, MA, USA) was used. Nitric acid, suprapur grade from Merck (Darmstadt, Germany), was further purified using a quartz sub-boiling distillation unit. Both the water purification system and the sub-boiling distillation unit were operated in a clean room class 1000. Suprapur grade reagents from Merck (Darmstadt, Germany) as NaCl and NaHCO$_3$ were used to clean materials, prepare solutions and preserve and analyse samples. Natural element standards were obtained from CPI international (Amsterdam, The Netherlands) as 1000 µg·mL^{-1} stock standard solutions. The working standard solutions were prepared gravimetrically in 1% v/v sub-boiled HNO$_3$ using serial dilutions.

4.2. Samples

Nuclear spent fuel leachates analogues were prepared in a buffer of NaCl 19 mmol L^{-1} and NaHCO$_3$ 1 mmol L^{-1} pH 8.06 with a uranium concentration from 1x10^{-7} to 10^{-5} mol L^{-1} and spiked with different concentrations of ^{90}Sr. Blanks with the same composition were also prepared.

A boiling water reactor (BWR) fuel with a local burn-up of 58 GWd(tHM)$^{-1}$ was used to perform the stability studies from irradiated fuel in contact with simplified groundwater in oxidizing conditions for geological disposal applications. The main characteristic of the UO$_2$ spent nuclear fuel are summarized in Table 3.

Table 3. UO$_2$ spent nuclear fuel sample characteristics.

Segment	BWR
Total Mass (g)	2.5957
Fuel Mass (g)	2.062
Sr Inventory (µg/g)	1307
Leaching solution	Carbonated water
Contact Time (days)	0.05–315

As leaching solution a carbonated solution was used. The solution (1 mmol L^{-1} NaHCO3 and 19 mmol L^{-1} NaCl), initial pH of 8.4 ± 0.1, was initially equilibrated with air under oxidizing conditions and a normal hot cell temperature (25 ± 5) °C, as described elsewhere.

At each sampling, unfiltered aliquots were acidified with 100 µL of concentrated HNO3 and diluted to prepare the diluted spent fuel leachates samples.

4.3. Flow Injection Set-Up

An automated system SC-2 DX SeaFAST (ESI, Omaha, NE, USA) was equipped as recommended by the manufacturer (Figure 7). The autosampler SC-2 DX was equipped with a sample probe with 1.0 mm inner diameter (ID), a vacuum pump for sample aspiration and two independent rinse pumps supplying the rinsing station with 1% HNO3 from 4 L reservoirs. The reagents were distributed by a syringe system (S400V) consisting of an ethylene chlorotrifluoroethylene (CTFE) valve with PFA rotor and four syringes: one 12 mL CTFE/polytetrafluoroethylene (PTFE) syringe (S1), and three 3 mL quartz/PFA syringes (S2, 3, and 4). The reagent flow paths were controlled by a valve module (FAST DX 3) with two 11-port CTFE valves (V1 and V2) and one 5-port CTFE valve (V3), all three with a PFA rotor. The sample coils utilized in this study had a volume of 0.6 and 2 mL. All tubing connecting the valves were made of PFA. Valve 3 was connected to the ICP-MS nebulizer. The eluent used for loading the sample was a solution of 4 mol L^{-1} HNO3. Samples were eluted with Milli-Q water. This system utilizes two different columns, a Nobias PA1 resin columns (200 µL) for pre-cleaning the loading eluent and a Sr-resin® (200 µL) as a pre-concentration column. The system was used in both direct and pre-concentration modes.

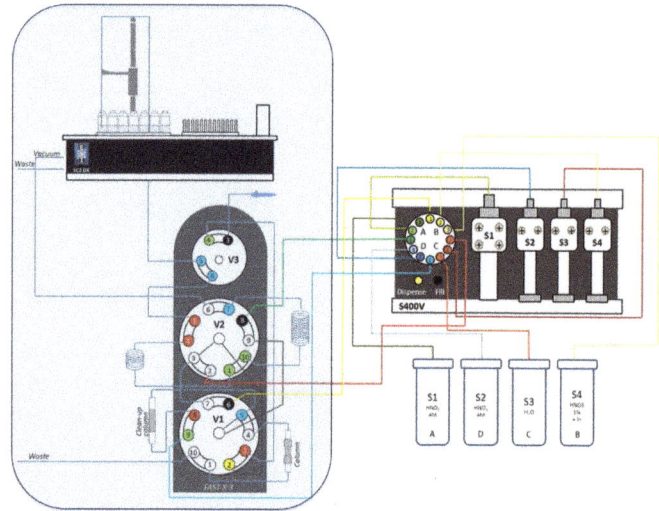

Figure 7. Schematic set-up of the SeaFAST system including the autosampler unit (SC2 DX), the syringe system (S400V), and valve module (FAST DX 3). Blue filled rectangle corresponds to the set-up components placed inside the glove box.

4.4. ICP-MS

A nuclearized NexION 300S ICP-MS (PerkinElmer, Inc., Shelton, CT, USA) equipped with a sample introduction system consisting of a quartz cyclonic spray chamber, a type C0.5 concentric glass nebulizer and a 2 mm bore quartz injector. The ICP-MS includes a dual-channel Universal Cell and uses Dynamic Reaction Cell™ (DRC™, PerkinElmer, Waltham, MA, USA) technology. In this study ultrapure O$_2$ (>99.9999%) was used as reaction gas in the collision-reaction cell (DRC). 'Syngistix™

for ICP-MS' software was used for defining the method set-up and data acquisition parameters. Instrument settings and conditions were optimized daily with typical values listed in Table 4. Prior to sample analysis, the NexION 300S ICP-MS was conditioned and tuned for maximum signal intensity and stability each day using a 1 µg L^{-1} standard tuning solution (PerkinElmer, Waltham, MA, USA).

Table 4. NexION® 300S ICP-MS instrumental conditions.

Nebulizer	Concentric
Tripe Cone material	Pt
Spray Chamber	Cyclonic
Torch and Injector	Quartz Torch and Quartz 2.0 mm bore injector
Power (W)	1300
Plasma Gas (L·min^{-1})	16
Aux Gas (L·min^{-1})	1.2
Neb Gas (L·min^{-1})	1.03
Sample Uptake Rate (mL·min^{-1})	0.2
Isotopes	^{86}Sr, ^{90}Sr, ^{91}Zr, ^{92}Zr, ^{115}In
Sweeps/Reading	5
Readings/Replicate	450
Replicates	1
Dwell Time (µs)	50
Measuring Time (s)/reading	1.6
Cell Gas	O$_2$
Cell Gas Flow (mL·min^{-1})	0.5
Potential of the cell rods (RPq)	0.45

Following calibration, the ICP-DRC-MS was purged with 1% prior to initial blank and sample measurements. Blanks were systematically intercalated to avoid memory effect between samples. ^{90}Sr standards of 20 and 40 pg g^{-1} were analyzed after each sample sequence to verify the calibration. External calibration was used for ^{90}Sr quantitation.

4.5. SeaFAST Procedure

The SC-2 DX SeaFAST device was controlled using a modified protocol of the default NexION SeaFAST methods. The complete operational sequence for strontium separation, pre-concentration and elution is detailed in Table 5. Briefly, at $t = 0$, the sample is loaded onto two separate loops and injected into the system. The ICP-MS read delay (Table 5 A) is the time it takes to load the loops and for the direct mode signal to stabilize. During this time, the pre-concentration column is loaded with sample from its sample loop and the matrix ions are washed from the column to the waste. In direct mode (Table 5 B), sample is diluted online approximately 2x, mixed with In solution as internal standard to monitor correct filling of the sample loop and elution. Then it is introduced directly to the ICP-MS. After the direct mode is finished, the pre-concentrated ^{90}Sr is eluted into the ICP-MS and the pre-concentration mode is acquired (Table 5 C). Following the pre-concentration elution, the column is cleaned and conditioned (Table 5 D) to prepare the column for the next sample. Although the system was optimized with a cleaning and pre-conditioning between samples, blank samples were always intercalated to avoid sample-to-sample memory effects.

Table 5. SeaFAST set-up with the NexION 300S DRC ICP-MS detection.

Step	Valve Position	Syringe	Description
Load sample into the coils (A)	V1: Load V2: Load V3: Load		Sample is load into the coils. Direct measurements is 600 µL and pre-concentration in the column is 2 mL.
Direct measurement and sample load into the column (B)	V1: Load V2: Inject V3: Load	S3 180 µL·min^{-1} S4 20 µL·min^{-1} S1 2500 µL·min^{-1}	Sample is eluted into the ICP-MS system. Sample is loaded on the Sr-resin®
Elution (C)	V1: Inject V2: Load V3: Load	S3 1000 µL·min^{-1}	Pre-concentrated ^{90}Sr is eluted backpressure into the ICP-MS system for online determination.
Preconditioning (D)	V1: Load V2: Load V3: Load	S3 2000 µL·min^{-1} S1 2500 µL·min^{-1}	The column is cleaned and pre-conditioned

4.6. Liquid Scintillation Counting

Liquid scintillation counting was performed using a Quantulus 1220 (PerkinElmer, Inc., Shelton, CT, USA) on ^{90}Sr standard solutions, analogues and in diluted spent nuclear fuel leachates. ^{90}Sr was separated on small extraction chromatography columns (Bio-rad, Hercules, CA, USA) packed with Sr-resin® (50–100 µm). 0.6 mL (3 free column volumes, FCV) of the sample (in 4 mol L^{-1} HNO$_3$) was gravimetrically added and the ^{90}Y was fully eluted out using 1.4 mL (7 FCV) 3 mol L^{-1} HNO$_3$ + 0.05 mol L^{-1} oxalic acid yielding the Y fraction. The column was further washed with 1 mL (5 FCV) 3 mol L^{-1} HNO$_3$ before the Sr fraction was eluted with 2 mL 0.05 mol L^{-1} HNO$_3$. A Cerenkov sample was prepared from the complete Y fraction by adding 10 mL Milli-Q H$_2$O. The ^{90}Sr fraction was properly mixed and then split equally into a 1 mL LSC and a 1 mL Cerenkov sample. The samples were gravimetrically added to a standard LSC counting vial and 10 mL Aqua Light scintillation liquid (LSC) or 10 mL H$_2$O (Cerenkov) was added to a total filling volume of 12 mL. This sample geometry was calibrated using a ^{90}Sr reference standard solution (2046 Bq g^{-1} ± 3.2%, k = 2, Eckert and Ziegler) to an efficiency of 98.8% for ^{90}Sr and ^{90}Y (LSC) and 66% for ^{90}Y (Cerenkov). The samples were repeatedly measured with a measurement time of 20 min to roughly 92% of full ^{90}Y ingrowth/decay. ^{90}Sr was quantified by several methods including direct LSC and by ingrowth/decay curves of ^{90}Y. The recovery of ^{90}Sr was measured repeating the exact separation procedure with a spiked ^{137}Cs, ^{85}Sr feed. The recovery yield of ^{90}Sr was determined to be 92% by direct HPGe gamma measurements of the Sr columns before and after ^{90}Sr elution.

Author Contributions: All authors listed have contributed sufficiently to the project to be included as authors. LSC measurements V.V.V.; leachates preparation, L.I.P.; ICP-MS measurements, S.M. and M.S.; analysis, writing–review and editing, S.V.W., D.S.P. and L.A.d.l.H.; project administration and supervision, D.S.P. and L.A.d.l.H. All authors have read and agreed to the published version of the manuscript.

Funding: This research received no external funding.

Acknowledgments: The authors would also like to thank the technical staff of JRC Karlsruhe.

Conflicts of Interest: The authors declare no conflict of interest.

References

1. Jeong, J.; Lee, Y.-M.; Kim, J.-W.; Cho, D.-K.; Ko, N.Y.; Baik, M.-H. Progress of the long-term safety assessment of a reference disposal system for high level wastes in Korea. *Prog. Nucl. Energy* **2016**, *90*, 37–45. [CrossRef]
2. Serrano-Purroy, D.; Clarens, F.; González-Robles, E.; Glatz, J.; Wegen, D.; De Pablo, J.; Casas, I.; Giménez, J.; Martínez-Esparza, A. Instant release fraction and matrix release of high burn-up UO2 spent nuclear fuel: Effect of high burn-up structure and leaching solution composition. *J. Nucl. Mater.* **2012**, *427*, 249–258. [CrossRef]

3. Martínez-Torrents, A.; Serrano-Purroy, D.; Casas, I.; De Pablo, J. Influence of the interpellet space to the Instant Release Fraction determination of a commercial UO2 Boiling Water Reactor Spent Nuclear Fuel. *J. Nucl. Mater.* **2018**, *499*, 9–17. [CrossRef]
4. Johnson, L.; Schneider, J.; Zuidema, P.; Gribi, P.; Mayer, G.; Smith, P. Demonstration of disposal feasibility for spent fuel, vitrified high-level waste and long-lived intermediate-level waste. In *NAGRA NTB 02-05 Project Opalinus Clay Safety Report*; Nagra: Wettingen, Switzerland, 2002.
5. Croudace, I.W.; Russell, B.C.; Warwick, P.W. Plasma source mass spectrometry for radioactive waste characterisation in support of nuclear decommissioning: A review. *J. Anal. At. Spectrom.* **2017**, *32*, 494–526. [CrossRef]
6. Hou, X.; Roos, P. Critical comparison of radiometric and mass spectrometric methods for the determination of radionuclides in environmental, biological and nuclear waste samples. *Anal. Chim. Acta* **2008**, *608*, 105–139. [CrossRef]
7. Bu, W.; Ni, Y.; Steinhauser, G.; Zheng, W.; Zheng, J.; Furuta, N. The role of mass spectrometry in radioactive contamination assessment after the Fukushima nuclear accident. *J. Anal. At. Spectrom.* **2018**, *33*, 519–546. [CrossRef]
8. Hou, X. Radioanalysis of ultra-low level radionuclides for environmental tracer studies and decommissioning of nuclear facilities. *J. Radioanal. Nucl. Chem.* **2019**, *322*, 1217–1245. [CrossRef]
9. Hou, X.; Zhang, W.; Wang, Y. Determination of Femtogram-Level Plutonium Isotopes in Environmental and Forensic Samples with High-Level Uranium Using Chemical Separation and ICP-MS/MS Measurement. *Anal. Chem.* **2019**, *91*, 11553–11561. [CrossRef]
10. Kavasi, N.; Sahoo, S.K.; Arae, H.; Aono, T.; Palacz, Z. Accurate and precise determination of 90Sr at femtogram level in IAEA proficiency test using Thermal Ionization Mass Spectrometry. *Sci. Rep.* **2019**, *9*, 16532. [CrossRef]
11. Wang, Z.; Huang, Z.; Xie, Y.; Tan, Z. Method for determination of Pu isotopes in soil and sediment samples by inductively coupled plasma mass spectrometry after simple chemical separation using TK200 resin. *Anal. Chim. Acta* **2019**, *1090*, 151–158. [CrossRef]
12. González-Robles, E.; Serrano-Purroy, D.; Sureda, R.; Casas, I.; De Pablo, J. Dissolution experiments of commercial PWR (52 MWd/kgU) and BWR (53 MWd/kgU) spent nuclear fuel cladded segments in bicarbonate water under oxidizing conditions. Experimental determination of matrix and instant release fraction. *J. Nucl. Mater.* **2015**, *465*, 63–70. [CrossRef]
13. Martínez-Torrents, A.; Serrano-Purroy, D.; Sureda, R.; Casas, I.; De Pablo, J. Instant release fraction corrosion studies of commercial UO_2 BWR spent nuclear fuel. *J. Nucl. Mater.* **2017**, *488*, 302–313. [CrossRef]
14. Feuerstein, J.; Boulyga, S.; Galler, P.; Stingeder, G.; Prohaska, T. Determination of 90Sr in soil samples using inductively coupled plasma mass spectrometry equipped with dynamic reaction cell (ICP-DRC-MS). *J. Environ. Radioact.* **2008**, *99*, 1764–1769. [CrossRef] [PubMed]
15. Kołacińska, K.; Chajduk, E.; Dudek, J.; Samczyński, Z.; Łokas, E.; Bojanowska-Czajka, A.; Trojanowicz, M. Automation of sample processing for ICP-MS determination of 90 Sr radionuclide at ppq level for nuclear technology and environmental purposes. *Talanta* **2017**, *169*, 216–226. [CrossRef]
16. Rodríguez, R.; Avivar, J.; Leal, L.O.; Cerdà, V.; Ferrer, L. Strategies for automating solid-phase extraction and liquid-liquid extraction in radiochemical analysis. *TrAC Trends Anal. Chem.* **2016**, *76*, 145–152. [CrossRef]
17. Rodríguez, R.; Borras, A.; Leal, L.O.; Cerdà, V.; Ferrer, L. MSFIA-LOV system for [226]Ra isolation and pre-concentration from water samples previous radiometric detection. *Anal. Chim. Acta* **2016**, *911*, 75–81. [CrossRef]
18. Rodríguez, R.; Leal, L.O.; Miranda, S.; Ferrer, L.; Avivar, J.; Garcia, A.; Cerdà, V. Automation of 99Tc extraction by LOV prior ICP-MS detection: Application to environmental samples. *Talanta* **2015**, *133*, 88–93. [CrossRef]
19. Trojanowicz, M.; Kolacinska, K. Recent advances in flow injection analysis. *Analyst* **2016**, *141*, 2085–2139. [CrossRef]
20. Lagerström, M.E.; Field, M.; Séguret, M.; Fischer, L.; Hann, S.; Sherrell, R. Automated on-line flow-injection ICP-MS determination of trace metals (Mn, Fe, Co, Ni, Cu and Zn) in open ocean seawater: Application to the GEOTRACES program. *Mar. Chem.* **2013**, *155*, 71–80. [CrossRef]
21. Rapp, I.; Schlosser, C.; Rusiecka, D.; Gledhill, M.; Achterberg, E.P. Automated preconcentration of Fe, Zn, Cu, Ni, Cd, Pb, Co, and Mn in seawater with analysis using high-resolution sector field inductively-coupled plasma mass spectrometry. *Anal. Chim. Acta* **2017**, *976*, 1–13. [CrossRef]

22. Wuttig, K.; Townsend, A.T.; Merwe, P.; Gault-Ringold, M.; Holmes, T.; Schallenberg, C.; Latour, P.; Tonnard, M.; Rijkenberg, M.J.; Lannuzel, D.; et al. Critical evaluation of a seaFAST system for the analysis of trace metals in marine samples. *Talanta* **2019**, *197*, 653–668. [CrossRef] [PubMed]
23. Currie, L.A. Limits for qualitative detection and quantitative determination. Application to radiochemistry. *Anal. Chem.* **1968**, *40*, 586–593. [CrossRef]
24. Anicich, V.G. *An Index of the Literature for Bimolecular Gas Phase Cation-Molecule Reaction Kinetics*; National Aeronautics and Space Administration (N.A.A.S. Administration), Ed.; California Institute of Technology: Pasadena, CA, USA, 2003.
25. Takagai, Y.; Furukawa, M.; Kameo, Y.; Suzuki, K. Sequential inductively coupled plasma quadrupole mass-spectrometric quantification of radioactive strontium-90 incorporating cascade separation steps for radioactive contamination rapid survey. *Anal. Methods* **2014**, *6*, 355–362. [CrossRef]
26. Ezure, H. Validation of ORIGEN Computer Code by Measurements on Nuclear Fuels of JPDR-1. *J. Nucl. Sci. Technol.* **1989**, *26*, 777–786. [CrossRef]
27. Grate, J.W.; Strebin, R.; Janata, J.; Egorov, O.; Ruzicka, J. Automated Analysis of Radionuclides in Nuclear Waste: Rapid Determination of 90Sr by Sequential Injection Analysis. *Anal. Chem.* **1996**, *68*, 333–340. [CrossRef]
28. Desmartin, P.; Kopajtic, Z.; Haerdi, W. Radiostrontium-90 (90Sr) Ultra-Traces Measurements by Coupling Ionic Chromatography (HPIC) and on Line Liquid Scintillation Counting (OLLSC). *Environ. Monit. Assess.* **1997**, *44*, 413–423. [CrossRef]
29. Fanghänel, T.; Rondinella, V.V.; Glatz, J.-P.; Wiss, T.; Wegen, D.H.; Gouder, T.; Carbol, P.; Serrano-Purroy, D.; Papaioannou, D. Reducing Uncertainties Affecting the Assessment of the Long-Term Corrosion Behavior of Spent Nuclear Fuel. *Inorg. Chem.* **2013**, *52*, 3491–3509. [CrossRef]
30. Asai, S.; Hanzawa, Y.; Okumura, K.; Shinohara, N.; Inagawa, J.; Hotoku, S.; Suzuki, K.; Kaneko, S. Determination of 79Se and 135Cs in Spent Nuclear Fuel for Inventory Estimation of High-Level Radioactive Wastes. *J. Nucl. Sci. Technol.* **2011**, *48*, 851–854. [CrossRef]
31. Alonso, J.I.G.; Sena, F.; Arbore, P.; Betti, M.; Koch, L. Determination of fission products and actinides in spent nuclear fuels by isotope dilution ion chromatography inductively coupled plasma mass spectrometry. *J. Anal. At. Spectrom.* **1995**, *10*, 381. [CrossRef]
32. Bienvenu, P.; Ferreux, L.; Andreoletti, G.; Arnal, N.; Lépy, M.C.; Bé, M.M. Determination of ^{126}Sn half-life from ICP-MS and gamma spectrometry measurements. *Radiochim. Acta* **2009**, *97*, 687–694. [CrossRef]
33. Catlow, S.A.; Troyer, G.L.; Hansen, D.R.; Jones, R.A. Half-life measurement of 126Sn isolated from Hanford nuclear defense waste. *J. Radioanal. Nucl. Chem.* **2005**, *263*, 599–603. [CrossRef]
34. Oberli, F.; Gartenmann, P.; Meier, M.; Kutschera, W.; Suter, M.; Winkler, G. The half-life of 126Sn refined by thermal ionization mass spectrometry measurements. *Int. J. Mass Spectrom.* **1999**, *184*, 145–152. [CrossRef]
35. Asai, S.; Toshimitsu, M.; Hanzawa, Y.; Suzuki, H.; Shinohara, N.; Inagawa, J.; Okumura, K.; Hotoku, S.; Kimura, T.; Suzuki, K.; et al. Isotope dilution inductively coupled plasma mass spectrometry for determination of ^{126}Sn content in spent nuclear fuel sample. *J. Nucl. Sci. Technol.* **2013**, *50*, 556–562. [CrossRef]
36. Rahman, M.M.; Macdonald, C.; Cornett, R.J. Separation of tin from tellurium: Performance of different extraction chromatographic materials. *Sep. Sci. Technol.* **2018**, *53*, 2055–2063. [CrossRef]

Sample Availability: Samples of the compounds are not available from the authors.

© 2020 by the authors. Licensee MDPI, Basel, Switzerland. This article is an open access article distributed under the terms and conditions of the Creative Commons Attribution (CC BY) license (http://creativecommons.org/licenses/by/4.0/).

Article

The Automation Technique Lab-In-Syringe: A Practical Guide

Burkhard Horstkotte * and Petr Solich

Department of Analytical Chemistry, Charles University, Faculty of Pharmacy, Akademika Heyrovského 1203, 500 05 Hradec Králové, Czech Republic; Petr.Solich@faf.cuni.cz
* Correspondence: Horstkob@faf.cuni.cz; Tel.: +420-495-067-507; Fax: +420-495-067-164

Academic Editor: Pawel Koscielniak
Received: 6 March 2020; Accepted: 28 March 2020; Published: 1 April 2020

Abstract: About eight years ago, a new automation approach and flow technique called "Lab-In-Syringe" was proposed. It was derived from previous flow techniques, all based on handling reagent and sample solutions in a flow manifold. To date Lab-In-Syringe has evidently gained the interest of researchers in many countries, with new modifications, operation modes, and technical improvements still popping up. It has proven to be a versatile tool for the automation of sample preparation, particularly, liquid-phase microextraction approaches. This article aims to assist newcomers to this technique in system planning and setup by overviewing the different options for configurations, limitations, and feasible operations. This includes syringe orientation, in-syringe stirring modes, in-syringe detection, additional inlets, and addable features. The authors give also a chronological overview of technical milestones and a critical explanation on the potentials and shortcomings of this technique, calculations of characteristics, and tips and tricks on method development. Moreover, a comprehensive overview of the different operation modes of Lab-In-Syringe automated sample pretreatment is given focusing on the technical aspects and challenges of the related operations. We further deal with possibilities on how to fabricate required or useful system components, in particular by 3D printing technology, with over 20 different elements exemplarily shown. Finally, a short discussion on shortcomings and required improvements is given.

Keywords: Lab-In-Syringe; automation of sample pretreatment; potentials and troubles; system setup and operation modes; tips and tricks in method development; 3D printing of instrument elements

---

Academic Editor: Pawel Koscielniak

1. Introduction

The term "Lab-In-Syringe" ambiguously describes two analytical techniques that have been developed within the last decade and given the same name. Both take advantage of a syringe to carry out sample preparation procedures. In the first approach, one or two, generally disposable, syringes are used to facilitate manual procedures while the second one refers to an automation technique, i.e., employing an automatic computer-controlled syringe pump. In both cases, the benefit is taken from the fundamental characteristics of a syringe: (i) size-adaptability, (ii) a steadily sealed compartment, (iii) ability to measure and transfer fluids, (iv) and the ease of emptying and wiping action of the piston on the barrel walls. In both cases, the syringe voids are used as mixing, reaction, and extraction chambers, most often to carry out liquid-phase microextraction (LPME) procedures.

In this article, we deal exclusively with the automation technique "Lab-In-Syringe" and present a tutorial on the system setup, operation modes, and required materials as well ... with the as a critical overview of its applicability, potential, and shortcomings.

Lab-In-Syringe (LIS) derives from the far better-known flow technique Sequential Injection Analysis (SIA) described already in 1990 by Růžička and Marshall [1,2]. Flow techniques (FTs) are versatile automation tools for laboratory procedures consisting of, e.g., solution metering, mixing, dilution, and analyte conversion into detectable forms. As the name suggests, these unit operations are carried out in flow within a tubing manifold and modulated by means of pumps and valves and generally by the aid of a liquid carrier that transports the injected sample and reagent solutions.

Usually, flow systems integrate appropriate flow-through detectors (optical, electrochemical, ...), achieving specific advantages in terms of analysis speed, reproducibility, and reliability as well as for analyte discrimination, e.g., via chemometrics. These are enabled by precise reaction timing, reproducible detector feed, controlled detector passage, and precise timing, so that, e.g., contact times of the analyte with immobilized enzymes or electrode surfaces are strictly reproducible. Alternatively, FT can be coupled to external instrumentation, e.g., HPLC or atomic spectrometric techniques, and used for preceding automated sample preparation by matrix elimination and analyte preconcentration. Determination selectivity of FT is achieved by tactical employment of selective reagents and enzymes, by kinetic differentiation, or by the hyphenated selective detection or separation techniques. For the latter two, FT can contribute significant advantages as a versatile tool for sample processing, by carrying out required analyte conversion, derivatization, or enrichment.

Typical chemical problems solved by FT are the automation of analytical assays applied to all kinds of liquid samples such as waters, beverages, body fluids, or fuels. Most often, FT are applied to inorganic analytes, such as nutrients, metal contaminants, or total indices (total phenols, MBAS, DIC, ...). For organic compounds, analyte selective enzymes (glucose—glucose oxidase), selective chromogenic reactions (formaldehyde with acetylacetone), or simple matrix elimination (gas diffusion for volatiles) are explored as well as taking advantage of more or less unique analyte's reactivity (e.g., ascorbic acid) or physical properties (e.g., a fluorescence activity). FT are ideal for the automation of reactions of fast kinetics, e.g., chemiluminescence assays, and versatile tools for monitoring of a sample stream originating from technical, environmental, or biological processes. One major trend is the automation of SPE or LLE procedures for sample cleanup and analyte enrichment for online-coupled instrumental techniques ICP, GFAAS, or HPLC, among others.

FT automation generally yields a significant gain in repeatability, sample throughput, and assay reliability and allows miniaturization compared to manual processing and reduction of sample and reagent volumes. Processing in a closed tubing system also implies a low risk of user's exposure to harmful reagents or sample contamination, as well as automatic system cleaning by the carrier flow.

The various FT and proposed methodologies differ in manifold configurations, operation characteristics and flow patterns, arrangement and types of valves and pumps, flow segmentation, modes of sample insertion, and obtained transient signals [3,4]. It is impossible here to give a picture of the many facets of FT with justified thoroughness. Therefore, earlier treatises and tutorials on the matter are recommended [5–8]. To introduce to Lab-In-Syringe, it is, however, necessary to elucidate at first two earlier FTs: SIA and the Flow-Batch concept.

SIA is based on a simple computer-controlled system consisting of a single bidirectional pump, typically an automatic syringe pump, and a selection valve with all required solutions and the detector connected on its lateral ports. A long tube of a dead volume that often exceeds the volume of the employed syringe, the so-denoted holding coil, connects the common port of the multiposition valve to the syringe pump. It serves to intermediately contain the subsequently aspirated sample and reagents upon which the solution zones penetrate and mix partly. The stacked solutions undergo further mixing and reaction during flow reversal and on their way towards the detector yielding a tailing peak. This way, chromogenic reactions can easily be automated and miniaturized. Procedural parameters (volumes, flow rates, timing, pausing) are easily adapted, enabling high operational flexibility. Of special interest can be steps with stopped-flow for prolonging reaction times or monitoring only the progress of a chromogenic reaction or formation of concentration gradients that are enabled by the incomplete mixing of the stacked solution zones undergoing further dispersion in-flow. On the other hand, homogenous mixing of widely different volumes of solutions, reactions requiring multiple reagents and steps, or maintaining mixing patterns in case of varying sample viscosity are less feasible to accomplish. Here, air-bubble segmentation of a confined zone of stacked solutions can be used to achieve homogenous mixing of small solution volumes.

Another option is the return to what has been denoted "beaker-chemistry" [3], i.e., the placement of a mixing chamber, such as the barrel of a disposable plastic syringe, on one lateral port of the selection valve. This enables the stepwise addition and mixing of solutions as needed, in other words, "batch-processing" but operated via the sequential injection analyzer, leading to "flow-batch analysis" [9]. The chamber, particularly if constantly agitated, enables homogenous mixing, widely independently from the initial sample viscosity, or open port sampling, e.g., using a pipette to introduce the sample to the flow analyzer. However, in the lack of a carrier flow, there is also a higher risk of carry-over effects and, using an atmospherically open mixing chamber, of contamination. In consequence, tedious cleaning is required often, if not even after each analysis. Chambers and the flow-batch concept have been implemented not only in SIA but also in combination with other FT, and for other purposes. A comprehensive review on this topic is recommended [10].

Flow-batch is one step towards discrete analyzers, i.e., liquid handling on versatile autosamplers or, more and more, by robotic arms that allows mimicking "hu-man-ual" handling of samples [3,11]. For the automation of very fast reactions, e.g., using chemiluminescence as detection principle, the study of reaction kinetics, membrane separations, online digestions, or continuous monitoring 24/7 and fieldwork, FT are ideally suited tools and superior to batch automation. However, a critical comparison of both automation concepts is apposite considering sample preparation. In batch and flow-batch automation, the preparative procedure, e.g., an LLE approach, can essentially be broken down into steps of stepwise addition of solutions, homogenous mixing, and liquid withdraw. Consequently, the workflow and procedural parameters are far more predictable than if mixing is based on dispersion-based flow patterns. It seems fair and reasonable to aim for combining both flow and batch-processing if automation of sample pretreatment procedures is intended, would it not be for the addressed shortcomings of flow-batch analysis.

Removing the holding coil from an SIA system violates a fundamental principle of this FT as the solutions will enter the syringe void from where they cannot be flushed out completely as they undergo turbulent mixing. Either severe carry-over or repeated syringe cleaning must be therefore accepted. However, intending to use the syringe as a controlled mixing chamber, flow-batch analysis is feasible in a simplified and more compact SIA system and, as it will be discussed, with increased efficiency while omitting several of its inherent weaknesses. Through the marriage of flow-batch and SIA concept, a new and versatile automation approach was created that enables new operation modes and suits ideally for accomplishing liquid-phase microextraction approaches and that has been now widely recognized by the term "Lab-In-Syringe" [12,13].

2. Lab-In-Syringe—Technical Milestones

To the best of our knowledge, it was Maya et al. who "misused" in 2012 for the first time the void of an automatic syringe pump to automate an analytical procedure, in concrete, dispersive liquid-liquid microextraction (DLLME) of benzo(α)pyrene into octanol [12]. In contrary to manually performed DLLME, the procedures started by aspiration of a mixture of water-immiscible extraction solvent and miscible dispersion solvent into the syringe followed by a very fast aspiration of the aqueous sample. The turbulence led to solvent dispersion and formation of small droplets observed as the characteristic cloudy state. After phase separation by droplet floatation and spontaneous coalescence, the analyte-enriched organic phase was subjected to low-pressure chromatography.

In the following, steps of analyte derivatization and complexation were added to determine copper and aluminum in water samples as hexanol-soluble complexes with bathocuproine and lumogallion, respectively [14,15]. As in flow-batch, a mixing chamber was added to the flow manifold to enable homogeneous mixing of the sample with the required reagents before carrying out in-syringe DLLME that consequently required intermediate cleaning.

Therefore, an important step in LIS development was the option of in-syringe homogenous mixing by using a magnetic stirring bar inside the syringe void. Possible modes of how to induce stirring are summarized in Figure 1.

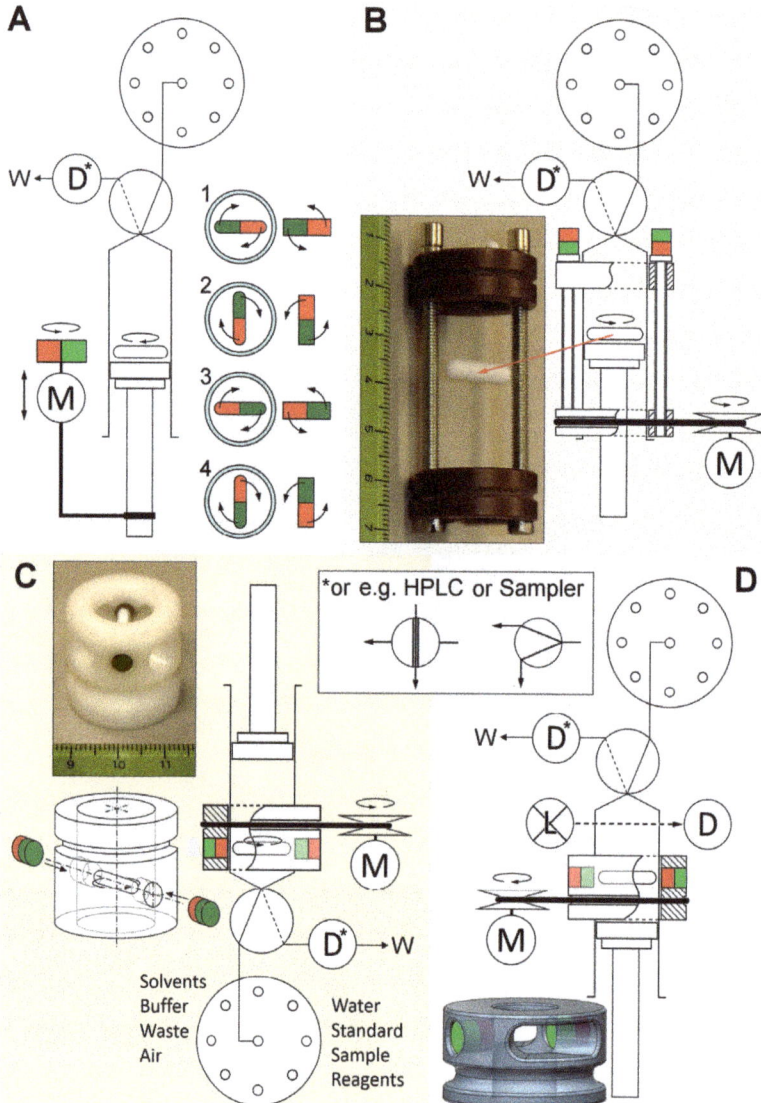

Figure 1. Modalities for system setup, syringe orientation, and stirring arrangements. Stir-drivers are shown in photographs with a levitated magnetic stirring bar. (**A,B,D**): Syringe in upright orientation, (**C**) Syringe used upside down. (**A**) Stirring induced by a closely placed motor with an attached bar magnet, (**B**) Stir-driver for the full stroke length of the syringe, (**C,D**) Stir-driver for fixed position. Upon turning the stir-drivers via a rubber ring and motor, they generate a rotating magnetic field around the syringe barrel. Abbreviations D—Detector, L—Light source, M—Motor, W—Waste.

In the first work reporting this approach, the stirring bar was forced to spin by a stir-driver as shown in Figure 1B [16]. It integrated two oppositely magnetized iron rods that were placed onto the syringe barrel to create an external rotating magnetic field along the stroke-length of the syringe on demand by the aid of a relay-controlled motor. The field was strong enough for rotation speeds of 1000 rpm. The superiority of this approach over using a mixing chamber was proven by reoptimizing the earlier extraction of aluminum-lumogallion but omitting the dispersion solvent and disrupting the extractant into fine droplets by the kinetic energy of the stirring bar instead. In effect, a lower solvent consumption and detection limit were achieved in a shorter time [15,16] and instantaneous homogenous mixing for differently viscose samples was verified. Hereafter, the suitability of LIS for (i) online standard preparation by adaptable in-syringe stock dilution and (ii) automation of a procedure involving 12-steps and 10 solutions for the derivatization and DLLME of chromate as Cr(III)-diphenyl carbazone complex was shown [17].

The resulting inability to empty the syringe completely was solved by simply turning the syringe upside-down. By this, air remained inside the syringe above the liquid level and expelled all solution content from the syringe during emptying (Figure 1C). The cost of this advantage is that each liquid transfer is delayed by the compressible air cushion that necessarily remains inside the syringe. Nonetheless, this arrangement enabled the reproducible performance of automated DLLME procedures into solvents denser than water while previous works had all used floating solvents. The classical assay of total cationic surfactants, requiring chloroform, was miniaturized and automated this way [18]. Noteworthy is that it was possible to use a much simpler stir-driver since the magnetic bar was not displaced with the syringe piston but remained at the bottom, near the syringe inlet. This enabled a focused magnetic field and higher rotation speeds of up to 3000 rpm that exceeded the typical stirring velocities in manual stirring-assisted DLLME severalfold.

Using the syringe itself as a detection cell by the aid of a fiber optic adaptor (see Figure 1D and Figure 4) was yet another milestone set by Maya et al. in 2012 [13] that justified the denomination "Lab-In-Syringe". This potentially increases the system simplicity and compactness, but a limitation is given by the requirement of using visible and near-UV wavelengths which can pass through a glass syringe.

Another lift in technical versatility was featuring a secondary entrance to the syringe void by drilling of a longitudinal channel through the syringe piston that enabled accessing the headspace created inside the syringe [19,20] or using the syringe as flow-through reactor [21].

Finally, it was by Anthemidis' working group to incorporate for the first time a secondary syringe pump into a LIS system and to use syringe heaters to enhance both liquid-gas transfer and reaction kinetics in the determination of ammonia with *ortho*-phthaldialdehyde [22].

3. Lab-In-Syringe—Characteristics, Potentials, and Limitations

In terms of technical versatility as automation- and flow technique, the following features and advantages of using a syringe as mixing, reaction, and extraction chamber are noteworthy:

(1) The syringe in a LIS system is a permanently sealed compartment, thus user exposure to reagents or sample contaminations during processing are avoided similarly as by using a tubing manifold.
(2) The syringe represents a size-adaptable void allowing both transferring and containing liquids, gases, and even sorbent suspension that makes it an ideal tool for the automation of laboratory procedures that require pipettes, burettes, and flasks. Consequently, the syringe now executes these functions for acting both as the pump and the mixing chamber in a very compact system.
(3) Size adaptation and the wiping action of the syringe piston on the inner wall of the syringe barrel also implies that the small remaining space of the emptied syringe requires cleaning. Therefore, this step can be done more efficiently and faster than in SIA-operated flow-batch either by fast aspiration of the cleaning solution with mixing by resulting turbulence or by in-syringe magnetic stirring. Using SIA for flow-batch operation, the entire chamber must be filled with the cleaning solution. Moreover, it should be stressed out that each solution transfer requires, in fact, two

steps, e.g., one for sample aspiration into the holding coil and a second for sample transfer into the mixing chamber.

(4) The size adaptability allows changing of the pressure inside the syringe if the syringe's head valve is turned to a permanently closed position. This can promote analyte evaporation at reduced pressure [19,22–24] or to solvate a gaseous compound at increased pressure [22].

(5) Placing a magnetic stirring bar inside the syringe allows using a far higher stirring rate than in an open chamber where the liquid content would be lost at vigorous stirring by splashing. Moreover, in-syringe stirring can be performed at any stage of the procedure and prolonged as needed. Also, presence of air bubbles or immiscible phases, i.e., extraction solvents, varying sample viscosity, or mixing of very small with very large solution volumes are without difficulty while in tubing-based FT, each item alone can present a significant challenge. Finally, in-syringe stirring allows for efficient void cleaning between analyses. Summing up, we consider it a waste of potential if choosing LIS for procedural automation but without in-syringe stirring. In this context, we draw attention to Figure 1 summarizing the modes to induce in-syringe stirring.

(6) The usual glass syringes are transparent, allowing its use as a cuvette for photometric detections. Even if detection is done outside the syringe, mixing the solutions homogeneously means that the Schlieren effect can be avoided, which is often troublesome in dispersion-based FT [25].

(7) Finally, a drilled-through piston enables (i) a second entrance to the syringe void [19], (ii), linkage to hyphenated instrumentation [21], and (iii) using the syringe as a flow-through reactor [21].

Summing up, LIS allows, in contrast to tube-based FT, reproducible treatment of samples of several milliliters in a sealed, size-adaptable reactor that enables the addition of further reagents, on-demand homogenous mixing, gas-liquid separation, in-situ detection, pressure adaption, and is more efficient in terms of system cleaning compared to the flow-batch approaches. Despite this versatility, there are still certain disadvantages against FT based on tubing manifolds:

(1) The main disadvantage is the large dead volume of the syringe that requires cleaning after each analysis. This step can take > 1 min or up to 30% of the total procedural time while most FT are "self-cleaning" by the action of a carrier flow. Performing this step efficiently is discussed in the previous listing, item 3 and section 5.12.

(2) Reproducible formation of a concentration gradient "in-space" is impossible while it is easily formed in the holding coil of an SIA system just by aspiration of different solutions. This tool can be used to determine two analytes of distinct reaction kinetics depending on the chemical milieu or to compensate for matrix effects. On the other hand, there is the possibility to aspirate solutions either stepwise or gradually into the syringe so that a gradient "in-time" is still feasible.

(3) Solution cooling and heating are more efficient in a thin tubing manifold than as for the bulk solution inside the syringe yet syringe heating to promote analyte volatilization has been reported recently using a resistance wire wrapped around the syringe barrel [22]. Earlier, we integrated an efficient heater into the short holding coil to enhance reaction kinetics [14].

(4) Similarly, sample digestion has been, to our knowledge, not yet accomplished inside the void of a LIS system. Classical pretreatment techniques based on membranes, i.e., gas-diffusion, are equally not feasible in-syringe yet the suitability for automation of head-space extraction can partly compensate for that.

(5) The main advantages of the flow-batch techniques, i.e., FT based on using an open mixing chamber, over LIS is that a sensing element, e.g., an electrode, can easily be integrated into the system (placement into the chamber), that open-port sampling is enabled, i.e., interfacing the analyzer with manual pipetting or other instrumentation, and that reactions that generate gases cannot induce critical overpressure.

4. Modes of Operation and Automated Methodologies

In the following, we give an overview of the possible operation modes and methodologies that have been automated using Lab-In-Syringe focusing rather on technical challenges than aiming for a comprehensive review of previous applications. Coupling LIS to other analytical instrumentation will be also addressed.

In the first place, the simplicity of automation of standard laboratory procedures, i.e., chromogenic assays, by LIS should be pointed out. In contrast to most FTs, mixing a large sample volume with step-by-step added reagents is straightforward and does not require such optimization as studying solution stacking and zone dispersion. Mere downscaling of an existing protocol and using the pre-optimized reagents is likely to work from the start. As the only complication, in-syringe solution heating could be quoted but was reported howbeit [14,22]. Since nearly instantaneous homogenous mixing can be achieved using a stir-driver, measurement before reaching the reaction steady-state can be considered possible, similarly as in another FT.

Only a few papers report on the automation of only chromogenic assays, all abstaining from in-syringe stirring but using an additional mixing chamber, stepwise aspiration of solutions, or air for solution homogenization inside the syringe. Using an almost identical system as in early LIS works, Ma's group reported on the analysis of ammonium and chromate in waters [26,27]. While the LIS concept was applied and cited, the authors invented their own flowery expression of "integrated syringe-pump-based environmental-water analyzer (iSEA)". In two works from Koscielniak's group, in-syringe mixing was integrated into FT analyzers for flexible preparation of sample solutions for iron determination studying chemometric tools for reduction of matrix effects [28,29]. Finally, the LIS-automated determination of ester content in biodiesel was reported where ethanol was used for in-syringe preparation of a homogenous phase of the hydrophobic sample and aqueous reagents to accomplish a chromogenic reaction [30].

Most of the developed analytical applications based on LIS have focused on the automation of various sample pretreatment procedures, which are summarized schematically in Figure 2. We also refer to commented videos on Youtube channel "In-Syringe Analysis/Lab-In-Syringe".

It is out of scope to represent and discuss all operation modes in a tutorial so that we chose those which added technical or procedural novelty no aiming for a comprehensive listing. Three review articles on automated liquid phase sample preparation are recommended instead [31–33].

Among all LIS applications, automation of LPME approaches and, in particular, DLLME, has been reported most often. Solvents both lighter and denser than water have been used following the schemes indicated in Figure 2A,B, respectively. Ideally, the phase to be used for detection or injection into an online coupled instrumentation is expulsed first to avoid carry-over so that an upright syringe orientation is advantageous for using a floating solvent and vice versa. Such systems have been used repeatedly ones combining LIS-DLLME either with mixing chambers or in-syringe stirring. Spectrophotometric detection has been used mostly taking advantage of analyte-selective chromogenic reactions or ion-pair formation [17,18] including utilization of liquid waveguide capillary cells [14], in-syringe detection [13], or fluorescence measurement [15,16].

Moreover, the online coupling of LIS-DLLME of phthalates and UV-filters to GC-MS via a micro-injection valve was reported including in-syringe silylation inside the extractant [34,35]. DLLM-extract collection for posterior scintillation counting of the enriched ^{99}Tc was reported where LIS automation significantly improved the time efficiency of an otherwise tedious sample cleanup [36].

Generally, phase separation is based on droplet floatation/sedimentation and spontaneous coalescence. However, this approach works best and fastest for solvents of low viscosity and in the absence of surfactants, which can cause the formation of stable emulsions. In this sense, Maya and co-workers demonstrated an alternative approach and collected 1-dodecanol droplets by in-flow solidifying in a 3D printed Peltier cooled phase separator that was handled by simple robotics [37].

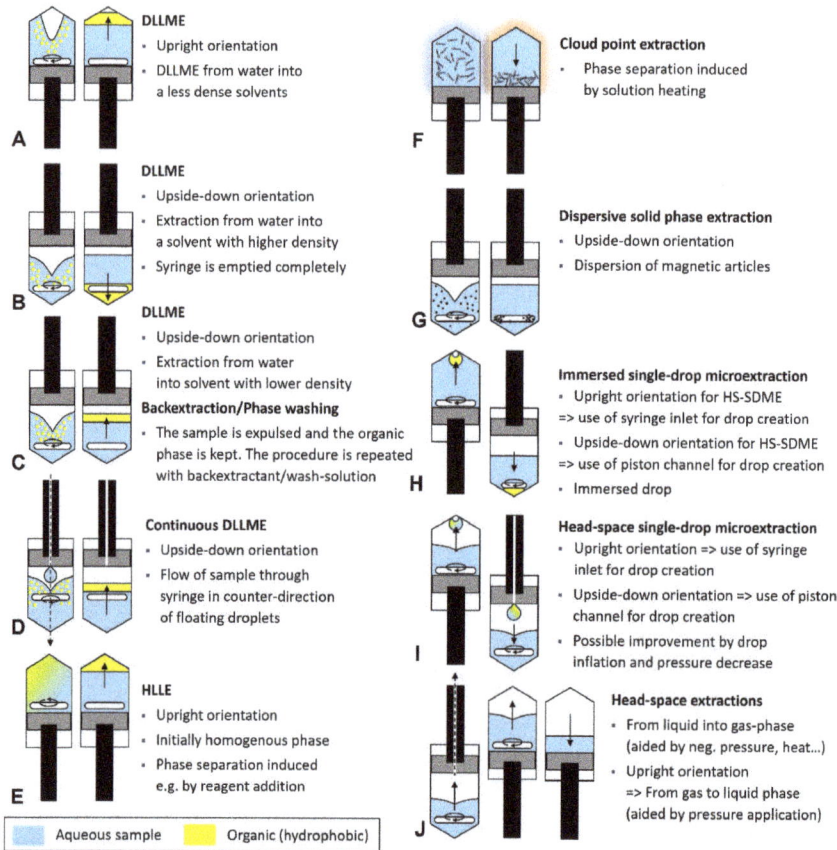

Figure 2. Overview of sample pretreatment procedures automated by Lab-In-Syringe with in-syringe stirring. In **A–G**, the left scheme indicates the extraction phase while the right one shows phase separation, in **H–J**, different modes of a similar methodology or operation are shown. Black arrows indicate the direction of analyte transfer, dotted back arrow indicate flow.

LIS-DLLME has been further coupled online to atomic spectrometric techniques ETAAS [38] and ICP-AES [39] achieving stirring-assisted solvent without any stir-driver (see Figure 1A). To achieve compatibility of extractant xylene with ICP-AES, a heated nebulization chamber had to be used [39]. An alternative to the direct injection of the organic extract and a mode that circumvents compatibility problems of the coupled instrumental technique is analyte back-extraction equally automated in-syringe. Using a floating solvent, an upside-down orientation of the syringe allows keeping the organic phase inside while discharging the sample. Hereafter, an appropriate reagent can be aspirated for dispersive back-extraction (Figure 2C) with posterior photometric detection [21,40] or coupling to ICP-AES [41]. For a solvent denser than water, the organic phase can be "parked" in the connected manifold to enable the complete discharge of the sample or, consistently, the syringe is used upright. A washing step for the elimination of remains of the sample matrix can be carried out between sample discharge and back-extractant aspiration. Here, mixing by turbulence suffices while renewed stirring and consequent solvent dispersion will prolong the procedural time by pausing for phase separation. This step can also be used for solvent washing for which a lower stirring rate can be advantageous to avoid long phase separation times [18].

Recently, we demonstrated DLLME with back-extraction converting the syringe into a flow-through chamber that was enabled by a channel in the syringe piston (Figure 2D). The sample was entering the syringe from above and flowing out below while a loss of solvent was hindered by keeping the velocity of the sample flow lower than the floatation speed of the solvent droplets [21]. Concerning alternatives to classical liquid-liquid extraction and related solvents, ionic liquids and deep eutectic solvents have been used in LIS-automated LPME methodologies. For instance, UV-filters were extracted into an ionic liquid and separated by coupled HPLC. An additional syringe was used in this case for extractant dilution prior to injection [42]. Elsewhere, a deep eutectic solvent was used as disperser as a modification of earlier reported chromate extraction as diphenylcarbazone-Cr(III) complex [17,43].

Bulatov's group has reported on LIS-automated homogenous liquid-liquid extraction (HLLE, Figure 2E). HLLE starts with a homogenous phase of the sample and a water-miscible solvent that exhibits better compatibility with liquid chromatography than classical extraction solvents. Phase separation is induced, e.g., by the addition of salt, change of pH, or addition of a tertiary solvent [31]. An important advantage of HLLE is that emulsion formation and loss of extraction solvent on matrix components such as particles are neglectable compared to the relatively large volume of used solvent and even only moderately hydrophobic compounds can be extracted efficiently. In a first work, phase separation of acetonitrile-water was induced by the addition of a concentrated glucose solution that enabled the extraction of pesticides from juices and separation by online-coupled HPLC [44]. In the second work, a switchable solvent (DEHPA) was used that formed a homogenous phase with the sample under alkaline condition (DEHPA ionized). Phase separation was then induced by aspiration of acid into the syringe to extract antimicrobial drugs before HPLC separation [45].

Moreover, in-syringe automated cloud point extraction (CPE) has been reported twice using Triton X-114 as the surfactant (Figure 2F) for the extraction of colored analyte derivates of antimony and epinephrine with subsequent spectrophotometric detection. In the first work, required solution heating to yield phase separation and enrichment of antimony as iodide complex was achieved by in-syringe dilution of high concentrated sulfuric acid [46]. A heating block was used in another work reporting CPE of epinephrine where the syringe was used for phase separation and abscission [47].

Maya and co-workers reported on LIS-automated dispersive solid-phase microextraction on model analyte malachite green and estrogens. They used a magnetic nanostructured sorbent (metal-organic framework) inside the upside-down oriented syringe for this purpose (Figure 2G). The sorbent particles were dispersed by in-syringe stirring while stopping the stirring, the magnetic sorbent was attracted to the stirring bar. The sample was easily exchanged first by a washing agent to remove any sample remains and then by a suitable eluent, all without risking a loss of sorbent particles [48,49].

Two works report on LIS-automated extraction into a drop of extraction solvent that is in direct contact with the sample. Advantages are a simpler setup since fast stirring was not required and omission of emulsion formation as low stirring speed suffices and does not impair drop integrity. A drop of solvent, that simply floated on the aqueous sample and was stabilized by the stirring vortex, enabled extracting silver as thiocarbamate complex with subsequent analysis by online connected ETAAS [50]. In another work, lead was determined by spectrophotometry after being extracted as dithizone complex comparing two approaches for drop formation (Figure 2H) [51]. With the syringe upright, a drop of floating solvent clang on the syringe inlet and was stabilized by an air bubble inside. For the second approach, the syringe was used upside-down and a drop of chloroform was used with a lentil-shaped stirring cross turning slowly inside the drop. The new approach of "in-drop stirring" was superior: Both drop stabilization and surface movement that enhanced analyte transfer were achieved and a stirring time of only 150 s was sufficient for quantitative extraction.

An interesting feature of LIS is the gas-tightness and size adaptability of the syringe, which enabled the automation of various sample pretreatment methodologies involving gas-liquid transfer as well as positive and negative pressure applications. In-syringe head-space single-drop microextraction (HS-SDME, Figure 2I) was first applied to the selective determination of ethanol in wines using a mixture of sulfuric acid and chromate as chromogenic drop reagent that turned green at the reaction with the evaporated ethanol from the sample inside the syringe. The process was supported by negative pressure application and inflating the drop surface by a small bubble inside [23]. Adding a second syringe pump, the precise formation of a drop of palladium nanoparticle suspension was done by Anthemidis' group to extract elementary, thus volatile, mercury from the sample with subsequent transfer of the enriched drop of Pd nanoparticles to ETAAS [24]. In a third work, contact of the drop with the head-valve manifold was strictly avoided. The syringe pump was placed upside-down and a drop of indicator solution was generated inside the headspace by pushing the reagent with a second pump through a drilled channel in the first syringe piston. The color change of the drop was then measured online by fiber optics allowing the determination of ammonium in surface waters [19].

Finally, headspace extraction, gas transfer, and capturing gaseous compounds into an appropriate reagent is feasible (Figure 2J). Online coupling of a simple LIS system to GC-FID was reported that allowed the transfer of 80% of the gaseous phase after in-syringe headspace enrichment with the volatile analytes (benzene, toluene, ethylbenzene, xylene). This simple procedure yielded similar sensitivity than former reports requiring a secondary analyte enrichment or trapping [20].

In a LIS system combing two syringes, ammonia was forced to volatilize in syringe 1 that was assisted by heating and negative pressure application. Subsequently, the analyte was captured and let react with the fluorogenic reagent *ortho*-phthaldialdehyde in syringe 2 that was assisted equally by heating and by compression of the transferred gas [22]. Finally, for the determination of dissolved inorganic carbon, in-syringe capturing of released CO_2 was done by an indicator solution [52].

5. Tips and Tricks for System Setup, Method Development, Optimization, and Characterization

After this overview that demonstrates the high versatility of Lab-In-Syringe, in this section we give tips and tricks that should assist in the instrumental setup and use of the LIS technique. Setups of Lab-In-Syringe systems including 3D printed system elements are shown in Figures 3–5.

Figure 3. Lab-In-Syringe system from a syringe pump (SP) with 2 port head valve (HV) with an upright oriented syringe (S). 3D printed elements are shown in blue color and disassembled, required screws are not shown. A simple stir-driver (5) is connected to a motor (M), featured from a computer fan moving, with a removable top-attachment with pulley wheel (3) via a rubber band. On the piston carriage (PC), a universal adapter (1) allows fixing a holder for the motor (2) and a lifting lug (4) to move stir-driver and motor upwards with the syringe piston. An SLA-printed 5 cm detection flow cell (6) is mounted on top of the syringe pump using a universal adaptor (7). The photo shows the SLA-3D printed adaptor (item 1) for easy mounting additional elements to the piston carriage.

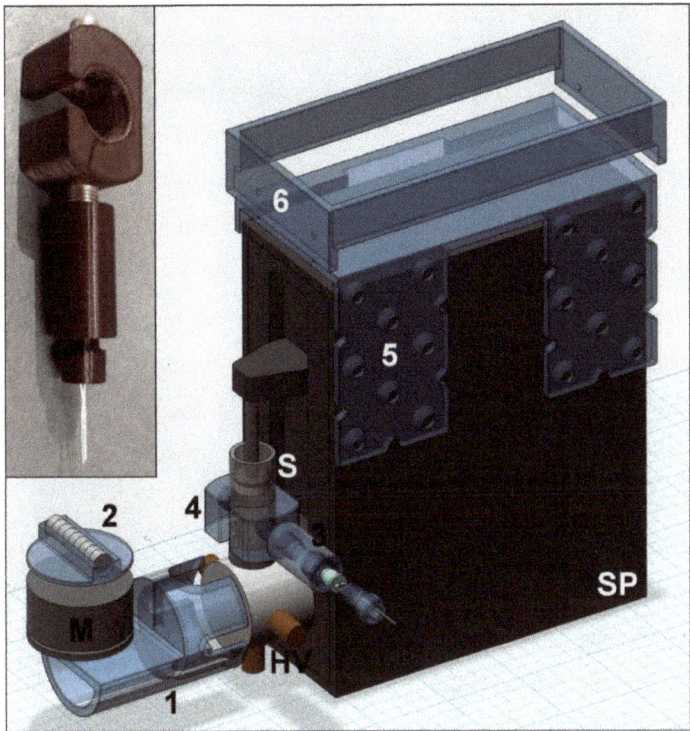

Figure 4. Lab-In-Syringe system from a syringe pump (SP) with 3 port head valve (HV) with an upside-down orientated syringe (S). 3D printed elements are shown in blue color and disassembled, required screws are not shown. A removable adaptor for the head valve (1) allows the positioning of the motor (M) close to the syringe. A holder for NdFeB magnets (2) is glued on-top that induce stirring bar rotation inside the syringe at the motor start. A LED holder (3) is screwed onto a fiber-optic adaptor (4) that is placed onto the glass barrel for in-syringe photometric measurements (Photometer and optical fibers not shown). Elements 5 and 6 compose a tray and universal support for deposition of tools (fittings, screwdriver, etc.) and to attach further items, e.g., the relay board for motor control. The photo shows the 3D printed elements 3 and 4 for in-syringe spectrophotometry.

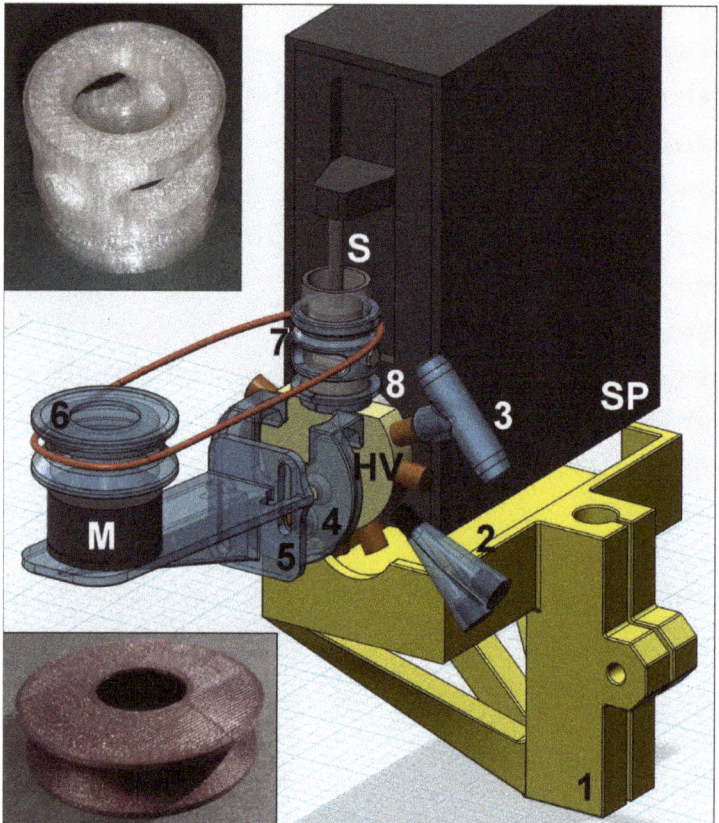

Figure 5. Lab-In-Syringe system from a syringe pump (SP) with nine port head valve (HV) with an upside-down orientated syringe (S). 3D printed elements are shown in blue color and disassembled, required screws are not shown. The syringe is situated on a laboratory stand compatible support (1). On the head valve, SLA-printed elements add the functionalities of a flow-through port (2) and connecting a drain tube to waste as well as air aspiration (3), respectively. Elements 4 allow mounting further elements on the head valve, here a support (5) for positioning and leveling of the stirring motor (M). The motor is equipped with a pulley-wheel (6, photography below) connecting the motor via a rubber band to a simple stir-driver (7, photograph above) that is held in place by a supporting ring (8).

5.1. Syringe Orientation

One of the first steps surely is to decide on how to position the syringe. Upright syringe orientation or the use of a secondary syringe are advantageous if highly precise liquid handling or pressure application is required or for such DLLME procedures using a solvent with lower density than water (Figure 2A). For DLLME using a floating solvent, it has been also proven most helpful to aspirate air into the syringe to create an open liquid surface. This enables vortex formation during in-syringe stirring and facilitates the dispersion of the extraction solvent.

Vice-versa, upside-down syringe orientation will be more suitable for dispersing a solvent that is denser than water (ionic liquids, halogenated solvents, deep eutectic solvents, etc.) as the stirring bar will be situated on the bottom of the syringe and close or inside the solvent and promote dispersion (Figure 2B). This setup has also proven useful if a floating solvent is used for a procedure involving back-extraction so that the sample must be expulsed first (Figure 2C).

Often, we found the upside-down syringe orientation to be the better choice as it allows emptying the syringe nearly completely and the syringe piston is not in contact with the sample. This can be advantageous in the case of particulate sample matrix or hydrophobic compounds sticking to PTFE. To minimize friction and eventually mechanical wearing of the piston in this arrangement, add a drop of glycerol as lubricant into the upward opening of the syringe barrel.

In upside-down orientation, the user must plan that each liquid propulsion must be followed by at least 1–2 s of pausing due to the compressibility of the air cushion inside the syringe that delays any liquid movement. In consequence, this setup is less suitable for handling viscose liquids or for small volumes. Moreover, tubing of larger or equal diameters than 0.8 mm i.d. should be used in combination with a 2.5 mL syringe to reduce the formation of negative pressure at aspiration at higher speeds and solution degassing.

Before system and method setup, preliminary experiments on the expected extraction conditions (pH, ion-pairing reagent, etc.) and, foremost, to select possible extraction solvents (density, stickiness, etc.) is surely advantageous as this will decide on syringe orientation. The optimization is then started using the same solutions and volumetric ratios as used in the standard procedure or tested.

5.2. Solvents and Chemicals

There are hardly any limitations on usable solvent or reagents as contact surfaces of the syringes are typically glass and inert plastics. However, the syringe piston will eventually be attacked by the solutions inside the syringe so that regular cleaning of the syringe piston is recommended. In case of high ionic strength of sample or reagents, incrustation of salt must be prevented/removed as it will damage the syringe piston with time. In principle, the same rules apply as for seal wash in liquid chromatography. Low viscosity, low water solubility, high vapor pressure, and a significant difference in density towards the aqueous sample are beneficial features of a solvent intended for DLLME procedures to achieve efficient solvent dispersion, phase separation, and droplet coalescence and to avoid degassing or solvent loss by dissolution in the aqueous phase. If stable emulsions are formed, nondispersive approaches such as DI-SDME will be more suitable. One imitation to be highlighted is that hydrophobic solvents tend to stick to the PTFE surfaces of syringe piston and inlet. Here, cleaning with moderate hydrophobic solvents such as isopropanol, diminishing the inlet diameter (see Figure S1), and an upside-down syringe orientation are helpful.

5.3. Syringe Model, Holding Coil, and Head Valve

We have worked for a long time with Cavro® automatic syringe pumps of 3 cm and 6 cm stroke length with rotary head valves (Tecan Trading AG, Switzerland) but there should be no limitations regarding the syringe pump model. Pressure stability > 2 bar of the syringe head valve is highly desirable to avoid problems related to limited robustness and all materials in contact with liquid should be chemically inert (PTFE, glass, PEEK, etc.).

The holding coil is a system component that will be required only occasionally and should generally be short to maximize the volume that can actually be used for the in-syringe accomplished procedure. Complete omission of a holding coil is possible using syringe pump models that feature a multiposition head valve (six or nine ports). It is indeed astonishing how much these pumps have proven fit-for-purpose so that it remains an open question whether they were originally intended for some in-syringe automation.

However, even on a 9-port head valve, the lateral ports are rapidly occupied for waste disposal, water, cleaning solution, sample, a detector or secondary instrument, e.g., HPLC, and possibly standard solution for in-syringe addition, buffers, solvents, etc. To build up positive or negative pressure inside the syringe, a permanently closed position is further required. Finally, aspiration of air might be needed and is highly useful to enable vortex formation inside the syringe, gas-liquid separations, or aspiration of all liquid from the holding coil into the syringe void.

The required position for waste and air can be combined in particular if the syringe is used in an upside-down position. For this, a PEEK tube of only a few centimeters is connected to the selection valve and pierces on its other side a wide tube (several millimeters) that acts as a drain to waste. This way, the dead volume required for waste disposal is minimal and, considering that in this syringe orientation, the air is the last thing to be expulsed from the syringe at content discharge, aspiration of any undesired solution with air is neglectable. An adapter for this purpose is shown in Figure 5.

5.4. Syringe Dimensions

The length-to-diameter ratio of the syringe can be of some importance. For example, a slender syringe will facilitate the separation and removing the organic from the aqueous phase while exhibiting a small surface area in case of gas-liquid transfers. In the case of headspace extraction, a wider diameter at equal syringe volume will be advantageous. In respect to the syringe size, one important argument for choosing a smaller one is this allows building up a higher pressure with the available force of the syringe pump, which might be a limiting factor in some applications. Decision criteria can further include the available sample size or the aimed preconcentration factor. Finally, the shape and material at the transition zone between syringe inlet and barrel should be contemplated; a smooth outflow of both liquid and bubbles at emptying the syringe is highly desirable. Moreover, we would recommend a syringe diameter so wide that in-syringe stirring is feasible and air bubbles cannot segment the liquid column inside the syringe.

5.5. Syringe Inlet

To minimize the dead volume in the syringe inlet we have proposed to insert a piece of 0.8 m tubing as shown in Figure S1B that also increases the flow velocity and turbulence for stirring-less in-syringe DLLME [14]. To increase the hydrophilicity of the syringe inlet we have used a glass capillary to stabilize a liquid drop in the syringe inlet for headspace single-drop microextraction [23].

5.6. Motors

Slow-turning motors can be taken from old VCR devices. For faster speeds, we have good experience featuring motors from computer fans as shown in Figure S2A–D. Models with pulse width control have proven more reliable in terms of speed stability and starting power at low speed. Moreover, they are easily regulated and brushless, i.e., they show a far longer lifetime than cheap hobby motors which will fail after a few weeks. Higher priced models are suitable for high spinning speeds and show generally a greater momentum than the ones from computer fans. Several analog control circuits have been described [16,18,41]. In the long term, power control by a microcontroller as recently proposed [37] is surely the way how to improve the reliability of motor operation, a key issue of LIS. In Figures 3–5 and Figure S2E we show 3D printed motor supports as well as motor attachments that either facilitate fixation of magnets on top of the employed motor or that serve as a pulley wheel if a stir-bar driver is utilized and must be forced to rotate via a rubber band.

5.7. Stirring Without Stir-Driver

We can only recommend using the approach of in-syringe stirring since we consider the gain in versatility higher than its backsides, i.e., a higher dead-volume and required system extension. In all cases, we used NdFeB magnets and commercial stirring bars for this purpose with positive results.

The simplest option to accomplish in-syringe stirring is to place a motor of low to a moderate velocity close to the syringe with a pile of disk magnets fixed on top. Adding or removing magnets then allows adapting the magnetic force (see Figures 1A and 4). In consequence, the stirring bar inside the syringe will follow the rotation speed of the motor. For simple mixing at low spinning speed, there is no need for computer control so that this option is ideal for the automation of bare chromogenic assays or headspace extractions [20,23]. To avoid tumbling or bouncing of the stirring bar, a short stirring bar or a stirring cross (Figure S1D) that is a fair approximation to a round disk are ideal. The exact placement of the motor using such stirring bars will be of low importance.

If higher speeds are desired, we found that the force of the stirring bar and of the magnetic pile must be higher and the motor must be placed closer to the syringe. However, in such a case it is crucial that the motor exhibits a high momentum at its start. If too weak, the magnetic pile and thus the motor are retained by the attraction from the stir bar inside the syringe. We recommend using a stirring bar that fits smoothly into the barrel and to start at low rotation speed before accelerating [18,41]. The motor must also be lifted with the piston to assure leveling of the magnetic pile and bar magnet and avoid stir bar bouncing. Optimization will take some effort yet stirring fast enough for DLLME are feasible.

5.8. Stir-Driver

Alternatively, an element that attracts and consequently aligns the stirring bar from its both sides can be used, which we have denoted, in lack of a better term, "stir-driver". It is forced to rotate around the syringe axis by the motor via a simple rubber ring (See Figure 1B–D, Figures 3 and 5) purchasable at any stationery store. Evenness, close-to-round profile, and > 15 cm of circumference are fit. These rubber rings degrade in laboratory air and must be exchanged at times.

Two kinds of stir-drivers have been proposed so far that can be produced with little effort in a tool shop or from hardware store material. Using the syringe in an upright position, the stirring bar must move up and down with the piston it is laying on so that the creation of a magnetic field along the full stroke length of the syringe is required. As shown in Figure 1B, only two plastic rings, one with an annual groove for the rubber band, that fits smoothly over the barrel, two long iron screws, and two smaller NdFeB magnets are needed. A thorough description can be found elsewhere [16].

A simpler stir-driver can be used when the stirring bar remains in the same position, typically when the syringe is used upside-down [18]. Here, a single plastic ring, e.g., featured from a piece of PVC water pipe, with two holes for the two, NdFeB magnets opposing each other and an annular groove is required as shown in Figure 1C. An improved version is easily produced by 3D printing that allows adding cut-outs for better observation of the inner part of the syringe as shown in Figure 1D. As the magnetic force is more focused in this design, stirring rates up to 3000 rpm are possible without that the stirring bar "loses track" but rotates synchronously with the driver. At such high velocities, the use of a dispersion solvent should not be required anymore. Moreover, the NdFeB magnets levitate the stirring bar inside the syringe so that friction is minimal and lifting the stir-driver will allow dislocating the stirring bar to the perfect position, e.g., the boundary between organic and aqueous phases as previously described [21].

5.9. In-Syringe Detection

The simple stir-driver or a driver-less system setup leave space for using an optical-fiber adaptor (see Figure 1D and Figure S1E) to enable in-syringe spectrophotometric detection [13,19]. The advantage is that there is no need for solution transfer that might cause baseline alterations by Schlieren formation. In any case, it is recommended to use a detection cell from material that is easily wettable by the phase of interest, i.e., glass for the aqueous, polymers for the organic one. A 3D printed element is shown in Figure 4.

5.10. Piston Channel

Accessing the syringe via a secondary inlet can be of use for phase separation, converting the syringe into a flow-through reactor, or creation a hanging drop inside the syringe void [13,19,20]. After drilling the piston longitudinally, a short PEEK tube can be glued in with silicone or special adhesive (Figure S1C). If not required, the piston channel can be easily closed with a short piece of silicone rubber tube as a connector and a simple plug.

5.11. Worn Out Piston Head

With time, the piston head of each syringe wears out by friction, abrasion by particles or salt crystals, and it will no longer seal sufficiently. As it is generally made of PE or PTFE that deform under pressure, a certain improvement can be achieved by pressing the piston head on a flat and smooth surface with a slightly circling motion that will bell out the piston head to a slightly wider diameter and achieve tighter sealing.

5.12. Syringe Cleaning

This step is easily done by repeated aspiration of 20–25% of the syringe volume of an appropriate cleaning solution with activated in-syringe stirring and expulsion to waste immediately or after a few seconds. As a universal approach we recommend cleaning subsequently with a pure or diluted miscible solvent (e.g., with acid or base, possibly mixed in-syringe), and water. If the syringe is in an upright orientation, i.e., it cannot be emptied completely, the final step can be performed with the intended sample if sufficient in amount to minimize undesired sample dilution. Using the syringe upside-down, additional aspiration of air and stopping the stirring during emptying is recommended to efficiently empty the syringe.

5.13. Optimization

The order of aspiration and placement of solutions on the selection valve should be carefully contemplated to minimize carry-overs and turning times. The less important solution or the largest volume should be aspirated last so that any aspiration error causes a relatively small variation in the analytical result.

During in-syringe mixing of chlorinated solvents and aqueous sample, we have observed the formation of a slight overpressure that builds up due to spontaneous degassing. In upside-down orientation and solution heating, e.g., at the addition of methanol to water, the headspace will expand and equally increase the pressure inside the syringe slightly. This can deteriorate the reproducibility of the aspirated volume in this setup. The problem can be minimized, e.g., by slow aspiration of air as the final step, applying negative pressure with the head valve turned to a permanently closed position, or saturate the solvent with water.

Expectable dependencies of the signal intensity on selected experimental parameters for a LIS-automated DLLME procedure are shown in Figure S3 so to help newcomers in respect of what to expect during method optimization. Extraction and back-extraction times follow simple saturation behavior while a stepwise increase will be observed for the stirring speed where it becomes fast enough to achieve solvent dispersion. Time of phase separation can be visually controlled and so that generally no further optimization study is needed. While phase separation is typically achieved in 20-60 seconds depending on the solvent density, spontaneous droplet coalescence will vary with the solvent viscosity and can require additional time. It must be also taken into account that a part of the solvent will dissolve in the aqueous phase and depending on the ionic strength that will differ between standard and samples (Figure S3). To achieve high reproducibility, low solvent solubility, increasing the ionic strength by salt addition, or the use of a solvent volume that surpasses the solubility in water several times, is therefore required.

We would refrain from optimizing all parameters by experimental design and at once. The user must also rely on his own experience and reasoning to decide which parameters' behaviors and interactions are difficult to foresee, thus requiring such optimization, while other parameters can be fixed first to higher, "save-side" values and evaluated later, possibly in univariant studies, e.g., the stirring rate. The choice of the final conditions should not only aim for the highest performance but also the high reliability, robustness, and comparability between standard and matrix-loaded samples.

In method development it must be considered that any magnetic stirrer inside the syringe will decrease the applicable stroke length by a few millimeters, i.e., the usable void volume will decrease by a few hundred microliters. If using the syringe in upside-down orientation, this volume is then occupied by the air remaining in the syringe.

5.14. Head-Space Extractions

In the automation of HS-SDME, it should be kept in mind that even using the syringe upside-down, cleaning the syringe by stirring will also wet the piston head and can imply undesired contamination of the channel inlet, requiring e.g., a slower stirring rate. On the other hand, creating a drop in the normal entrance of the syringe [19–21] implies an even larger risk of drop contamination. The use of a secondary syringe pump of much smaller dimensions for precise drop formation has been, therefore, a significant advantage [21,24].

5.15. Determination of Performance

Preconcentration factors and extraction efficiency can be calculated from Equations (1), (2), and (3). We would like to highlight that the effective preconcentration cannot be calculated from the analytical signals but rather the analyte concentrations found in the sample and the extract. For their calculation utilizing the analytical signal, the detection sensitivity for the analyte in both sample and extraction medium must be taken into account.

$$\text{Potential preconcentration factor} = V_{Extract}/V_{Sample} \quad (1)$$

$$\text{Effective preconcentration factor} = c_{Extract}/c_{Sample} = \frac{Signal_{Extract} \times Sensitivity_{Sample}}{Signal_{Sample} \times Sensitivity_{Extract}} \quad (2)$$

$$\text{Extraction efficiency} = c_{Extract} \times V_{Extract}/c_{Sample} \times V_{Sample} \quad (3)$$

5.16. Determination of Stirring Rate

We recommend using a laser tachometer that can be purchased for less than 30 Euro. If a stir-driver is used, the reflective foil is fixed on the stir-driver or the ratio of the diameters of the pulley wheel and the stir-bar driver must be taken into account. For low stirring rates, the use of a bicycle tachometer is possible as it uses a hall sensor so that measurement is based on moving magnets. However, the high rotation speeds often used for in-syringe stirring are likely to be outside the typical application range of such devices so that the results might be unreliable.

5.17. Determination of Dead Volumes

For characterization of a system as given in Figure 1B (syringe is upright) it can be useful knowing the dead volumes of the holding coil and the space inside the syringe that cannot be emptied due to the stirring bar. For this, we recommend the following:

(1) The absorbance of a solution of a highly water-soluble, stable, and colored substance (e.g., 10 mmol/L potassium dichromate) is measured in a dilution of 1:10 offline in a glass cuvette, yielding Abs1. If the LIS integrates spectrophotometric detection, the syringe is cleaned several times with the solution that is then finally passed through the flow cell.
(2) A 1:10 mixture of dye solution and water is then prepared in-syringe. The syringe and holding coil must be cleaned and initially filled with water and the syringe piston at its highest position. First, the dye solution is aspirated from one lateral port of the selection valve and then the volume of water from a second port. After mixing, the syringe content is pushed out through an empty channel, collected and measured likewise, yielding Abs2. The volume of water that was not aspirated into the syringe corresponds to the volume that has been originally inside the holding coil so that an additional dilution is caused only by the water that was already inside the syringe (the dead volume to be evaluated).

(3) The procedure is then repeated but starting with an empty holding coil, yielding Abs3. If correctly performed, the absorbance values should decrease in the order Abs3 > Abs1 > Abs2. From these values, the dead volumes can be calculated by Equations (4), (5), and (6) abstaining here from any detailed mathematical derivation:

$$V_{syringe} = (Abs1 - Abs2)/Abs2 \times Dilution \times V_{Dye} \text{ here, Dilution} = 10 \tag{4}$$

$$V_{total} = V_{Water} \cdot 1/(1/(Abs3/Abs1 \times V_{Dye}/(V_{Dye} + V_{Water}) - 1)) \tag{5}$$

$$V_{holding\ coil} = V_{total} - V_{syringe} \tag{6}$$

6. Contributions to Lab-In-Syringe by 3D Printing

Lab-In-Syringe is perhaps the latest offspring of FT and finding commercial support is thus difficult. On the other hand, there are certain elements required for system setup, first and foremost the stir-driver, that appear to require a specialized tool shop. However, fabrication of the required elements that include supports for motor and syringe pump, stir-drivers, motor attachments, e.g., a pulley-wheel for forcing the stir-driver to rotate via a rubber band or a magnet holder, as well as elements in contact with solutions can easily be accomplished by 3D printing.

The two most commonly used and economic 3D printing techniques that fully comply with all needs are stereolithography (SLA) and fused deposition modeling (FDM). Printers based on both techniques are available for about 300 Euro. Now, there are so many users that the likelihood is high that a printer can be found in the circle of acquaintances. Besides, the printing material is cheap and adequate designing software for element design can be downloaded for free or used online.

In SLA, the object is created layer-by-layer from a monomer solution that polymerizes/hardens where it is illuminated by a laser or an LED matrix. The technique yields high surface quality, high special resolution, and liquid-tight sealed structures, making SLA useful for elements in liquid contact. However, the choice of available polymers is limited, solvent resistivity is rather low, and only smaller objects can typically be printed.

FDM deposits molten material, provided as a filament, in thin strings that are laid out so close to each other that a solid object is created, row-by-row and layer-by-layer. The selection possibilities of materials and printable sizes are far larger but obtaining liquid-tight layer-bonds and smooth surfaces require some optimization or post-print treatment, e.g., solvent vapor smoothing. The technique is ideal for printing supports, motor attachment, and stir-drivers. In fact, we found that SLA printed stir-drivers had a velvety surface that exhibited too much friction for proper operation. On the other hand, FDM prints with PLA showed the required slippery, smooth, and hard surface.

In Figures 3–5, we show and explain three exemplary LIS setups with about 20 elements that were 3D printed using FDM technology (if not indicated otherwise) as well as successfully tested and used for LIS applications. Selected 3D printed elements are shown as photos in Figures 3–5 and in Figure S2E with specific functions indicated in the respective captions. The elements include supports for motors, e.g., that can be easily attached to the syringe pump head valve, a lifting stand for a syringe pump (of interest if the pump is turned upside-down), motor attachments, stir-drivers, motor heads, flow adaptors to increase the functionality of the head-valve, and detection related elements. By this, we aim for an overview of how to use 3D printing for fit-for-purpose imperative as well as useful system components. Moreover, in addition to Figures 1 and 2, they show a more visual impression of possible setups of Lab-In-Syringe systems.

7. Conclusions

Lab-In-Syringe can be considered as a hybrid of flow-batch analysis and SIA. It diverts the typical setup of an SIA system from its intended use and breaks the unwritten rules of its parental technique. On the other hand, it is indisputable that LIS is a versatile alternative to tubing-based FT, that it allows automation of many sample pretreatment techniques, and shows specific advantages, most importantly, the ease to treat milliliter sample volumes. In this article, we have aimed for bringing LIS closer to newcomers and users of other FT by showing the different ways how this flexible technique can be used with high benefit. We hope that with time the LIS technique will spread even more and become, as SIA, recognized not only as "Tools of Inventive Science" but as standard laboratory automation technique.

The similarity in sample volumes and required instrumentation (a syringe pump and a solution selector) between Lab-In-Syringe and autosampler systems allows the straightforward merging of both techniques, i.e., using autosampler instrumentation for LIS automation.

In this work, we have shown the different modes of operation and examples with simple but useful additions and devices for LIS analyzer systems enabled by 3D printing. We believe that the LIS technique will undergo further diversification and that new developments will follow that also include this prototyping technique.

Supplementary Materials: The following items are available online, Figure S1: Details to LIS system adaptation; Figure S2: Preparation of a brush-less motor for in-syringe stirring; Figure S3: Typical signal dependencies in the optimization of DLLME protocols automated by Lab-In-Syringe.

Author Contributions: Conceptualization, writing, graphic preparation, reviewing, and editing, B.H., project administration, and funding acquisition, P.S. All authors have read and agreed to the published version of the manuscript.

Funding: This research was supported by the project EFSA-CDN (No. CZ.02.1.01/0.0/0.0/16_019/0000841) co-funded by ERDF.

Acknowledgments: The authors thank I.H. Šrámková for proof-reading and helpful discussion.

Conflicts of Interest: The authors declare no conflict of interest.

Abbreviations

DLLME	Dispersive Liquid-Liquid Microextraction
CPE	Cloud Point Extraction
FDM	Fused deposition modeling
FT	Flow techniques
LIS	Lab-In-Syringe
LPME	Liquid-phase microextraction
NdFeB	Neodymium alloy as used for strong magnets
SIA	Sequential Injection Analysis
STA	Stereolithography

References

1. Růžička, J.; Marshall, G. Sequential injection: A new concept for chemical sensors, process analysis and laboratory assays. *Anal. Chim. Acta* **1990**, *237*, 329–343. [CrossRef]
2. Economou, A. Sequential-injection analysis (SIA): A useful tool for on-line sample handling and pre-treatment. *Trends Anal. Chem.* **2005**, *24*, 416–425. [CrossRef]
3. Horstkotte, B.; Miro, M.; Solich, P. Where are modern flow techniques heading to? *Anal. Bioanal. Chem.* **2018**, *410*, 6361–6370. [CrossRef] [PubMed]
4. Zagatto, E.A.G.; Rocha, F.R.P. The multiple facets of flow analysis. A tutorial. *Anal. Chim. Acta* **2020**, *1093*, 75–85. [CrossRef] [PubMed]
5. Trojanowicz, M.; Kołacińska, K. Recent advances in flow injection analysis. *Analyst* **2016**, *141*, 2085–2139. [CrossRef] [PubMed]
6. Kolev, S.D.; Mc Kelvie, I. *Advances in Flow Injection Analysis and Related Techniques*, 1st ed.; Elsevier: Amsterdam, The Netherlands, 2008; Volume 54.
7. Trojanowicz, M. Flow injection analysis. In *Instrumentation and Applications*, 1st ed; World Scientific Publishing: Singapore, 2000.
8. Hansen, E.H.; Růžička, J. Flow Injection Analysis. Tutorial & News on Flow-Based Micronalytical Techniques. Available online: http://www.flowinjectiontutorial.com (accessed on 5 March 2020).

9. Dias Diniz, P.H.G.; de Almeida, L.F.; Harding, D.P.; de Araújo, M.C.U. Flow-batch analysis. *Anal. Bioanal. Chem.* **2012**, *35*, 39–49. [CrossRef]
10. Zagatto, E.A.G.; Carneiro, J.M.T.; Vicente, S.; Fortes, P.R.; Santos, J.L.M.; Lima, J.L.F.C. Mixing chambers in flow analysis: A review. *J. Anal. Chem.* **2009**, *64*, 524–532. [CrossRef]
11. Prabhu, G.R.D.; Urban, P.L. The dawn of unmanned analytical laboratories. *Trends Anal. Chem.* **2017**, *88*, 41–52. [CrossRef]
12. Maya, F.; Estela, J.M.; Cerdà, V. Completely automated in-syringe dispersive liquid-liquid microextraction using solvents lighter than water. *Anal. Bioanal. Chem.* **2012**, *402*, 1383–1388. [CrossRef]
13. Maya, F.; Horstkotte, B.; Estela, J.M.; Cerdà, V. Lab in a syringe: Fully automated dispersive liquid-liquid microextraction with integrated spectrophotometric detection. *Anal. Bioanal. Chem.* **2012**, *404*, 909–917. [CrossRef]
14. Horstkotte, B.; Alexovič, M.; Maya, F.; Duarte, C.M.; Andruch, V.; Cerdà, V. Automatic determination of copper by in-syringe dispersive liquid–liquid microextraction of its bathocuproine-complex using long path-length spectrophotometric detection. *Talanta* **2012**, *99*, 349–356. [CrossRef] [PubMed]
15. Suárez, R.; Horstkotte, B.; Duarte, C.M.; Cerdà, V. Fully-automated fluorimetric determination of aluminum in seawater by in-syringe dispersive liquid-liquid microextraction using lumogallion. *Anal. Chem.* **2012**, *84*, 9462–9469. [CrossRef] [PubMed]
16. Horstkotte, B.; Suárez, R.; Solich, P.; Cerdà, V. In-syringe stirring: A novel approach for magnetic stirring-assisted dispersive liquid-liquid microextraction. *Anal. Chim. Acta* **2013**, *788*, 52–60. [CrossRef]
17. Henriquez, C.; Horstkotte, B.; Solich, P.; Cerdà, V. In-syringe magnetic-stirring-assisted liquid-liquid microextraction for the spectrophotometric determination of Cr(VI) in waters. *Anal. Bioanal. Chem.* **2013**, *405*, 6761–6769. [CrossRef]
18. Horstkotte, B.; Suárez, R.; Solich, P.; Cerdà, V. In-syringe magnetic stirring assisted dispersive liquid-liquid micro-extraction with solvent washing for fully automated determination of cationic surfactants. *Anal. Meth.* **2014**, *6*, 9601–9609. [CrossRef]
19. Šrámková, I.; Horstkotte, B.; Sklenářová, H.; Solich, P.; Kolev, S.D. A novel approach to Lab-In-Syringe Head-Space Single-Drop Microextraction and on-drop sensing of ammonia. *Anal. Chim. Acta* **2016**, *934*, 132–144. [CrossRef] [PubMed]
20. Horstkotte, B.; Lopez de los Mozos Atochero, N.; Solich, P. Lab-In-Syringe automation of stirring-assisted room-temperature headspace extraction coupled online to gas chromatography with flame ionization detection for determination of benzene, toluene, ethylbenzene, and xylenes in surface waters. *J. Chromatogr. A* **2018**, *1555*, 1–9. [PubMed]
21. Fikarová, K.; Horstkotte, B.; Sklenářová, H.; Švec, F.; Solich, P. Automated continuous-flow in-syringe dispersive liquid-liquid microextraction of mono-nitrophenols from large sample volumes using a novel approach to multivariate spectral analysis. *Talanta* **2019**, *202*, 11–20. [CrossRef] [PubMed]
22. Giakisikli, G.; Anthemidis, A.N. Automatic pressure-assisted dual-headspace gas-liquid microextraction. Lab-in-syringe platform for membraneless gas separation of ammonia coupled with fluorimetric sequential injection analysis. *Anal. Chim. Acta* **2018**, *1033*, 73–80. [CrossRef]
23. Šrámková, I.B.; Horstkotte, B.; Solich, P.; Sklenářová, H. Automated in-syringe single-drop head-space micro-extraction applied to the determination of ethanol in wine samples. *Anal. Chim. Acta* **2014**, *828*, 53–60. [CrossRef]
24. Mitani, C.; Kotzamanidou, A.; Anthemidis, A.N. Automated headspace single-drop microextraction via a lab-in-syringe platform for mercury electrothermal atomic absorption spectrometric determination after in situ vapor generation. *J. Anal. Atom. Spectr.* **2014**, *29*, 1491–1498. [CrossRef]
25. Dias, A.C.B.; Borges, E.P.; Zagatto, E.A.G.; Worsfold, P.J. A critical examination of the components of the Schlieren effect in flow analysis. *Talanta* **2006**, *68*, 1076–1082. [CrossRef] [PubMed]
26. Zhu, X.; Deng, Y.; Li, P.; Yuan, D.; Ma, J. Automated syringe-pump-based flow-batch analysis for spectrophotometric determination of trace hexavalent chromium in water samples. *Microchem. J.* **2019**, *145*, 1135–1142. [CrossRef]
27. Ma, J.; Li, P.; Chen, Z.; Lin, K.; Chen, N.; Jiang, Y.; Chen, J.; Huang, B.; Yuan, D. Development of an Integrated Syringe-Pump-Based Environmental-Water Analyzer (iSEA) and Application of It for Fully Automated Real-Time Determination of Ammonium in Fresh Water. *Anal. Chem.* **2018**, *90*, 6431–6435. [CrossRef] [PubMed]

28. Paluch, J.; Kozak, J.; Wieczorek, M.; Kozak, M.; Kochana, J.; Widurek, K.; Konieczna, M.; Koscielniak, P. Novel approach to two-component speciation analysis. Spectrophotometric flow-based determinations of Fe(II)/Fe(III) and Cr(III)/Cr(VI). *Talanta* **2017**, *171*, 275–282. [CrossRef] [PubMed]
29. Wieczorek, M.; Rengevicova, S.; Świt, P.; Woźniakiewicz, A.; Kozak, J.; Kościelniak, P. New approach to H-point standard addition method for detection and elimination of unspecific interferences in samples with unknown matrix. *Talanta* **2017**, *170*, 165–172. [CrossRef]
30. Soares, S.; Melchert, W.R.; Rocha, F.R.P. A flow-based procedure exploiting the lab-in-syringe approach for the determination of ester content in biodiesel and diesel/biodiesel blends. *Talanta* **2017**, *174*, 556–561. [CrossRef]
31. Alexovič, M.; Horstkotte, B.; Solich, P.; Sabo, J. Automation of dispersive liquid-liquid microextraction and related techniques. Approaches based on flow, batch, flow-batch and in syringe modes. *Trends Anal. Chem.* **2017**, *86*, 39–55.
32. Alexovič, M.; Horstkotte, B.; Solich, P.; Sabo, J. Automation of static and dynamic non-dispersive liquid phase microextraction. Part 1: Approaches based on extractant drop-, plug-, film- and microflow-formation. *Anal. Chim. Acta* **2016**, *906*, 22–40. [CrossRef]
33. Alexovič, M.; Horstkotte, B.; Solich, P.; Sabo, J. Automation of static and dynamic non-dispersive liquid phase microextraction. Part 2: Approaches based on impregnated membranes and porous supports. *Anal. Chim. Acta* **2016**, *907*, 18–30. [CrossRef]
34. Clavijo, S.; del Rosario Brunetto, M.; Cerdà, V. In-syringe-assisted dispersive liquid-liquid microextraction coupled to gas chromatography with mass spectrometry for the determination of six phthalates in water samples. *J. Sep. Sci.* **2014**, *37*, 974–981. [CrossRef] [PubMed]
35. Clavijo, S.; Avivar, J.; Suarez, R.; Cerdà, V. In-syringe magnetic stirring-assisted dispersive liquid-liquid microextraction and silylation prior gas chromatography-mass spectrometry for ultraviolet filters determination in environmental water samples. *J. Chromatogr. A* **2016**, *1443*, 26–34. [CrossRef] [PubMed]
36. Villar, M.; Avivar, J.; Ferrer, L.; Borràs, A.; Vega, F.; Cerdà, V. Automatic in-syringe dispersive liquid–liquid microextraction of 99Tc from biological samples and hospital residues prior to liquid scintillation counting. *Anal. Bioanal. Chem.* **2015**, *407*, 5571–5578. [CrossRef] [PubMed]
37. Vargas Medina, D.A.; Santos-Neto, Á.J.; Cerdà, V.; Maya, F. Automated dispersive liquid-liquid microextraction based on the solidification of the organic phase. *Talanta* **2018**, *189*, 241–248. [CrossRef]
38. Wang, X.; Xu, G.; Chen, P.; Sun, Y.; Yao, X.; Lv, Y.; Guo, W.; Wang, G. Fully-automated magnetic stirring-assisted lab-in-syringe dispersive liquid–liquid microextraction for the determination of arsenic species in rice samples. *RSC Advances* **2018**, *8*, 16858–16865. [CrossRef]
39. Sanchez, R.; Horstkotte, B.; Fikarová, K.; Sklenářová, H.; Maestre, S.; Miro, M.; Todoli, J.L. Fully Automatic In-Syringe Magnetic Stirring-Assisted Dispersive Liquid Liquid Microextraction Hyphenated to High-Temperature Torch Integrated Sample Introduction System-Inductively Coupled Plasma Spectrometer with Direct Injection of the Organic Phase. *Anal. Chem.* **2017**, *89*, 3787–3794. [CrossRef]
40. González, A.; Avivar, J.; Cerdà, V. Determination of priority phenolic pollutants exploiting an in-syringe dispersive liquid–liquid microextraction–multisyringe chromatography system. *Anal. Bioanal. Chem.* **2015**, *407*, 2013–2022. [CrossRef]
41. Horstkotte, B.; Fikarová, K.; Cocovi-Solberg, D.J.; Sklenářová39, H.; Solich, P.; Miro, M. Online coupling of fully automatic in-syringe dispersive liquid-liquid microextraction with oxidative back-extraction to inductively coupled plasma spectrometry for sample clean-up in elemental analysis: A proof of concept. *Talanta* **2017**, *173*, 79–87. [CrossRef]
42. Suárez, R.; Clavijo, S.; Avivar, J.; Cerdà, V. On-line in-syringe magnetic stirring assisted dispersive liquid-liquid microextraction HPLC–UV method for UV filters determination using 1-hexyl-3-methylimidazolium hexafluorophosphate as extractant. *Talanta* **2016**, *148*, 589–595. [CrossRef]
43. Shishov, A.; Terno, P.; Moskvin, L.; Bulatov, A. In-syringe dispersive liquid-liquid microextraction using deep eutectic solvent as disperser: Determination of chromium (VI) in beverages. *Talanta* **2020**, *206*, 120209. [CrossRef]
44. Timofeeva, I.; Shishov, A.; Kanashina, D.; Dzema, D.; Bulatov, A. On-line in-syringe sugaring-out liquid-liquid extraction coupled with HPLC-MS/MS for the determination of pesticides in fruit and berry juices. *Talanta* **2017**, *167*, 761–767. [CrossRef] [PubMed]

45. Pochivalov, A.; Vakh, C.; Garmonov, S.; Moskvin, L.; Bulatov, A. An automated in-syringe switchable hydrophilicity solvent-based microextraction. *Talanta* **2020**, *209*, 120587. [CrossRef] [PubMed]
46. Frizzarin, R.M.; Portugal, L.A.; Estela, J.M.; Rocha, F.R.P.; Cerdà, V. On-line lab-in-syringe cloud point extraction for the spectrophotometric determination of antimony. *Talanta* **2016**, *148*, 694–699. [CrossRef] [PubMed]
47. Davletbaeva, P.; Falkova, M.; Safonova, E.; Moskvin, L.; Bulatov, A. Flow method based on cloud point extraction for fluorometric determination of epinephrine in human urine. *Anal. Chim. Acta* **2016**, *911*, 69–74. [CrossRef] [PubMed]
48. Maya, F.; Cabello, C.P.; Estela, J.M.; Cerdà, V.; Palomino, G.T. Automatic In-Syringe Dispersive Microsolid Phase Extraction Using Magnetic-Metal Organic Frameworks. *Anal. Chem.* **2015**, *87*, 7545–7549. [CrossRef] [PubMed]
49. Gonzalez, A.; Avivar, J.; Maya, F.; Palomino Cabello, C.; Turnes Palomino, G.; Cerdà, V. In-syringe dispersive mu-SPE of estrogens using magnetic carbon microparticles obtained from zeolitic imidazolate frameworks. *Anal. Bioanal. Chem.* **2017**, *409*, 225–234. [CrossRef]
50. Giakisikli, G.; Anthemidis, A.N. An automatic stirring-assisted liquid-liquid microextraction system based on lab-in-syringe platform for on-line atomic spectrometric determination of trace metals. *Talanta* **2017**, *166*, 364–368. [CrossRef]
51. Šrámková, I.H.; Horstkotte, B.; Fikarová, K.; Sklenářová, H.; Solich, P. Direct-immersion single-drop microextraction and in-drop stirring microextraction for the determination of nanomolar concentrations of lead using automated Lab-In-Syringe technique. *Talanta* **2018**, *184*, 162–172. [CrossRef]
52. Sasaki, M.K.; Souza, P.A.F.; Kamogawa, M.Y.; Reis, B.F.; Rocha, F.R.P. A new strategy for membraneless gas-liquid separation in flow analysis: Determination of dissolved inorganic carbon in natural waters. *Microchem. J.* **2019**, *145*, 1218–1223. [CrossRef]

© 2020 by the authors. Licensee MDPI, Basel, Switzerland. This article is an open access article distributed under the terms and conditions of the Creative Commons Attribution (CC BY) license (http://creativecommons.org/licenses/by/4.0/).

Article

Fluorimetric Method for the Determination of Histidine in Random Human Urine Based on Zone Fluidics

Antonios Alevridis [1], Apostolia Tsiasioti [1], Constantinos K. Zacharis [2] and Paraskevas D. Tzanavaras [1,*]

[1] Laboratory of Analytical Chemistry, School of Chemistry, Faculty of Sciences, Aristotle University of Thessaloniki, GR-54124 Thessaloniki, Greece; alevridi@chem.auth.gr (A.A.); atsiasioti@gmail.com (A.T.)
[2] Laboratory of Pharmaceutical Analysis, Department of Pharmaceutical Technology, School of Pharmacy, Aristotle University of Thessaloniki, GR-54124 Thessaloniki, Greece; czacharis@pharm.auth.gr
* Correspondence: ptzanava@chem.auth.gr; Tel.: +30-2310997721; Fax: +30-2310997719

Academic Editor: Pawel Koscielniak
Received: 6 March 2020; Accepted: 2 April 2020; Published: 4 April 2020

Abstract: In the present study, the determination of histidine (HIS) by an on-line flow method based on the concept of zone fluidics is reported. HIS reacts fast with o-phthalaldehyde at a mildly basic medium (pH 7.5) and in the absence of additional nucleophilic compounds to yield a highly fluorescent derivative ($\lambda_{ex}/\lambda_{em}$ = 360/440 nm). The flow procedure was optimized and validated, paying special attention to its selectivity and sensitivity. The LOD was 31 nmol·L^{-1}, while the within-day and day-to-day precisions were better than 1.0% and 5.0%, respectively (n = 6). Random urine samples from adult volunteers (n = 7) were successfully analyzed without matrix effect (<1%). Endogenous HIS content ranged between 116 and 1527 µmol·L^{-1} with percentage recoveries in the range of 87.6%–95.4%.

Keywords: histidine; random human urine; zone fluidics; o-phthalaldehyde; derivatization; stopped-flow; fluorimetry

---

1. Introduction

L-Histidine (HIS) is an essential amino-acid that has proven to play unique roles in human organism [1,2]. Representative examples include: (i) when it is co-administered with iron, it helps towards the most rapid increase in its plasma levels in patients with anemia, especially to those with chronic kidney failure [3]; (ii) in vitro studies have demonstrated that due to its metal chelating properties HIS is the most effective hydroxyl radical scavenger compared to several amino acids that were examined [4]; (iii) it has significant anti-inflammatory properties, mainly due to the production on histamine through enzymatic decarboxylation [5]; and (iv) a recent study also revealed impressive improvement of the metabolic syndrome through supplementation with HIS [6,7]. The critical role of HIS along with arginine and tryptophan in the strengthening of the immune system particularly during cancer immune-therapies has very recently been reviewed [8]. Detection of histidine in urine samples is associated with the diagnosis of histidine metabolism disorders, particularly 'histidinemia' at elevated levels in physiological fluids (normal level: 130–2100 mM in urine) [9] or with the level of histamine secretion in allergic patients [10].

From an analytical chemistry point of view, a search through the scientific literature revealed a continuous interest in the development of novel methods for the determination of HIS. Many of the recently reported methods are indirect based on the well-known affinity and interaction of HIS with metal ions such as Cu(II) and Ni(II). For example, a colorimetric naked-eye chemosensor was based on a thiazolylazo dye–Ni(II)–HIS system visual color change from red to yellow. Although amino acids were reported not to interfere, no validation and application to real samples was included [11]. In a similar manner, HIS was found to reverse the quenching effect of Cu(II) ions on the fluorescence of semiconductor quantum dots (QD), enabling the analysis of HIS at the micromolar level [12]. According to the authors, although the optimal pH of the sensor favors the selectivity against –SH containing amino acids (e.g., cysteine), the endogenous Fe and Cu content of biological samples should always be taken into account for living cell imaging applications.

Automation of analytical methods through flow-based configurations is always attractive and up-to-date due to the unique advantages of high throughput, strict and precise control of experimental conditions, enhanced selectivity based on kinetic differentiations, etc. Among flow-based methods the concept of Zone Fluidics (ZF) offers some additional features including effective manipulation of reactants zones at the micro-liter level, adaptation of various chemistries through single channel configurations, minimization of waste generation, even efficient coupling to separation techniques as a front-end sample preparation platform [13,14]. ZF coupled to analytical derivatization have proven as an advantageous alternative to the development of robust and reliable methods for a variety of analytes such as hydrazine [15], adamantane derivatives [16], dopamine [17], creatinine [18], etc.

The goal of the present study was to develop, validate, and apply a reliable and fast method for the determination of HIS in random urine samples. To achieve this task, we have chosen to automate the reaction between HIS and *o*-phthalaldehyde (OPA) through a ZF flow platform. OPA has been reported to react with HIS at a basic pH to form a relatively stable and highly fluorescent derivative [19]. The above-mentioned reaction is promising since it proceeds without the need of nucleophilic compounds and is therefore selective against numerous primary amino-group containing analytes that react with OPA through the traditional mechanism [20]. Additionally, histamine that is structurally analogous form an unstable derivative at alkaline medium requiring acidification prior to detection [21], while glutathione reacts with OPA at a significantly more basic pH [22].

From a literature point of view, to the best of our knowledge, only one method for the determination of HIS has been published based on ZF [23]. It is based on chemiluminescence (CL) detection of a rather complicated chemical system—HIS increases the catalytic activity of Mn(II) salts in the CL reaction with luminol-hydrogen peroxide in the presence of dioximes in sodium borate medium. Although the authors report application in real beer samples, the selectivity of the method should be of concern in real world applications since—as could be expected—some common metal ions that are well known to act catalytically in luminol-based reactions (Fe(III), Cu(II), Co(II), etc.) and amino acids interfere at comparable of even lower levels to HIS.

2. Results and Discussion

2.1. Preliminary Experiments

Preliminary experiments were carried out in the batch mode to verify the reaction of OPA and HIS in the absence of nucleophilic compounds. A representative off-line fluorimetric detect (FL) spectrum is shown in Figure S1 (Supplementary Material). All subsequent experiments were performed at $\lambda_{ex} = 360$ nm and $\lambda_{em} = 440$ nm.

In a second series of preliminary studies, it was confirmed that the reaction can proceed under zone fluidic conditions and that was promising for further development and optimization. Initial ZF experiments were performed under the following starting variables: [OPA] = 10 mmol·L^{-1}, pH = 7–9, T = ambient temperature, V = 50 µL for all zones, and [HIS] = 10 µmol·L^{-1}. The order of mixing proved to have negligible impact on the sensitivity and the aspiration order of OPA/Buffer/Sample was adopted for all subsequent experiments.

2.2. Development of the ZF Method

The development of the ZF method involved experimental studies of the various chemical and instrumental parameters that are expected to affect the performance of the procedure, such as: the pH of the reaction (6.0–9.0), the reaction time (0–120 s), the reaction temperature (25–40 °C), the amount concentration of OPA (5–15 mmol·L^{-1}), and the volumes of the sample (50–125 µL) and the OPA/Buffer (25–75 µL).

As can be seen in Figure 1, the reaction is clearly favored at a mildly basic pH of 7.5. It should be noted that our findings under flow conditions are not in absolute agreement with the batch method, since Hakanson et al. have reported an optimum reaction pH of 11.2–11.5 [19]. Such variations are common in flow methods due to the kinetic character of the procedures, compared to the "thermodynamic" batch analogues. For example, the reaction can be favored kinetically at a mild alkaline pH, while the more alkaline pH reported in [17] might provide a more stable product at a lower reaction rate. Additionally, this is an important feature in terms of selectivity since most of the OPA reactions take place in significantly more alkaline pH regions. For example, Glutathione (GSH) reacts with OPA at pH values > 8.0 and the highest sensitivity is achieved at pH > 12.0.

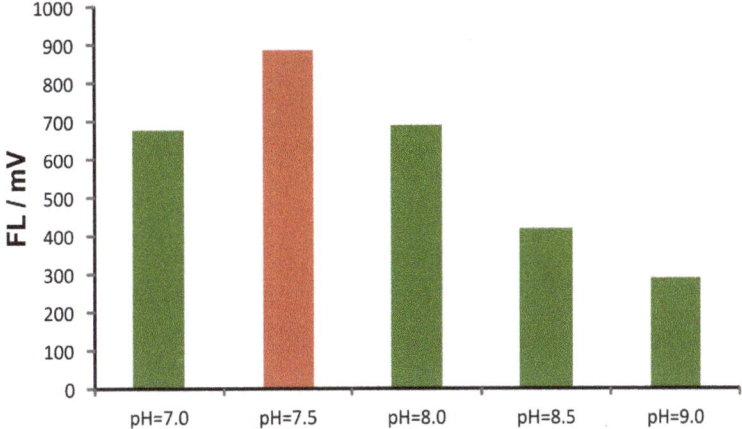

Figure 1. Effect of the pH on the fluorescence intensity of histidine–o-phthalaldehyde (HIS–OPA) derivative under flow conditions.

The effect of the reaction time was investigated under stopped-flow conditions. In brief, following aspiration of the zones in the holding coil (HC) of the ZF setup, the flow was reversed for 30 s at 0.6 mL·min^{-1}. The reaction mixture is "trapped" in the reaction coil (RC) and the flow is stopped for increasing time intervals. As can be seen in Figure 2, there is a non-linear increase in the range of 0–90 s, while the signals practically leveled-off thereafter. A stopped-flow reaction time of 60 s was selected as a compromise between sensitivity and sampling throughput. On the other hand, variation of the reaction temperature up to 60 °C offered a ca. 25% sensitivity enhancement and the latter value was adopted by thermostating the reaction coil using an HPLC column oven (see experimental section).

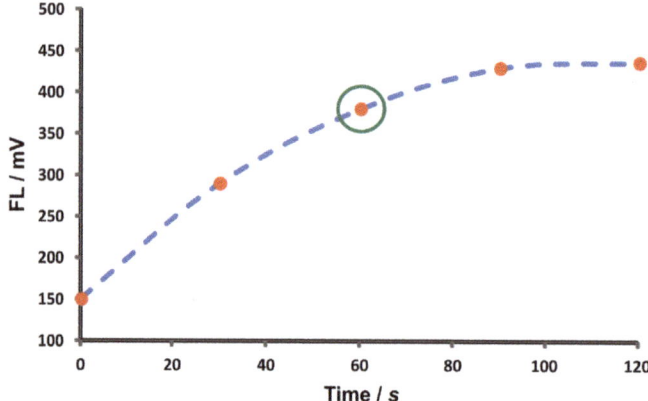

Figure 2. Effect of the reaction time under stopped-flow conditions; $t = 0$ corresponds to the delivery of the reaction product to the detector through the reaction coil without the stopped-flow step.

The volumes of the samples and reagents zones are an important variable that affects both the mixing efficiency and the dispersion/sensitivity. The sample injection volume vs. the fluorescence intensity had an almost linear profile in the range of 50–125 µL (at 2 µmol·L^{-1} HIS) and the signals practically doubled in this range. A sample injection volume of 100 µL was selected in terms of satisfactory sensitivity, sampling rate, and sample consumption. On the other hand, the volumes of the OPA reagent and the buffer had a rather moderate effect on the method. In both cases, 50 µL were selected taking into account the sensitivity and the consumption of the reagents.

Finally, the amount concentration of the reagent was investigated in the range of 5–15 mmol·L^{-1}. The criteria were the sufficient excess of the reagent and its effect on the kinetics of the reaction (sensitivity). The experimental results confirmed that the method was unaffected at OPA amount concentrations >10 mmol·L^{-1}, where the signals reached a plateau. The latter value was therefore selected for subsequent validation experiments.

An overview of the studied instrumental and chemical variables, the investigated range and the selected values is included in Table 1.

Table 1. Overview of the chemical and instrumental variables.

Variable	Studied Range	Selected Value
pH	7.0–9.0	7.5
Reaction time (stopped-flow, s)	0–120	60
Temperature (°C)	25–60	60
Sample volume (µL)	50–125	100
OPA volume (µL)	25–75	50
Buffer volume (µL)	25–75	50
OPA concentration (mmol·L^{-1})	5–15	10

2.3. Validation of the ZF Method

The proposed method has been validated for linearity, limits of detection (LOD) and quantification (LOQ), precision, selectivity, matrix effect, and accuracy.

2.3.1. Linearity, LOD and LOQ

The method proved to offer satisfactory linearity in the range of 125–2000 nmol·L^{-1} (20–310 µg·L^{-1}) HIS. The regression equation was obtained in a "cumulative" way by incorporating the results from more than 90 standard solutions analyzed in different working days ($n = 8$). In this way, the calibration curve is more representative including potential day-to-day variations. The following regression Equation (1) was obtained:

$$F = 420.1\ (\pm 3.4)\ [\text{HIS}] + 63.1\ (\pm 3.9),\ r^2 = 0.994 \tag{1}$$

where F is the fluorescence intensity as measured by the detector. Linearity was further validated using the back-calculated concentrations (residuals). The percent residuals were distributed randomly around the "zero axis" and ranged between −13.2% and +9.5%. It should be noted that the highest positive and negative values correspond to residuals at the lowest point of the calibration curve close to the LOQ of the method.

The LOD and LOQ was estimated based on the following Equations (2):

$$\text{LOD} = 3.3 \times SD_b/s \text{ and } \text{LOQ} = 10 \times SD_b/s \tag{2}$$

where SD_b is the standard deviation of the intercept and s is the slope of the respective regression lines. The calculated LOD/LOQ for the analyte was 31 and 93 nmol·L^{-1}, respectively, corresponding to 4.8 and 14.4 µg·L^{-1} HIS.

2.3.2. Within and between Day Precisions

The within-day precision was validated at the 1.0 µmol·L^{-1} level by repetitive injections ($n = 8$). The relative standard deviation (RSD) was 0.5 % (see Figure S2 in supplementary section).

The day-to-day precision was evaluated by independent calibration curves obtained at different working days ($n = 8$). As can be seen in Table S1 (supplementary section), the RSD of the regression slopes was 4.2%, verifying the repeatability of the flow procedure.

2.3.3. Selectivity and Matrix Effect

The selectivity/matrix effect of the proposed method was evaluated towards three axes:

(i) against amino-acids and biogenic amines that can potentially react with OPA;
(ii) using an artificial urine matrix spiked with the analyte;
(iii) using a pooled human urine sample also spiked with the analyte.

As can be seen in Table 2, several amino-acids and biogenic amines were examined on the basis of their reaction with OPA at the optimal values set for the determination of HIS. All compounds were set at 10.0 µmol·L^{-1} (10-fold excess compared to HIS) with the exception of Histamine and Glutathione that were examined at equimolar concentrations to HIS (1.0 µmol·L^{-1}). Based on the experimental results the selectivity factor (S.F.) for each potential interfering compound (INT) was calculated according to the following Equation (3):

$$\text{S.F.} = \frac{FL(\text{HIS})}{FL(\text{INT})} \times \frac{c(\text{INT})}{c(\text{HIS})} \tag{3}$$

Table 2. Selectivity of the proposed method.

Examined Compound	Amount Concentration (µmol·L^{-1})	FL (mV)	Selectivity Factor (S.F.)
Glycine	10	30	148
Glutamate	10	140	32
Alanine	10	NR [a]	N/A
Lysine	10	NR	N/A
Threonine	10	NR	N/A
Cysteine	10	55	81
Serine	10	NR	N/A
Tyrosine	10	NR	N/A
Histamine	1	110	4.1
Glutathione	1	132	3.4

[a] NR: not reacted.

The results can be categorized as follows:

(i) Alanine, Lysine, Threonine, Serine, and Tyrosine do not seem to react with OPA at the experimental conditions of the ZF method since the obtained signals were not different compared to the blank values.

(ii) Glycine, Cysteine, and Glutamate react with OPA, but the selectivity factors are high, ranging between 32–148.

(iii) Histamine and Glutathione seem to cause the most serious interference at equimolar levels to HIS. Both compounds are known to react with the tagging reagent in the absence of nucleophilic compounds, yielding fluorescent derivatives [22,24]. However, the selectivity factors of ca. four are quite satisfactory taking into account the fact that HIS is in great excess in the real samples compared to the above compounds. The derivative of Histamine is more stable and with higher fluorescence at acidic medium, while the reaction of Glutathione is favored kinetically in more alkaline pH.

To evaluate the matrix effect, a widely accepted artificial urine sample has been prepared as described in the experimental section. As can be seen in Table 3, the matrix effect was examined at several dilution factors of the artificial urine at HIS concentrations covering the linearity range of the method. The slopes of the matrix matched curves were compared against the aqueous regression line and the matrix effect was evaluated as the relative error (%). The experimental results confirmed significant negative matrix effect (signal suppression) at 1:5 and 1:10 dilutions of the artificial urine ranging between −35% and −55%. On the other hand, the matrix effect was minimized (<5%) for dilution factors higher than 1:100. The obtained results were quite promising since the sensitivity of the method is high and real urine dilutions of >500-fold are expected to be necessary for quantification of HIS based on its reported levels in the literature [25].

Table 3. Matrix effect using artificial urine (250–2000 nmol·L^{-1}).

	Dilution	Slope	Matrix Effect (%)
Aqueous curve	-	420.1	-
Artificial urine	1:5	185.3	−55.9
Artificial urine	1:10	271.8	−35.3
Artificial urine	1:100	441.3	+5.0
Artificial urine	1:250	428.5	+2.0

Further evaluation of the matrix effect was carried out using a real pooled urine sample ($n = 8$). In brief, 500 µL of each urine sub-sample were transferred in a centrifuge tube and mixed. An equal volume of ice-cold acetonitrile (4 mL) was added to precipitate proteins followed by centrifugation.

The obtained solution was diluted 250-fold and spiked with HIS in the range of 250–1500 nmol·L^{-1}. The experimental regression Equation (4) was:

$$F = 417.5\ (\pm 3.1)\ [\text{HIS}] + 242.8\ (\pm 2.9),\ r^2 = 0.999 \tag{4}$$

offering a <1% matrix effect compared to the cumulative aqueous calibration curve and enabling its practical use for analyzing the individual real urine samples.

2.3.4. Accuracy of the ZF Method

The accuracy of the method was validated at three concentration levels (500, 1000, and 1500 nmol·L^{-1}) preparing two independent series of samples in the pooled urine matrix. Quantification was carried out using the cumulative aqueous calibration curve described in Section 2.3.1. The experimental results are presented in Table 4 and are quite satisfactory for this type of analysis with percent recoveries ranging between 87.6% and 95.4%.

Table 4. Accuracy of the developed method.

Added (nmol·L^{-1})	Found (nmol·L^{-1})	Recovery (%)
500	438 (±21)	87.6
500	443 (±15)	88.6
1000	926 (±30)	92.6
1000	943 (±38)	94.3
1500	1431 (±51)	95.4
1500	1412 (±47)	94.1

2.4. Applications of the ZF Method

Seven random urine samples from male and female volunteers were analyzed according to the optimized and validated method described above. Samples were preliminarily screened to determine the necessary dilution factor in order to fall within the range of the calibration curve. In all cases, the dilution factors were either 500 or 1000 being consistent with the validation of the method in terms of potential matrix effects. The experimental findings are tabulated in Table 5. The concentration of HIS ranged between 115.8 and 1527 µmol·L^{-1} and are in accordance with reference values for adults (>18 years) obtained in an extensive study (>800 samples) that has been reported in the literature [25].

Table 5. Analysis of random urine samples.

Sample	Histidine (µmol·L^{-1})	S.D. (n = 3)
Urine-A	384	12
Urine-B	1293	42
Urine-C	116	6
Urine-D	1527	50
Urine-E	629	18
Urine-F	445	21
Urine-G	873	34

3. Materials and Methods

3.1. Instrumentation

The ZF setup was consisted of the following parts: a Minipuls3 peristaltic pump (Gilson, Middleton, WI, USA); a low-pressure micro-electrically actuated 10-port valve (Valco, Brockville, ON, Canada); a RF-551 flow-through spectrofluorimetric detector operated at high sensitivity (Shimadzu, Kyoto, Japan); PTFE tubing was used for the connections of the flow configuration (0.5 or 0.7 mm i.d.); and Tygon tubing was used in the peristaltic pump. An HPLC column heater (Jones Chromatography,

Hengoed, Mid Glamorgan, UK) was employed for the temperature control (60 ± 0.5 °C) of the reaction coil (100 cm/0.5 mm i.d.); the latter was tightly wrapped around the stainless-steel body of an old HPLC column (4.6 mm i.d.).

Control of the ZF system was performed through a LabVIEW (National Instruments, Austin, TX, USA) based program developed in house; while data acquisition (peak heights) was carried out through the Clarity® software (version 4.0.3, DataApex, Prague, Czech Republic). Off-line spectra were recorded using a RF-5301PC batch spectrofluorophotometer (Shimadzu, Kyoto, Japan).

3.2. Reagents and Solutions

Histidine (HIS, 99%) was purchased by Sigma (St. Louis, MO, USA); o-phthalaldehyde (OPA, Fluka, Munich, Germany), KH_2PO_4 (Merck, Darmstadt, Germany), NaOH (Merck) and HCl (Sigma) were all of analytical grade. Doubly de-ionized water was produced by a Milli-Q system (Millipore, Bedford, MA, USA).

The standard stock solution of the analyte was prepared daily at the 1000 µmol·L^{-1} level in water. Working solutions were prepared by serial dilutions in water. The derivatizing reagent (OPA) was prepared at an amount concentration of 10 mmol·L^{-1} by firstly dissolving in 0.5 mL methanol and subsequently adding 9.5 mL water [22]. This solution was stable for a practical period of 3–4 working days at 4 °C in an amber glass vial. Phosphate buffer (100 mmol·L^{-1}) was also prepared daily and regulated to the desired pH value (pH = 7.5) by drop-wise addition of NaOH (1 mol·L^{-1}).

Synthetic urine (200 mL in water) was prepared according to the literature [26] and was consisted of the following (analytical grade): lactic acid (1.1 mmol·L^{-1}), citric acid (2.0 mmol·L^{-1}), sodium bicarbonate (25 mmol·L^{-1}), urea (170 mmol·L^{-1}), calcium chloride (2.5 mmol·L^{-1}), sodium chloride (90 mmol·L^{-1}), magnesium sulfate (2.0 mmol·L^{-1}), sodium sulfate (10 mmol·L^{-1}), potassium dihydrogen phosphate (7.0 mmol·L^{-1}), di-potassium hydrogen phosphate (7.0 mmol·L^{-1}), and ammonium chloride (25 mmol·L^{-1}). The pH of the solution was adjusted to 6.0 by addition of 1.0 mol·L^{-1} HCl.

All other amino-acids and biogenic amines employed in the selectivity studies were of analytical grade and were supplied by Sigma. All solutions were prepared in water at the levels mentioned in the respective section.

3.3. ZF Procedure

The experimental conditions of the optimized ZF sequence for the determination of HIS can found in Table 6 and are also depicted graphically in Figure 3. In brief, OPA (50 µL, 10 mmol·L^{-1}), buffer (50 µL, 100 mmol·L^{-1} phosphate/pH = 7.5) and sample/standards (100 µL) were sequentially aspirated in the holding coil (HC).

Table 6. ZF steps for the determination of Histidine.

Time (s)	Pump Action	Flow Rate (mL min^{-1})	Volume (µL)	Valve Position	Action Description
0	Off	-	-	1	Selection of OPA reagent port
5	Aspirate	0.6	50	1	Aspiration of OPA in the holding coil
1	Off	-	-	2	Selection of buffer port
5	Aspirate	0.6	50	2	Aspiration of buffer in the holding coil
1	Off	-	-	3	Selection of sample port
10	Aspirate	0.6	100	3	Aspiration of sample in the holding coil
1	Off	-	-	4	Selection of detector port
30	Deliver	0.6	300	4	Propulsion of reaction mixture to reaction coil
60	Off	-	-	4	Stopped-flow step
120	Deliver	0.6	1200	4	Detection of derivative/end of measuring cycle

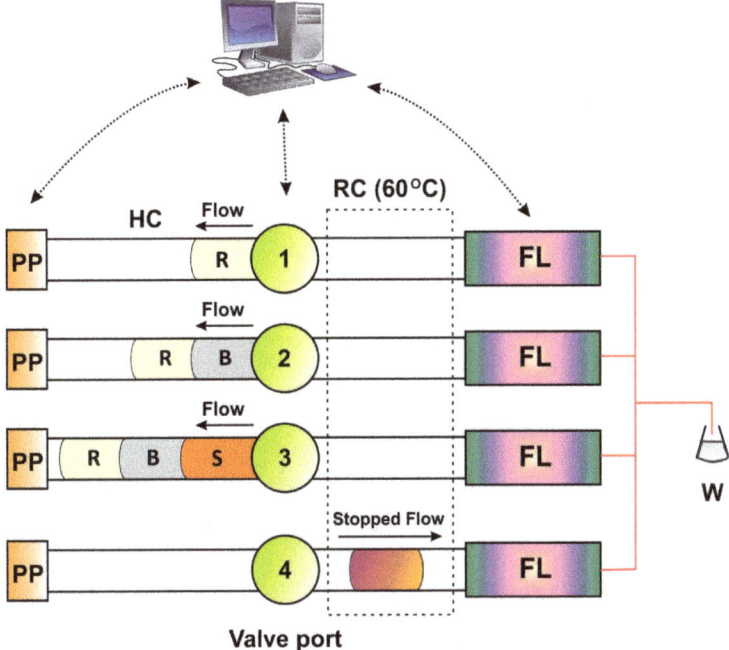

Figure 3. ZF sequence for the determination of HIS: PP = peristaltic pump; R = OPA reagent (10 mmol·L^{-1}); B = buffer (100 mmol·L^{-1} phosphate, pH = 7.5); S = sample; HC = holding coil (300 cm/0.7 mm i.d.); RC = reaction coil (100 cm/0.5 mm i.d.); FL = fluorimetric detector ($\lambda_{ex}/\lambda_{em}$ = 360/440 nm); W = waste.

Upon flow reversal, the stacked reaction zones were propelled towards the thermostated reaction coil (RC, 100 cm/60 °C) at a flow rate of 0.6 mL min^{-1} and the reaction was allowed to develop for 60 s under stopped-flow conditions. Downstream fluorimetric detection was carried out at $\lambda_{ex}/\lambda_{em}$ = 360/440 nm. The sampling throughput was 16 h^{-1}.

3.4. Preparation of Urine Samples

Random urine samples were kindly donated voluntarily (no ethical approval was required) by male and female members of the laboratory. Samples where either processed immediately or stored at −20 °C until analysis [25].

Due to the selectivity and sensitivity of the developed method, sample preparation included the following simple and rapid steps:

(i) protein precipitation with addition of ice-cold acetonitrile (1 + 1);
(ii) centrifugation (4000 rpm, 10 min);
(iii) 500–1000-fold dilution depending on the levels of HIS in the real samples;
(iv) Analysis by the ZF method.

4. Conclusions

The developed on-line fluorimetric method for the determination of Histidine offers some interesting features:

1. It utilizes readily available reagents and due to the zone fluidics-based concept the consumption and generation of wastes is minimal compared to continuous flow techniques such as HPLC and Flow Injection Analysis.
2. The method is based on direct reaction and is advantageous compared to indirect methods based on inhibitory effects.
3. The high sensitivity of the method down to the nano-molar level enables the direct analysis of Histidine in human urine with minimum sample preparation.
4. The unique mechanism of the derivatization reaction excluded interference from most amino-acids and biogenic amines offering a highly selective platform for the determination of the analyte in the complicated samples without matrix effects.
5. Application in random urine samples was successful at a reasonable sampling frequency of 16 h^{-1}.

Supplementary Materials: The following are available online, Figure S1: Representative FL off-line spectra of the HIS-OPA derivative. Figure S2: Within day precision at 1.0 µmol·L^{-1} HIS (n = 8). Table S1: Day-to-day precision of the developed method.

Author Contributions: Conceptualization, P.D.T. and C.K.Z.; methodology, P.D.T. and C.K.Z.; validation, A.T., A.A., and C.K.Z.; data curation, P.D.T., A.T., and C.K.Z; writing—original draft preparation, A.T. and C.K.Z.; writing—review and editing, P.T. and C.K.Z. All authors have read and agreed to the published version of the manuscript.

Funding: This research received no external funding.

Conflicts of Interest: The authors declare no conflict of interest.

References

1. Vera-Aviles, M.; Vantana, E.; Kardinasari, E.; Koh, N.; Latunde-Dada, G. Protective Role of Histidine Supplementation Against Oxidative Stress Damage in the Management of Anemia of Chronic Kidney Disease. *Pharmaceuticals* **2018**, *11*, 111. [CrossRef] [PubMed]
2. El-Sayed, N.; Miyake, T.; Shirazi, A.; Park, S.; Clark, J.; Buchholz, S.; Parang, K.; Tiwari, R. Design, Synthesis, and Evaluation of Homochiral Peptides Containing Arginine and Histidine as Molecular Transporters. *Molecules* **2018**, *23*, 1590. [CrossRef] [PubMed]
3. Jontofsohn, R.; Heinze, V.; Katz, N.; Stuber, U.; Wilke, H.; Kluthe, R. Histidine and iron supplementation in dialysis and pre-dialysis patients. *Proc. Eur. Dial. Transplant Assoc.* **1975**, *11*, 391–397. [PubMed]
4. Nair, N.G.; Perry, G.; Smith, M.A.; Reddy, V.P. NMR Studies of Zinc, Copper, and Iron Binding to Histidine, the Principal Metal Ion Complexing Site of Amyloid-β Peptide. *J. Alzheimer's Dis.* **2010**, *20*, 57–66. [CrossRef] [PubMed]
5. Lee, Y.; Hsu, C.; Lin, M.; Liu, K.; Yin, M. Histidine and carnosine delay diabetic deterioration in mice and protect human low density lipoprotein against oxidation and glycation. *Eur. J. Pharmacol.* **2005**, *513*, 145–150. [CrossRef]
6. DiNicolantonio, J.J.; McCarty, M.F.; OKeefe, J.H. Role of dietary histidine in the prevention of obesity and metabolic syndrome. *Open Heart* **2018**, *5*, e000676. [CrossRef]
7. Feng, R.N.; Niu, Y.C.; Sun, X.W.; Li, Q.; Zhao, C.; Wang, C.; Guo, F.C.; Sun, C.H.; Li, Y. Histidine supplementation improves insulin resistance through suppressed inflammation in obese women with the metabolic syndrome: a randomised controlled trial. *Diabetologia* **2013**, *56*, 985–994. [CrossRef]
8. Tantawy, A.A.; Naguib, D.M. Arginine, histidine and tryptophan: A new hope for cancer immunotherapy. *PharmaNutrition* **2019**, *8*, 100149. [CrossRef]
9. Sun, S.-K.; Tu, K.-X.; Yan, X.-P. An indicator-displacement assay for naked-eye detection and quantification of histidine in human urine. *Analyst* **2012**, *137*, 2124. [CrossRef]
10. Eaton, K.K.; Howard, M.; Hunnisett, A. Urinary Histidine Excretion in Patients with Classical Allergy (Type A Allergy), Food Intolerance (Type B Allergy), and Fungal-type Dysbiosis. *J. Nutri. Environ. Med.* **2004**, *14*, 157–164. [CrossRef]

11. Nagae, T.; Aikawa, S.; Inoue, K.; Fukushima, Y. Colorimetric detection of histidine in aqueous solution by Ni^{2+} complex of a thiazolylazo dye based on indicator displacement mechanism. *Tetrahedron Lett.* **2018**, *59*, 3988–3993. [CrossRef]
12. Chabok, A.; Shamsipur, M.; Yeganeh-Faal, A.; Molaabasi, F.; Molaei, K.; Sarparast, M. A highly selective semiconducting polymer dots-based "off–on" fluorescent nanoprobe for iron, copper and histidine detection and imaging in living cells. *Talanta* **2019**, *194*, 752–762. [CrossRef] [PubMed]
13. Marshall, G.; Wolcott, D.; Olson, D. Zone fluidics in flow analysis: potentialities and applications. *Anal. Chim. Acta* **2003**, *499*, 29–40. [CrossRef]
14. Tzanavaras, P.D.; Zacharis, C.K.; Karakosta, T.D.; Zotou, A.; Themelis, D.G. High-Throughput Determination of Quinine in Beverages and Soft Drinks Based on Zone-Fluidics Coupled to Monolithic Liquid Chromatography. *Anal. Lett.* **2013**, *46*, 1718–1731. [CrossRef]
15. Karakosta, T.D.; Christophoridis, C.; Fytianos, K.; Tzanavaras, P.D. Micelles Mediated Zone Fluidics Method for Hydrazine Determination in Environmental Samples. *Molecules* **2019**, *25*, 174. [CrossRef]
16. Tzanavaras, P.D.; Papadimitriou, S.; Zacharis, C.K. Automated Stopped-Flow Fluorimetric Sensor for Biologically Active Adamantane Derivatives Based on Zone Fluidics. *Molecules* **2019**, *24*, 3975. [CrossRef]
17. van Staden, J.K.F.; State, R. Determination of Dopamine Using the Alkaline Luminol–Hydrogen Peroxide System for Sequential Injection–Zone Fluidics Analysis. *Anal. Lett.* **2016**, *49*, 2783–2792. [CrossRef]
18. Ohira, S.-I.; Kirk, A.B.; Dasgupta, P.K. Automated measurement of urinary creatinine by multichannel kinetic spectrophotometry. *Anal. Biochem.* **2009**, *384*, 238–244. [CrossRef]
19. Håkanson, R.; Rönnberg, A.L.; Sjölund, K. Improved fluorometric assay of histidine and peptides having NH2-terminal histidine using o-phthalaldehyde. *Anal. Biochem.* **1974**, *59*, 98–109. [CrossRef]
20. Hanczkó, R.; Molnár-Perl, I. Derivatization, stability and chromatographic behavior of o-phthaldialdehyde amino acid and amine derivatives: o-Phthaldialdehyde/ 2-mercaptoethanol reagent. *Chromatographia* **2003**, *57*, S103–S113. [CrossRef]
21. Yoshimura, T.; Kamataki, T.; Miura, T. Difference between histidine and histamine in the mechanistic pathway of the fluorescence reaction with ortho-phthalaldehyde. *Anal. Biochem.* **1990**, *188*, 132–135. [CrossRef]
22. Tsiasioti, A.; Iakovidou, I.; Zacharis, C.K.; Tzanavaras, P.D. Automated fluorimetric sensor for glutathione based on zone fluidics. *Spectrochim. Acta Part A Mol. Biomol. Spectrosc.* **2019**, *229*, 117963. [CrossRef] [PubMed]
23. Gao, C.; Fan, S. Influence of Zone Stacking Sequences on CL Intensity and Determination of Histidine in Sequential Injection Analysis. *Anal. Lett.* **2008**, *41*, 1335–1347. [CrossRef]
24. Tzanavaras, P.D.; Deda, O.; Karakosta, T.D.; Themelis, D.G. Selective fluorimetric method for the determination of histamine in seafood samples based on the concept of zone fluidics. *Anal. Chim. Acta* **2013**, *778*, 48–53. [CrossRef]
25. Amino Acids, Quantitative, Random, Urine. Available online: https://neurology.testcatalog.org/show/AAPD (accessed on 2 March 2020).
26. Brooks, T.; Keevil, C.W. A simple artificial urine for the growth of urinary pathogens. *Lett. Appl. Microbiol.* **1997**, *24*, 203–236. [CrossRef]

Sample Availability: Samples of the compounds mentioned in the text are not available from the authors but they can be purchased from the manufacturers mentioned in the "reagents and solutions" section.

© 2020 by the authors. Licensee MDPI, Basel, Switzerland. This article is an open access article distributed under the terms and conditions of the Creative Commons Attribution (CC BY) license (http://creativecommons.org/licenses/by/4.0/).

Article

Enzymatic Reactions in a Lab-on-Valve System: Cholesterol Evaluations

Jucineide S. Barbosa [1,2], Marieta L.C. Passos [1,*], M. das Graças A. Korn [2,3] and M. Lúcia M.F.S. Saraiva [1,*]

1. LAQV, REQUIMTE, Department of Chemical Sciences, Laboratory of Applied Chemistry, Faculty of Pharmacy, Porto University, Rua de Jorge Viterbo Ferreira, 228, 4050-313 Porto, Portugal
2. Instituto de Química, Departamento de Química Analítica, Campus Universitário de Ondina, Universidade Federal da Bahia, 40170-115 Salvador, Bahia, Brazil
3. INCT de Energia e Ambiente–Instituto de Química, Campus Universitário de Ondina, Universidade Federal da Bahia, 40170-115 Salvador, Bahia, Brazil
* Correspondence: marietapassos@gmail.com (M.L.C.P.); lsaraiva@ff.up.pt (M.L.M.F.S.S.); Tel.: +351-220-428-643 (M.L.C.P.); +351-220-428-674 (M.L.M.F.S.S.)

Academic Editor: Pawel Koscielniak
Received: 19 July 2019; Accepted: 7 August 2019; Published: 9 August 2019

Abstract: The micro sequential injection analysis / lab-on-valve (μSIA-LOV) system is a miniaturized SIA system resulting from the implementation of a lab-on-valve (LOV) atop of the selection valve. It integrates the detection cell and the sample processing channels into the same device, promoting the reduction of reagent consumption and waste generation, the improvement of the versatility, and the reduction of the time of analysis. All of these characteristics are really relevant to the implementation of enzymatic reactions. Additionally, the evaluation of cholesterol in serum samples is widely relevant in clinical diagnosis, since higher values of cholesterol in human blood are actually an important risk factor for cardiovascular problems. An automatic methodology was developed based on the μSIA-LOV system in order to evaluate its advantages in the implementation of enzymatic reactions performed by cholesterol esterase, cholesterol oxidase and peroxidase. Considering these reactions, the developed methodology was also used for the evaluation of cholesterol in human serum samples, showing reliable and accurate results. The developed methodology presented detection and quantification limits of 1.36 and 4.53 mg dL^{-1} and a linear range up to 40 mg dL^{-1}. This work confirmed that this μSIA-LOV system is a simple, rapid, versatile, and robust analytical tool for the automatic implementation of enzymatic reactions performed by cholesterol esterase, cholesterol oxidase, and peroxidase. It is also a useful alternative methodology for the routine determinations of cholesterol in real samples, even when compared with other automatic methodologies.

Keywords: cholesterol; serum samples; lab-on-valve; automation; enzymatic reaction

1. Introduction

The popularity of biocatalysis has been increasing in the last decades, due to the high specificity and selectivity of enzymes. At the same time, they have been used to substitute some hazardous reagents, and due to their biodegradability, being in accordance with the principles of Green Chemistry. On the other hand, their use can present some drawbacks, such as their dependence on some factors such as temperature, ionic strength, pH and cofactors. Additionally, they can suffer inhibition by some compounds, and their lifetime is limited [1]. In order to overcome these limitations, enzyme has been used in flow-based methodologies. Flow systems allow efficient control of the reaction conditions, maximizing the enzyme activity [1,2], and increasing the reproducibility.

Additionally, using a flow system, the measurements can be done in non-equilibrium conditions increasing the number of analyses per time, and decreasing the amounts of reagents, leading to decreasing costs per analysis, and to a reduction of waste generation. Thus, a large number of enzymatic reactions have been implemented in flow systems, and in particular in sequential injection systems (SIA) [1]. The operation principle of SIA systems is the sequential aspiration of reagent and sample to a holding coil. After that, the product zone is propelled, by the reversal flow, to the detector [3].

The µSIA-LOV [4], whose representation is presented in Section 3.2, is basically a miniaturization of a SIA system, and works on the same SIA principle. The miniaturization results from the implementation of a lab-on-valve (LOV) atop of the selection valve. The LOV integrates the detection cell, leading to an analytical flow path with an internal volume of microliters [4]. Sample processing channels were also integrated into the same device, promoting the proximity between the injection port and the flow cell. Additionally, the possibility of integrating the optical fibers, responsible for the detection, in different positions, allows the implementation of different kinds of detection (absorbance, fluorescence, reflectance) [4], making the system very versatile. The sample processing achieved by reversed flow and by random access provided by the multiposition valve is also very flexible [4]. Thus, this miniaturized system allows, even when compared with SIA, the reduction of reagent consumption (important for enzymatic reactions due to the high reagents costs) and waste generation, improve the versatility of the system and reduce the time of analysis [5]. Additionally, the reduction of sample and reagents volumes promote an improvement in the overlapping of sample and reagents zones, improving the assays repeatability [6]. All of these characteristics are really relevant to the implementation of enzymatic reactions. Additionally, it is also possible to implement the bead injection concept, using different kinds of beads for the sample treatment (preconcentration or matrix removal) [4], or also for enzymes immobilization [1,4]. Thus these previous characteristics turn a µSIA-LOV system into an interesting and challenging alternative to implement enzymatic reactions. Thus, some works about the implementation of enzymatic reactions in µSIA-LOV systems have been reported [6–12].

In this work it is intended to evaluate, for the first time, the µSIA-LOV potentialities in the implementation of three reactions catalyzed by cholesterol oxidase, cholesterol esterase and peroxidase, following described in Scheme 1.

$$\text{Cholesteryl acetate} + H_2O \xrightarrow{\text{Cholesterol esterase}} \text{Cholesterol} + \text{Acetic acid}$$

$$\text{Cholesterol} + O_2 \xrightarrow{\text{Cholesterol oxidase}} \Delta\text{-Cholestenone} + H_2O_2$$

$$2H_2O + 4\text{-Aminoantipyrine} + \text{Phenol} \xrightarrow{\text{Peroxidase}} \text{Quinoneimine} + 4H_2O$$

Scheme 1. Enzymatic reactions used for the quantification of cholesterol.

In these reactions, cholesterol ester is hydrolyzed by the cholesterol esterase, to free cholesterol. This free cholesterol is then oxidized by cholesterol oxidase to cholest-4-en-3-one simultaneously with the reduction of O_2 to hydrogen peroxide. The hydrogen peroxide can react with different chromogenic indicator systems such as ABTS, 3-dimethylaminobenzoic acid (DMAB)-aminoantipyrine (AAP), or 3,5-dichloro-2-hydroxybenzene sulfonic acid (DCHBS)-AAP to be detected spectrophotometrically [13]. In this work, phenol with 4-AAP was selected as the chromogenic system, and the measurement of the formed quinoneimine is proportional to the total concentration of cholesterol [14,15].

Simultaneously with the evaluation of the potentialities of the implementation of the enzymatic reactions promoted by cholesterol oxidase, cholesterol esterase and peroxidase in a µSIA-LOV system, it is also intended to use the resulted automatic methodology to evaluate total cholesterol levels in real samples.

For adults, cholesterol levels less than 200 mg dL^{-1} are desirable. High levels of cholesterol in the blood (hypercholesterolemia) resulted from a non-balanced diet [16], a sedentary lifestyle, and from some genetic factors [16], and all of this can result in cholesterol accumulation in the arterial walls, leading to atherosclerosis (hardening, thinning and chronic inflammation) [17]. Patients with these kinds of problems have an increased risk of stroke [18], ischemic heart disease [19,20] and peripheral vascular disease [21]. Thus, the evaluation of the cholesterol levels in the blood is essential in order to identify the risk of illness in the general population. The development of enzymatic methodologies with this objective has already been reported both in batch [22–28] and flow analysis mode [13,29–36] (Table 1).

Despite the fact that some methodologies for the enzymatic determinations of cholesterol have already been developed based on flow analysis, from the best of our knowledge, this is the first time that a µSIA-LOV system is used with this goal.

Thus, it is intended to combine the study of the potentialities of the use of µSIA-LOV for the implementation of enzymatic reactions promoted by cholesterol esterase, cholesterol oxidase and peroxidase, with a development of an automatic, robust, reliable, economic and fast methodology for the quantification of cholesterol in serum samples.

Table 1. Enzymatic methodologies based on conventional batch and flow-based procedures for the determination of cholesterol levels.

Mode	Methodology	Enzymes	Matrix	Analytes	Detection Mode	Sample Treatment	Linear Range	Detection Limit	Reference
Batch		CO, CE, POD	Serum	Total cholesterol	Amperometry	Dilution with ethanol and triton X-100	100–400 mg·dL^{-1}	n.a.	[22]
		CO, CE, POD	Serum	Total cholesterol	Chemiluminescence	Dilution with phosphate buffer solution	n.a.	n.a.	[23]
		CO, CE, POD	Serum	Free and total cholesterol	Chemiluminescence	Dilution with triton X-100	0.4–40 mg·dL^{-1}	0.2 mg·dL^{-1}	[24]
		CO	-	Free cholesterol	Voltammetry	Dilution with isopropanol and triton X-100	0.2–60.0 nmol·L^{-1}	0.05 nmol·L^{-1}	[25]
		CO, CE	Serum	Total cholesterol	Amperometry	Dilution with triton X-100	10–700 mg·dL^{-1}	0.1 mg·dL^{-1}	[26]
		CO, CE	Serum	Total cholesterol	UV-Vis spectrophotometry	Dilution with triton X-100	10–100 μmol·L^{-1}	2.9 μmol·L^{-1}	[27]
		CO, CE	-	Total cholesterol	Voltammetry	Dilution with isopropanol and triton X-100	5–5000 μg·mL^{-1}	3.0 μg·mL^{-1}	[28]
Flow analysis	FIA	CO, CE, POD	-	Total cholesterol	UV-Vis spectrophotometry and fluorimetry	Dilution with isopropanol and triton X-100	0.02–0.20 g·L^{-1} 0.005–0.05 g·L^{-1}	0.0020 g·L^{-1} 0.0004 g·L^{-1}	[29]
	FIA	CO, CE, POD	Serum	Free and total cholesterol	Potentiometry	Dilution with triton X-100	Up to around 10^{-3} mol·L^{-1}	3.0 × 10^{-3} mol·L^{-1}	[30]
	FIA	CO, CE, POD	Serum	Free and total cholesterol	UV-Vis spectrophotometry	Dilution with isopropanol and triton X-100	0.5–0.8 mmol·L^{-1}	n. a.	[31]
	FIA	CO, CE, POD	Serum	Total cholesterol	UV-Vis spectrophotometry	Dilution with triton X-100	0.11–0.86 mmol·L^{-1}	n. a.	[13]
	FIA	CO, CE, POD	Serum	Total cholesterol	Potentiometry	Dilution with isopropanol and triton X-100	0.05–3.0 mmol·L^{-1}	0.01 mmol·L^{-1}	[32]
	MFA	CO, CE, POD	Serum	Total cholesterol	UV-Vis spectrophotometry	Dilution with phenol and triton X-100	Up to 10.3 mmol·L^{-1}	n. a.	[33]
	MCFIA	CO, CE, POD	Serum	Total cholesterol	Chemiluminescence	Dilution with isopropanol and triton X-100	25–125 mg·L^{-1}	3.7 mg·L^{-1}	[34]
	AF4-PFRD	CO, CE, POD	Serum	Cholesterol and triglycerides	UV-Vis spectrophotometry	-	10–250 mg·dL^{-1}	n. a	[35]
	FIA microfluidic chip	CO	-	Free cholesterol	Amperometry	Dilution with triton X-100	50–400 mg·dL^{-1}	10 mg·dL^{-1}	[36]

CO, cholesterol oxidase; CE, cholesterol esterase; POD, peroxidase; FIA, flow injection analysis; MFA, mono-segmented flow analysis; MCFIA, multi-commutated flow injection analysis; AF4-PFRD, Asymmetrical flow field-flow fractionation with on-line, dual post-fractionation reaction detection; n. a., not available.

2. Results and Discussion

In this work, the objective is to implement some enzymatic reactions promoted by cholesterol esterase, cholesterol oxidase and peroxidase, in a μSIA-LOV system and at the same time to use this developed methodology for the quantification of total cholesterol in human serum samples.

2.1. Preliminary Studies

Before the μSIA-LOV system optimization, and since the cholesterol has low solubility in aqueous solutions, the effect of different concentrations of Triton X-100 in the solubilization of this compound were studied. Concentrations in a range between 1.25 and 10% were tested. It was verified that the lowest concentration of Triton X-100 that allows the solubilization of cholesterol was 1.5%. Thus this was the chosen concentration to be used for the preparation of standard solutions of cholesterol acetate.

2.2. μSIA-LOV System Optimization

In order to obtain the best performance with the developed μSIA-LOV system, parameters such as reagent concentrations and volumes, reaction time and other physical factors of the system, were studied. The optimization started by the evaluation of reagents concentrations, in order to certify that any reagent is a limiting reagent for the enzymatic reactions. Regarding the chromogenic reagent, composed by peroxidase enzyme, 4-aminoantipyrine (4-AAP) and phenol, the study started with the concentration of peroxidase. Concentrations between 18 and 146 $U·mL^{-1}$ were tested. Despite the increase of the analytical signals with the increasing concentration of enzyme, the concentration of 36.5 $U·mL^{-1}$ was chosen as a compromise between the sensitivity and the low consumption of the enzyme. The chosen concentration of 4-AAP was 0.10% in a tested interval between 0.02 and 2.00%. The sensitivity was increasing until this concentration (around 91% compared with the concentration of 0.02%), remaining stable for higher concentrations. The concentration of phenol was established in 6% due to its solubility and the intention to spend the least required amount of this toxic compound. For the enzymes, it was considered a concentration of 5.00 $U·mL^{-1}$ for the cholesterol esterase and 2.00 $U·mL^{-1}$ for the cholesterol oxidase (Figure 1) after testing concentrations between 2.00 and 12.5 $U·mL^{-1}$, and 0.50 and 4.00 $U·mL^{-1}$, respectively.

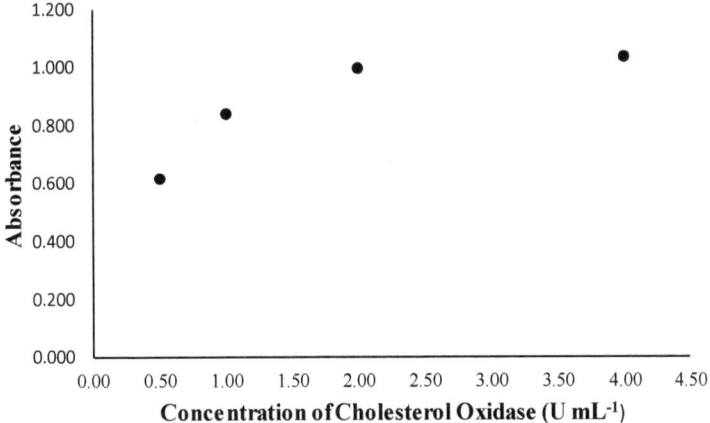

Figure 1. Effect of cholesterol oxidase in the analytical signals.

They were the lowest concentrations that lead to the highest sensitivities (around 62 and 24% higher when compared with concentrations of 2.00 and 0.050 $U·mL^{-1}$, for cholesterol oxidase and cholesterol esterase, respectively). The established volumes were 20 μL for the chromogenic reagent and 10 µL for the sample and for the enzymes. However, it was verified that dividing each aliquot into two parts and intercalating them, the mixture was improved, avoiding the formation of double peaks which occurred with the aspiration of the entire aliquots. It happens due to the proximity of the sample injection port to the detection cell. It promotes the use of reduced volumes and this reduction promotes the increase of the overlapping reagents zones [4]. Not only the aliquots division, but also the reagents order of aspiration is widely related to the interdispersion of the solutions inside the system. Thus, the effect of three different orders was studied. The first one was chromogenic reagent (CR)/cholesterol oxidase (CO)/cholesterol esterase (CE)/sample/CE/sample/CO/CR, the second was CR/CO/sample/CO/CE/sample/CE/CR, and the third was CR/CE/CO/sample/CO/CE/CR. The first order was the most efficient, since it leads to a higher sensitivity (around 1%, and 120% higher than with order 2 and order 3, respectively). The flow rate also intervenes in the interdispersion of the reagents, and at the same time influences the time of reaction. This last parameter can also be adjusted using a stopped-flow period, promoting the contact of enzymes, sample and chromogenic reagent, without the dispersion effect. Thus, it was verified that using a flow rate of 1 mL min^{-1} and a stopped-flow period of 6 min, the sensitivity increased around 48% when compared with the use of 5 min period. In the opposite, it was decreased around 15% when compared with a stopped flow period of 10 min, but at the same time, it promoted an increase of 40% in the determination frequency. The flow rate of 1 mL min^{-1} and a stopped-flow period of 6 min were selected as a compromise between sensitivity and the time spent per analysis. The determination frequency is usually higher when a µSIA-LOV system is used, due to the short course of the sample, and the reactants between injection ports and detection cell [4].

After the establishment of the previous conditions for the developed µSIA-LOV system, there was performed a calibration curve using cholesterol acetate standard solutions. This calibration curve (Figure 2) is represented by the equation AU = (1.96 ± 0.12) × 10^{-2} Conc. (mg·dL^{-1}) − (1.13 ± 2.99) × 10^{-2} (with confidence limits of 95%, for the intercept and the slope) with an R of 0.9995.

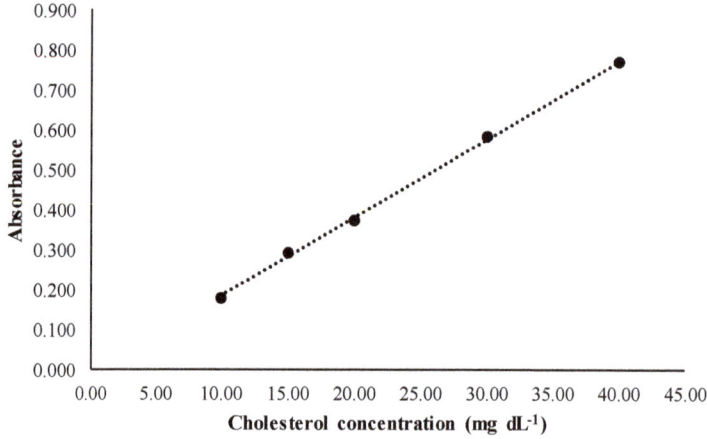

Figure 2. Calibration curve using cholesterol acetate standard solutions.

The detection and quantification limits are the concentrations obtained for the intercept, plus three and ten times Sy/x, respectively [37]. Thus, the obtained detection and quantification limits were 1.36 and 4.53 mg dL^{-1}, respectively.

A relative standard deviation (RSD) of 3.6% was obtained using 10 consecutive injections of 20 mg dL^{-1} of cholesteryl acetate standard solution, and this proved the repeatability of the developed methodology.

Taking into account the time needed to complete an analytical cycle (including the stopped-flow period and the aspiration and propulsion periods) a determination frequency of around 8 determinations h^{-1} was obtained.

2.3. Evaluation of Total Cholesterol Concentrations in Serum Samples Using the Developed µSIA-LOV Methodology

In order to attest the efficiency of the developed µSIA-LOV methodology in the quantification of total cholesterol in real samples, some reference samples of human serum (ABX Penta N Control) and some real human serum samples were selected and evaluated. All the samples were diluted in order to obtain different concentrations included in the linear concentration range from the calibration curve. The obtained results were presented in Table 2.

The developed methodology showed to be accurate since the obtained errors between the reference and the obtained concentrations for the total cholesterol in the samples were lower than 5.5%.

Additionally, the absence of statistical differences between real and obtained results was verified, since according to the t-test [37], the tabulated value of 2.16 was higher than the obtained t value of −0.82 (for 95% of confidence level), showing the accuracy of the results.

Table 2. Obtained results for the determination of total cholesterol in reference samples of human serum, using the proposed µSIA-LOV methodology.

Kind of Sample	Sample	Standard Concentrations (mg·dL^{-1})	Calculated Concentrations (mg·dL^{-1})	Error (%)
Reference samples	1	109	113.9 ± 0.7	4.5
	2	109	105.5 ± 0.6	−3.2
	3	109	108.6 ± 1.5	−0.3
	4	104	109.3 ± 0.2	5.1
	5	104	105.0 ± 0.0	1.0
Real samples	6	223	228.1 ± 0.1	2.3
	7	283	283.3 ± 1.3	0.1
	8	279	263.6 ± 3.7	−5.5
	9	272	269.2 ± 1.8	−1.0
	10	225	214.4 ± 0.2	−4.7
	11	272	257.4 ± 2.0	−5.4
	12	279	274.7 ± 3.3	−1.5
	13	211	213.5 ± 0.5	1.2
	14	283	292.3 ± 1.1	3.3

2.4. Comparison between the Developed Methodology and Other Flow-based Methodologies Used for Cholesterol Evaluation

As expected, the developed methodology, since it is a flow-based methodology, presents some advantages when compared with the batch procedures, namely promoting the reduction of sample and reagents consumption. The proposed methodology also presents some other advantages when compared with other automatic methodologies based on flow analysis (referred in Table 1). When compared with the methodology that refers to the development of a microfluidic chip [36], it presents a lower detection limit and lower sample consumption. In addition, the chip-based flow injection system was more laborious, since it needs the fabrication of functionalized carbon nanotubes working, silver reference and platinum counter electrode layers on the chip.

When compared with the asymmetrical flow field-flow fractionation (AF4) methodology [35], our manifold was much simpler, and the sample consumption was also smaller (50%).

Comparing with the multi-commutated flow procedure [34], the developed methodology allows a reduction in the detection limit, and at the same time, the sample preparation is more "green", since it does not use isopropanol as a solvent.

The mono-segmented flow analysis [33], as it is supposed by the principle of this methodology, uses air bubbles between samples. This forces the removal of the air from the stream before detection, turning the process more complex and the methodology more irreproducible. Additionally, the consumption of reagents is considerably higher, since with this methodology the reagents are propelled continuously (a mixture of cholate, Triton X-100, phenol and buffer is used as the carrier, and a mixture of three enzymes and 4-aminoantipyrine, as a reagent). The same situation of the continuous propulsion of the reagents happens with flow injection analysis-based methodologies [13,29,30,32]. Additionally, the FIA methodologies proposed by Situmorang et al. [32], Fernandez-Romero et al. [29], and Krug et al. [31] use isopropanol (an organic toxic solvent) in the sample preparation. In four [13,29,31,32] of the five referred FIA methodologies, the sample volumes are higher (4, 7, 9, and 5 times, respectively) than the proposed methodology. Two [13,31] of the previously referred FIA methodologies use the ABTS instead of 4-aminoantipyrine and phenol for the quantification of hydrogen peroxide resulting from the enzymatic reactions. Despite the fact that the ABTS were apparently less harmful than the phenol, it presents some drawbacks, since it can inhibit the peroxidase enzyme [31], or can suffer auto-oxidation in the presence of samples prepared in Triton X-100 and isopropanol [13]. The preparation of the peroxidase electrode referred in one of these FIA methodologies [30] is also laborious, in opposition to the easy handled developed methodology.

3. Materials and Methods

3.1. Reagents and Solutions

High purity water (Millipore, Danvers, MA, USA) with a specific conductivity < 0.1 $\mu S\ cm^{-1}$ and analytical reagent grade were used for the preparation of all solutions.

A 0.1 $mol \cdot L^{-1}$ phosphate buffer solution with a pH of 7.0 was used as a carrier and also for the preparation of enzyme solutions. This buffer solution was prepared by mixing appropriate amounts of 0.20 $mol \cdot L^{-1}$ Na_2HPO_4 (Sigma Aldrich St. Louis, MO, USA) and 0.20 $mol \cdot L^{-1}$ NaH_2PO_4 (Sigma Aldrich) solutions, and diluting them twice with water.

A 2.00 $U \cdot mL^{-1}$ solution of cholesterol oxidase (from *E. coli*) solution was prepared by the dissolution of the solid enzyme (Sigma Aldrich EC 1.1.3.6, 40 $U \cdot mg^{-1}$ of solid) in a buffer solution.

A 5.00 $U \cdot mL^{-1}$ solution of cholesterol esterase (from a porcine pancreas) solution was prepared by a dissolution of the solid enzyme (Sigma Aldrich EC 3.1.1.13, 35 $U \cdot mg^{-1}$ of solid) in 10 $mmol \cdot L^{-1}$ sodium cholate (Sigma) previously prepared in a buffer solution. The sodium cholate acts as a cholesterol esterase activator [24,33].

A 36.5 $U \cdot mL^{-1}$ peroxidase solution was also prepared by the dissolution of an appropriate amount of solid enzyme (Sigma Aldrich EC1.11.1.7, type I, 146 $U \cdot mg^{-1}$ of solid) in a buffer solution. This solution of the enzyme was used in the preparation of the chromogenic reagent, as well as a 6% phenol (Merck) solution, and a 0.10% 4-aminoantipyrine (4-AAP) (Sigma Aldrich) solution. The mixture of this solution was done with the proportion of 1:0.75:1 (phenol:4-AAP:peroxidase).

In order to prepare the standard solutions of cholesterol ester, and since it has low solubility in water [24], Triton X-100 was used as a solvent [38]. A 100 $mg \cdot dL^{-1}$ of cholesteryl acetate solution was prepared in Triton X-100 1.5% (v/v). For that, 5 mg of cholesteryl acetate were added and stirred in 75 µL of Triton X-100 (Sigma Life Science), previously heated until 72 °C. After the complete dissolution, 3 mL of water also previously heated were added, and the temperature and the stirring were kept for 30 min. Then, more 1.5 mL of hot water was added, and the stirring and the temperature were kept for more 30 min. The resulted solution was cooled keeping the stirring and finally, the volume of 5 mL was completed with water at room temperature. Using this final standard solution, a different solution with different concentrations were prepared by appropriate dilutions.

A 1.5% Triton X-100 was also prepared according to the previously described procedure in order to be used for the blank assays.

For the evaluation of total cholesterol, certified samples of serum ABX Penta N Control from Horiba Medical, and real human serum samples were used.

3.2. Apparatus

The developed methodology was based in a µSIA-lab-on-valve (LOV) system. This manifold (Figure 3) was composed by a multi-syringe pump model Multi-burette 4S from Crison (Barcelona, Spain) equipped with a syringe of 2.5 mL and a 10-port selection valve (Valco, Vici C25-3180EMH, Houston, USA), where a customized LOV was placed.

The control system was made using a homemade program written in VisualBasic language. A computer with an OOIBase32TM software version 2.0.6.5 of Ocean Optics was used to record the analytical signals.

All of the components of the system were connected using a 0.8 mm i.d. PTFE tubing (Idex).

For the detection, a spectrophotometer USB4000 (from Ocean Optics Inc., Dunedin, FL, USA), a light source (a tungsten halogen lamp, LS-1-LL, from Ocean Optics Inc.) and a pair of optic fibers with 600 µm of core diameter, that connect the light source with the flow cell (integrated in the LOV) and that with the spectrophotometric detector, were used.

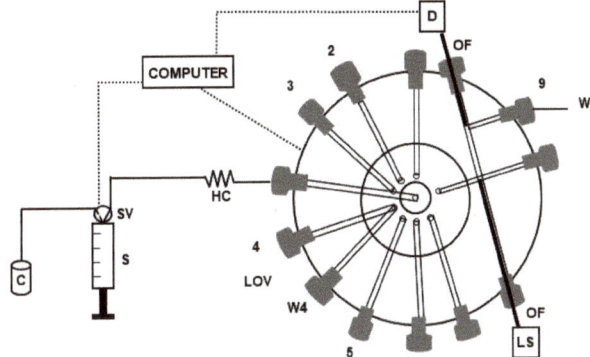

Figure 3. Developed µSIA-LOV system for the implementation of the enzymatic reactions used for the total cholesterol quantification. S, syringe; SV, solenoid valve; C, carrier (Phosphate buffer, pH 7.0); LOV, lab-on-valve; HC, holding coil (with 2 m length and a figures of eight configuration); D, spectrophotometric detector; OF, optical fiber; LS, light source; 2, chromogenic reagent; 3, cholesterol oxidase; 4, sample/standard solution; 5 cholesterol esterase; W, waste; W4, waste from port.

3.3. Micro Sequential Injection Procedure

The total cholesterol determinations based on the use of enzymatic reaction were performed using a µSIA-LOV. Before the start of the analytical cycle it was necessary to prime the system, filling the central tube with the carrier solution (phosphate buffer solution at pH of 7.0), and the lateral tubes of the selection valve with the reagents. The tubes from positions 2, 3, 4, and 5 were filled with the chromogenic reagent, cholesterol oxidase, sample/standard solutions of cholesteryl acetate and cholesterol esterase, respectively. After this preparation, the analytical cycle (Table 3) was performed by the aspiration of 10 µL of chromogenic reagents (step 1), 5 µL of cholesterol oxidase (step 2), 5 µL of cholesterol esterase (step 3) and 5 µL of sample (step 4). Then, the aspiration of 5 µL of cholesterol esterase (step 5), 5 µL of sample (step 6), 5 µL of cholesterol oxidase (step 7) and 10 µL of chromogenic reagent (step 8), was repeated. After the aspiration of all of these aliquots of reagents, the flow was stopped, the reagents remaining in the holding coil for a period of 6 min (step 9), in order to guarantee enough contact between the enzymes, the sample and the chromogenic reagent. Then, by a flow reversal, the fluids were propelled to the detection cell (a channel between the optic fibers, placed into the LOV) where the absorbance values were measured at 500 nm, and finally, this mixture was sent to the waste (step 10).

Table 3. Analytical cycle for the evaluation of total cholesterol.

Step	Position	Volume (µL)	Time (s)	Flow Rate (mL·min^{-1})	Direction	Event
1	2	10	4	0.15	Aspiration	Chromogenic reagent
2	3	5	2	0.15	Aspiration	Cholesterol oxidase
3	5	5	2	0.15	Aspiration	Cholesterol esterase
4	4	5	2	0.15	Aspiration	Sample
5	5	5	2	0.15	Aspiration	Cholesterol esterase
6	4	5	2	0.15	Aspiration	Sample
7	3	5	2	0.15	Aspiration	Cholesterol oxidase
8	2	10	4	0.15	Aspiration	Chromogenic reagent
9	9	-	360	0	Stopped flow	Stopped flow in the holding coil
10	9	10,000	75	1.25	Propulsion	Propulsion to the detector

3.4. Comparison Method

The accuracy of the obtained results with the developed methodology was evaluated in two ways. The first one was the use of commercial reference serum samples. The obtained results for these samples were compared with their theoretical concentration values. The second one was to use real samples of human serum, and compare the obtained concentration values with those obtained in a certified laboratory of clinical analysis, that uses a perfectly implemented methodology based on an enzymatic assay with spectrophotometric detection. These values are presented in Table 2, in the Section 2.3.

4. Conclusions

Concluding, a simple automatic methodology using a µSIA-LOV system was developed. It showed to be a useful tool for the application of enzymatic reactions promoted by cholesterol oxidase, cholesterol esterase, and peroxidase enzymes, and also for the evaluation of cholesterol in human serum samples. These results showed to be reliable and accurate, confirming the usefulness of this methodology for the implementation of enzymatic reactions, and at the same time its utility in the clinical field. The automation and the miniaturization leads to the reproducibility improvement, the reduction of reagents consumption, and waste generation. These advantages are relevant even when compared with other automatic methodologies based on the flow analysis. This showed that the developed methodology is in agreement with the principles of green analytical chemistry. Thus, it was verified that the developed µSIA-LOV methodology is a great alternative for the implementation of reactions catalyzed by cholesterol esterase, cholesterol oxidase and peroxidase and for the routine determinations of cholesterol.

Author Contributions: Conceptualization, M.L.C.P., M.G.A.K., and M.L.M.F.S.S.; formal analysis, J.S.B.; data curation, J.S.B., and M.L.C.P.; writing—original draft preparation, M.L.C.P.; writing—review and editing, J.S.B, M.G.A.K., and M.L.M.F.S.S.; supervision, M.L.C.P, M.G.A.K., and M.L.M.F.S.S.

Funding: This work was supported by the European Union (FEDER funds POCI/01/0145/FEDER/007265) and National Funds (FCT/MEC, Fundação para a Ciência e Tecnologia and Ministério da Educação e Ciência) under the Partnership Agreement PT2020 UID/QUI/50006/2019.

Acknowledgments: Authors acknowledge the financial support from FEDER Funds through the Operational Competitiveness Factors Program - COMPETE and by National Funds through FCT - within the scope of the project POCI-01-0145-FEDER-030163. Marieta Passos thanks FCT for the financial support. Jucineide Barbosa thanks CAPES Foundation, Ministry of Education of Brazil for the master fellowship 1554724. Authors also acknowledge Laura Pereira for the clinical analytical data obtained with the reference methodology.

Conflicts of Interest: The authors declare no conflict of interest.

References

1. Silvestre, C.I.C.; Pinto, P.; Segundo, M.A.; Saraiva, M.; Lima, J. Enzyme based assays in a sequential injection format: A review. *Anal. Chim. Acta* **2011**, *689*, 160–177. [CrossRef] [PubMed]
2. Araujo, A.N.; Lima, J.; Pinto, P.; Saraiva, M. Enzymatic determination of glucose in milk samples by sequential injection analysis. *Anal. Sci.* **2009**, *25*, 687–692. [CrossRef] [PubMed]
3. Ruzicka, J.; Marshall, G.D. Ssequential injection - A new concept for chemical sensors, process analysis and laboratory assays. *Anal. Chim. Acta* **1990**, *237*, 329–343. [CrossRef]
4. Ruzicka, J. Lab-on-valve: Universal microflow analyzer based on sequential and bead injection. *Analyst* **2000**, *125*, 1053–1060. [CrossRef]
5. Decuir, M.; Boden, H.; Carroll, A.; Ruzicka, J. Principles of micro sequential injection analysis in the lab-on-valve format andiIts introduction into a teaching laboratory. *J. Flow Injection Anal.* **2007**, *24*, 103.
6. Vidigal, S.; Toth, I.V.; Rangel, A. Sequential injection-LOV format for peak height and kinetic measurement modes in the spectrophotometric enzymatic determination of ethanol: Application to different alcoholic beverages. *Talanta* **2008**, *77*, 494–499. [CrossRef]
7. Chen, Y.; Ruzicka, J. Accelerated micro-sequential injection in lab-on-valve format, applied to enzymatic assays. *Analyst* **2004**, *129*, 597–601. [CrossRef] [PubMed]
8. Costa, D.; Passos, M.L.C.; Azevedo, A.M.O.; Saraiva, M. Automatic evaluation of peroxidase activity using different substrates under a micro sequential injection analysis/lab-on-valve (mu SIA-LOV) format. *Microchem. J.* **2017**, *134*, 98–103. [CrossRef]
9. Vidigal, S.; Toth, I.V.; Rangel, A. Sequential injection lab-on-valve system for the determination of the activity of peroxidase in vegetables. *J. Agric. Food Chem.* **2010**, *58*, 2071–2075. [CrossRef] [PubMed]
10. Vidigal, S.; Toth, I.V.; Rangel, A. Sequential injection lab-on-valve system for the on-line monitoring of hydrogen peroxide in lens care solutions. *Microchem. J.* **2009**, *91*, 197–201. [CrossRef]
11. Chen, Y.; Carroll, A.D.; Scampavia, L.; Ruzicka, J. Automated method, based on micro-sequential injection, for the study of enzyme kinetics and inhibition. *Anal. Sci.* **2006**, *22*, 9–14. [CrossRef] [PubMed]
12. Pavlicek, O.; Polasek, M.; Foltyn, M.; Cabal, J. Automated detection of organophosphate warfare gases (nerve agents) in air based on micro-SIA - lab-on-valve system. *J. Appl. Biomed.* **2013**, *11*, 27–32. [CrossRef]
13. Krug, A.; Gobel, R.; Kellner, R. Flow-injection analysis for total cholesterol with photometric detection. *Anal. Chim. Acta* **1994**, *287*, 59–64. [CrossRef]
14. Suman; Pundir, C.S. Co-immobilization of cholesterol esterase, cholesterol oxidase and peroxidase onto alkylamine glass beads for measurement of total cholesterol in serum. *Curr. Appl. Phys.* **2003**, *3*, 129–133.
15. Allain, C.C.; Poon, L.S.; Chan, C.S.G.; Richmond, W.; Fu, P.C. Enzymatic determination of total serum cholesterol. *Clin. Chem.* **1974**, *20*, 470–475. [PubMed]
16. Tan, M.H.; Dickinson, M.A.; Albers, J.J.; Havel, R.J.; Cheung, M.C.; Vigne, J.L. The effect of a high cholesterol and saturated fat diet on serum high-density lipoprotein-cholesterol, apoprotein-A-I, and apoprotein-E levels in normolipidemic humans. *Am. J. Clin. Nutr.* **1980**, *33*, 2559–2565. [CrossRef] [PubMed]
17. Stewart, A.J.; O'Reilly, E.J.; Moriarty, R.D.; Bertoncello, P.; Keyes, T.E.; Forster, R.J.; Dennany, L. A Cholesterol biosensor based on the NIR electrogenerated-chemiluminescence (ECL) of water-soluble CdSeTe/ZnS quantum dots. *Electrochim. Acta* **2015**, *157*, 8–14. [CrossRef]
18. Collins, R.; Armitage, J.; Parish, S.; Sleight, P.; Peto, R.; Heart Protect Study, C. Effects of cholesterol-lowering with simvastatin on stroke and other major vascular events in 20 536 people with cerebrovascular disease or other high-risk conditions. *Lancet* **2004**, *363*, 757–767.
19. Castelli, W.P.; Anderson, K. A population at risk - Prevalence of high cholesterol levels in hypertensive patients in the Framingham-study. *Am. J. Med.* **1986**, *80*, 23–32. [CrossRef]
20. Ezzati, M.; Lopez, A.D.; Rodgers, A.; Vander Hoorn, S.; Murray, C.J.L.; Comparative Risk Assessment Collaborating Group. Selected major risk factors and global and regional burden of disease. *Lancet* **2002**, *360*, 1347–1360. [CrossRef]
21. Emanuelsson, F.; Nordestgaard, B.G.; Benn, M. Familial Hypercholesterolemia and risk of peripheral arterial disease and chronic kidney disease. *J. Clin. Endocrinol. Metab.* **2018**, *103*, 4491–4500. [CrossRef] [PubMed]

22. Karube, I.; Hara, K.; Matsuoka, H.; Suzuki, S. Amperometric determination of total cholesterol in serum with use of immobilized cholesterol esterase and cholesterol oxidase. *Anal. Chim. Acta* **1982**, *139*, 127–132. [CrossRef]
23. Taniguchi, A.; Hayashi, Y.; Yuki, H. Determination of cholesterol with a laboratory-built chemiluminescence system. *Anal. Chim. Acta* **1986**, *188*, 95–100. [CrossRef]
24. Malavolti, N.L.; Pilosof, D.; Nieman, T.A. Determination of cholesterol with a microporous membrane chemiluminescence cell with cholesterol oxidase in solution. *Anal. Chim. Acta* **1985**, *170*, 199–207. [CrossRef]
25. Gupta, V.K.; Norouzi, P.; Ganjali, H.; Faridbod, F.; Ganjali, M.R. Flow injection analysis of cholesterol using FFT admittance voltammetric biosensor based on MWCNT-ZnO nanoparticles. *Electrochim. Acta* **2013**, *100*, 29–34. [CrossRef]
26. Aggarwal, V.; Malik, J.; Prashant, A.; Jaiwal, P.K.; Pundir, C.S. Amperometric determination of serum total cholesterol with nanoparticles of cholesterol esterase and cholesterol oxidase. *Anal. Biochem.* **2016**, *500*, 6–11. [CrossRef] [PubMed]
27. Zhang, Y.; Wang, Y.N.; Sun, X.T.; Chen, L.; Xu, Z.R. Boron nitride nanosheet/CuS nanocomposites as mimetic peroxidase for sensitive colorimetric detection of cholesterol. *Sens. Actuator B-Chem.* **2017**, *246*, 118–126. [CrossRef]
28. Huang, Y.; Cui, L.J.; Xue, Y.W.; Zhang, S.B.; Zhu, N.X.; Liang, J.T.; Li, G.Y. Ultrasensitive cholesterol biosensor based on enzymatic silver deposition on gold nanoparticles modified screen-printed carbon electrode. *Mater. Sci. Eng. C-Mater. Biol. Appl.* **2017**, *77*, 1–8. [CrossRef]
29. Fernandezromero, J.M.; Decastro, M.D.L.; Valcarcel, M. Enzymatic determination of total cholesterol in serum by flow-injection analysis. *J. Pharm. Biomed. Anal.* **1987**, *5*, 333–340. [CrossRef]
30. Yao, T.; Wasa, T. Flow-injection system for simultaneous assay of free and total cholesterol in blood-serum by use of immobilized enzymes. *Anal. Chim. Acta* **1988**, *207*, 319–323. [CrossRef]
31. Krug, A.; Suleiman, A.A.; Guilbault, G.G.; Kellner, R. Colorimetric determination of free and total cholesterol by flow-injection analysis with fiber optic detector. *Enzyme Microb. Technol.* **1992**, *14*, 313–316. [CrossRef]
32. Situmorang, M.; Alexander, P.W.; Hibbert, D.B. Flow injection potentiometry for enzymatic assay of cholesterol with a tungsten electrode sensor. *Talanta* **1999**, *49*, 639–649. [CrossRef]
33. Araujo, A.N.; Catita, J.A.M.; Lima, J. Monosegmented flow-analysis of serum cholesterol. *Farmaco* **1999**, *54*, 51–55. [CrossRef]
34. Pires, C.K.; Reis, B.F.; Galhardo, C.X.; Martelli, P.B. A multicommuted flow procedure for the determination of cholesterol in animal blood serum by chemiluminescence. *Anal. Lett.* **2003**, *36*, 3011–3024. [CrossRef]
35. Rambaldi, D.C.; Reschiglian, P.; Zattoni, A.; Johann, C. Enzymatic determination of cholesterol and triglycerides in serum lipoprotein profiles by asymmetrical flow field-flow fractionation with on-line, dual detection. *Anal. Chim. Acta* **2009**, *654*, 64–70. [CrossRef]
36. Wisitsoraat, A.; Sritongkham, P.; Karuwan, C.; Phokharatkul, D.; Maturos, T.; Tuantranont, A. Fast cholesterol detection using flow injection microfluidic device with functionalized carbon nanotubes based electrochemical sensor. *Biosens. Bioelectron.* **2010**, *26*, 1514–1520. [CrossRef]
37. Miller, J.N.; Miller, J.C. *Statistics and Chemometrics for Analytical Chemistry*, 6th ed.; Pearson Education Limited: Harlow, UK, 2010.
38. Richmond, W. Use of cholesterol oxidase for assay of total and free cholesterol in serum by continuous-flow analysis. *Clin. Chem.* **1976**, *22*, 1579–1588.

Sample Availability: Samples are not available from the authors.

© 2019 by the authors. Licensee MDPI, Basel, Switzerland. This article is an open access article distributed under the terms and conditions of the Creative Commons Attribution (CC BY) license (http://creativecommons.org/licenses/by/4.0/).

Article

Determination of Albumin, Glucose, and Creatinine Employing a Single Sequential Injection Lab-at-Valve with Mono-Segmented Flow System Enabling In-Line Dilution, In-Line Single-Standard Calibration, and In-Line Standard Addition

Kanokwan Kiwfo [1,2,3], Wasin Wongwilai [1,4], Tadao Sakai [3], Norio Teshima [3] and Kate Grudpan [1,2,4,*]

1. Center of Excellence for Innovation in Analytical Science and Technology, Chiang Mai University, Chiang Mai 50200, Thailand; k.kanokwan11@gmail.com (K.K.); wwongwilai@gmail.com (W.W.)
2. Department of Chemistry and Graduate programs in Chemistry, Faculty of Science, Chiang Mai University, Chiang Mai 50200, Thailand
3. Department of Applied Chemistry, Aichi Institute of Technology, 1247 Yachigusa, Yakusa-cho, Toyota 470-0392, Japan; tadsakai@octn.jp (T.S.); teshima@aitech.ac.jp (N.T.)
4. Science and Technology Research Institute, Chiang Mai University, Chiang Mai 50200, Thailand
* Correspondence: kgrudpan@gmail.com; Tel.: +66-5394-1917

Received: 25 October 2019; Accepted: 19 February 2020; Published: 4 April 2020

Abstract: A mono-segmented sequential injection lab-at-valve (SI-LAV) system for the determination of albumin, glucose, and creatinine, three key biomarkers in diabetes screening and diagnosis, was developed as a single system for multi-analyte analysis. The mono-segmentation technique was employed for in-line dilution, in-line single-standard calibration, and in-line standard addition. This made adjustments to the sample preparation step easy unlike the batch-wise method. The results showed that the system could be used for both fast reaction (albumin) and slow reaction (glucose with enzymatic reaction and creatinine). In the case of slow reaction, the analysis time could be shortened by using the reaction rate obtained with the SI-LAV system. This proposed system is for cost-effective and downscaling analysis, which would be applicable for small hospitals and clinics in remote places with a small number of samples but relatively fast screening would be needed.

Keywords: SI-LAV; mono-segmented flow; in-line dilution; in-line single-standard calibration; in-line standard addition; albumin; glucose; creatinine

1. Introduction

The determination of biomarkers in clinical samples is essential for screening for diagnosis and/or medical treatment for many diseases. Albumin, glucose, and creatinine are, for example, the key biomarkers for diabetes mellitus. In a large hospital, the three biomarkers are usually determined by using an automatic analyzer, so as to serve a large number of the clinical samples. In some, each of the three biomarkers may be operated by using an individual analyzer. However, in some small hospitals, a few samples may be encountered. In the latter cases, as the hospitals may be in remote places, a simple, cost-effective instrument may serve the requirement for clinical analysis of the three biomarkers, but the determination of the three biomarkers should be carried out in a short analysis period.

Recently, the mono-segmentation method has been employed in various flow-based techniques such as simultaneous multiple injection, multi-commutated flow system, micro-titration, and sequential injection. The mono-segmentation method involves creating a stack zone between air segments for eliminating dispersion and dilution effects from the carrier stream. It has been used in corporation with various analytical techniques, such as the electrochemical method, spectrophotometric method, and titration method [1–13], with different tasks in the chemical analysis steps from sample preparation to detection via in-line sample dilution, in-line single-standard calibration, and in-line standard addition, as summarized in Table 1. Although there have been a number of reports regarding mono-segmentation, most of the works engaged the development in applying mono-segmentation for one or two of the three tasks (in-line sample dilution, in-line single-standard calibration, and in-line standard addition). There was only one report using mono-segmentation for the three tasks together by employing only a one-instrument setup for multi-analyte determinations. That work was with electrochemical techniques [8]. It is of interest to apply mono-segmentation in a one-instrument setup for the three tasks in steps of chemical analysis for assays of albumin, glucose, and creatinine for some specific aims.

Table 1. Usages of mono-segmentation in sequential injection analysis.

No.	Analyte(s)/Sample (s)	Detection/Technique	Reagent	Sample Conditioning (Inline Dilution)	Role of Mono-Segmented Inline Single Std. Calibration	Inline Std. Addition	Ref.
1	Fe(II) in pharmaceutical preparations, Cr(VI) in natural water and domestic waste water samples	solution handing for spectrophotometric determination (Fe(II) and Cr(VI)	KMnO$_4$ for Fe(II) and diphenylcarbazide for Cr(VI)			X	[1]
2	Fe(II) in anti-anemic medicine	spectrophotometric determination of Fe(II)	1,10-phenanthroline	X		X	[2]
3	sulfide in waters	spectrophotometric detection	Fe(III) and N,N-dimethyl-␀-phenylene diamine hydrochloride		X	X	[3]
4	atrazine	voltammetric detection			X	X	[4]
5	picloram in natural waters	voltammetric detection			X	X	[5]
6	Mg, Ca in water sample	flame atomic absorption spectrometric detection			X	X	[6]
7	methyl parathion in water sample	voltammetric detection			X	X	[7]
8	Zn(II), Cd(II), Pb(II) and Cu(II) in water samples	voltammetric dectection		X	X	X	[8]
9	Al in water and beverage samples	tritrarion with spectrophotometric determination	sodium hydroxide as a titrant and phenolphthalein or thymolphthalein indicator		X		[9]
10	benzoic acid in a real beverage sample	amperometric detection	biosensor is based on the inhibition effect of benzoic acid on the biocatalytic activity of tyrosinase, polyphenol oxidase.	X	X	X	[10]
11	B in plants	spectrophotometric detection	azomethine-H		X		[11]
12	Se (IV) in raw Se-enriched yeast	spectrophotometric detection	o-phenylenediamine		X	X	[12]
13	Al in water and beverage samples.	spectrophotometric detection	Eriochrome cyanine R		X	X	[13]

175

Sequential injection analysis with lab-at-valve (SI-LAV) has been developed in order to offer a simple, cost-effective, alternative system for flow-based analysis. An additional component was attached to a multi-position selection valve, allowing various chemical reactions to be manipulated and monitored. With respect to the lab-on-valve (LOV) technique, the modification in LAV utilizes a device and tool available in laboratory, with no extra mechanical work required [14].

In this work, a simple, cost-effective SI-LAV system with mono-segmentation operation was proposed for downscaling chemical analysis, with the aim of developing a fully automated single system for the assays of the three biomarkers (albumin, glucose, and creatinine) for screening purposes in a small hospital or a clinic. With mono-segmentation, three operations were adopted from sample preparation to detection steps as follows: In-line sample dilution, in-line single-standard calibration, and in-line standard addition.

Chemical reactions used in the SI-LAV system involve the ion association of protein with tetra-bromophenolphthalein ethyl ester (TBPE) for albumin determination [15,16], an enzymatic reaction with p-anisidine as chromogenic reagent for glucose determination [16], and Jaffé reaction for creatinine determination [17–19].

2. Results and Discussion

2.1. Determination of Albumin

The detection reaction for albumin determination is based on the ion association of protein with tetra-bromophenolphthalein ethyl ester (TBPE) in the presence of Triton X-100 at pH 3.2 to form a blue product [15,16]. The albumin reaction was used as a model study of fast reaction.

In the sample preparation step, the mono-segmented technique was employed for in-line sample dilution (Table S1 and Figure S1A). The sample could be diluted 2–140-fold to achieve a concentration suitable for the detection range. Next, the mono-segmented technique was applied for in-line single-standard calibration (Figure S1B). In between air segments, the mono-segment of 200 µL was created. Each 100-µL mono-segment contained a sequence of R1: Albumin reagent (5.0×10^{-5} mol·L^{-1} TBPE, 0.02% Triton X-100, 0.04 mol·L^{-1} acetate buffer pH 3.2), W: deionized (DI) water, and SD: standard solution (human serum albumin (HSA)) at volumes of 65, 35 − Y and Y µL, respectively. Alternatively, the mono-segmented technique could be used in in-line standard addition (Figure S1C). A mono-segment of 200 µL was created, containing a sequence of R1, W, SD, R1, W, and S: sample at volumes of 65, 35 − X, X, 65, 35 − Y, and Y µL, respectively. During in-line standard addition, the dilution factor could be increased by aspirating the predilution sample from the port number 2 and adjusting the diluent volume accordingly. This flexible process made it more convenient for multi-dilution. In the previous study, it was suggested that the diluted sample could reduce the interference by at least 20-fold [15].

Due to the fast reaction rate of albumin with TBPE, the reaction could approach the steady state in a short time. The absorbance at 605 nm was monitored. It was found that the absorbance of product (A_{605}) was proportional to albumin content, as shown in Figure 1. The linear range was up to 3.5 µg HSA in 200 µL, with a calibration equation of absorbance = 0.0558 (HSA (µg) in 200 µL) + 0.0491, r^2 = 0.992, limit of detection (LOD) (3σ) [20] = 0.4 µg in 200 µL. From the µg amount obtained in 200 µL, the concentration (mg/dL) of albumin in the original can be evaluated.

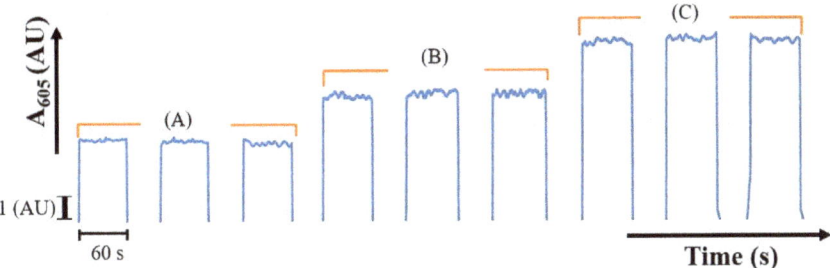

Figure 1. Analytical signal obtained from albumin content in 200 µL (**A**) 0, (**B**) 2.1, (**C**) 3.5 µg human serum albumin (HSA).

2.2. Determination of Glucose

The detection reaction for glucose determination involves glucose oxidase (GOD) promoting the oxidation of D-(+)-glucose. Glucono lactone and hydrogen peroxide are produced. The hydrogen peroxide obtained from the oxidation of glucose oxidizes p-anisidine to a red color compound in the presence of iron (II) used as a catalyst [16].

For in-line sample dilution (Table S2 and Figure S2A), the mono-segmented technique for a dilution of glucose could flexibly be carried out for 2–160 folds.

For in-line single-standard calibration (Figure S2B), the mono-segment of 200 µL was created, similar to the albumin assay but with the enzyme solution aspirated separately from other reagents to avoid enzyme denaturation. It consisted of repeated sequences of R2: Glucose reagent (0.04 mol·L^{-1} p-anisidine, 0.002 mol·L^{-1} iron (II) in 0.002 mol·L^{-1} H$_2$SO$_4$, 0.1 mol·L^{-1} acetate buffer, pH 4.5), R3: Glucose oxidase, W: DI water, and SD: standard glucose at volumes of 50, 10, 40 - Y, and Y µL, respectively.

In the case of in-line standard addition (Figure S2C), the mono-segment of 200 µL was created with a sequence of R2, R3, W, SD, R2, R3, W, and S, at volumes of 50, 10, 40 - Y, Y, 50, 10, 40 – X, and X µL, respectively. The dilution factor could be increased by adjusting the prediluted sample volume for suitable concentration, similar to the albumin system. Absorbance at 520 nm in the reaction chamber was monitored by using stop-flow mode. This showed that absorbance increased with time (Figure 2).

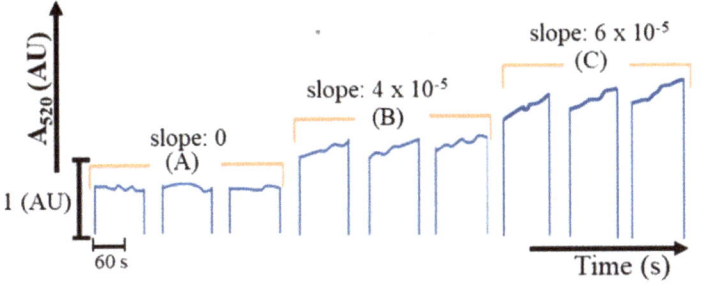

Figure 2. Analytical signal obtained from glucose content in 200 µL. (**A**) 0, (**B**) 2.4, and (**C**) 4 µg.

Unlike the reaction of albumin, the reaction of glucose could be monitored using SI-LAV in a presteady state due to its relatively slow reaction rate. The reaction rate was found to be proportional to the concentration of glucose, as it complied with the pseudo first-order reaction, as previously reported [21]. The reaction rate was calculated using a slope of absorbance at 520 nm recorded at reaction time 30–120 s in the presteady state. The increase in the reaction rate was proportional to glucose concentration, which is directly correlated to glucose content. A linear range was obtained up to 4 µg glucose in 200 µL, with a calibration equation of reaction rate = 1.52×10^{-5} (glucose (µg) in 200 µL) + 1.00×10^{-6}, r^2 = 0.986, LOD (3σ) [20] = 1.5 µg in 200 µL. From the µg amount obtained in 200 µL, the concentration (mg/dL) of glucose in the original can be evaluated.

In this study, the glucose oxidase in solution form was used for glucose determination, since it could exhibit stable activity across batch-wise preparation. It should be noted that the enzyme solution was readily homogenously mixed with the substrate, which helped shorten total analysis time by at least 10 times compared to the previous work using an enzyme-immobilized bead column [15]. Also, it was suggested that exploits of the standard addition method and reaction rate in calibration could reduce color interference, especially for urine samples.

2.3. Determination of Creatinine

The detection reaction for creatinine determination is based on the Jaffé reaction. Creatinine reacts with picrate in an alkaline medium to form a red-orange product [17–19].

First, the mono-segmented technique was applied for in-line creatinine sample dilution in the range of 2–100 folds (Table S3 and Figure S3A). It was reported previously that the creatinine sample had to be diluted by at least 80 fold in order to minimize the interference effect [19].

In the reaction and detection step (Table S4), the mono-segmented technique was applied in either in-line single-standard calibration or in-line standard addition. Unlike albumin and glucose assays, two mono-segments were created, as shown in Figure 3. First, the 100-µL mono-segment containing creatinine reagent was aspirated to into chamber (port number 10), followed by the dispensing of a 100-µL mono-segment of either creatinine standard or sample.

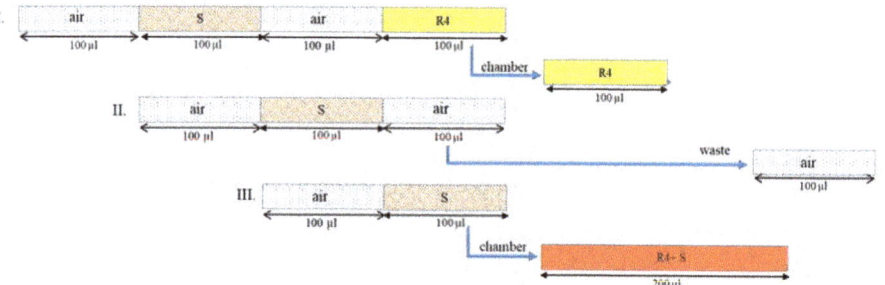

Figure 3. A sequence profile for creatinine determination (R4: Mixed reagent for creatinine determination (R4: 0.025 mol·L^{-1} sodium picrate, 0.75% NaOH, 0.03 mol·L^{-1} KH$_2$PO$_4$), S: Pretreated creatinine solution).

For in-line single-standard calibration, from above, another 100-µL mono-segment contained creatinine standard and DI water at volumes of X and 100 − X µL, respectively (Figure S3B). In the case of standard addition, another 100-µL mono-segment contained a pretreated sample (S), creatinine standard, and DI water (Figure S3C). The desired concentration could be achieved by adjusting the volume ratio. These mono-segments were aspirated into the LAV chamber successively, as shown in Figure 3.

The two mono-segments were injected into the chamber (see steps E to H in Table S4). Using stop-flow mode allowed the reaction to be followed for kinetics (slope of the signal refers to a reaction rate). In Figure 4, the linear range was up to 20 µg creatinine in 200 µL, with a calibration equation of reaction rate = 4.71×10^{-4} (creatinine (µg) in 200 µL) − 2.00×10^{-5}, r^2 = 0.996, LOD (3σ) [20] = 1.9 µg in 200 µL creatinine. From the µg amount obtained in 200 µL, the concentration (mg/dL) of creatinine in the original can be evaluated.

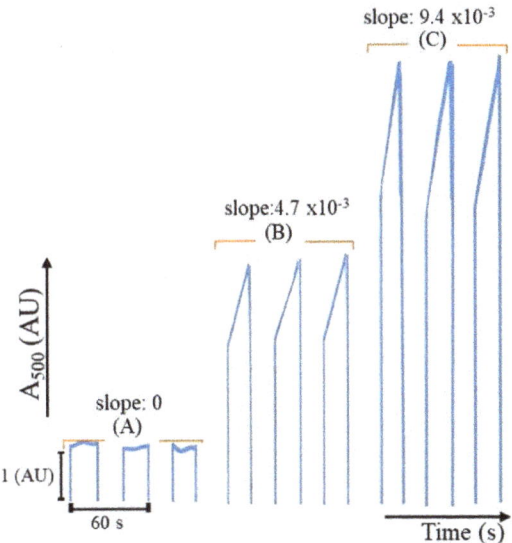

Figure 4. Analytical signal obtained from creatinine content in 200 μl (**A**) 0, (**B**) 10, and (**C**) 20 μg.

For glucose and creatinine determination, the results agreed with the previous study, showing that using kinetic data can reduce analysis time and could eliminate the effects of sample color and some interference in urine samples [19]. By using the mono-segmented SI-LAV system, the reaction rate of each reaction can be obtained. Those results from both reactions show that the use of the reaction rate to determine the concentration of analyte could be done in a shorter analytical time compared to those obtained at steady state. Therefore, the system could be useful, especially in the case of a slow reaction.

In all experiments, the in-line dilution using mono-segmentation allowed for the quick adjustment of the sample/standard volume to obtain the appropriate concentration within a linear range. If the first try with the dilution was unsuccessful, leading to the obtained signal being outside the linear range, the next attempt could be adjusted to be within the linear range. This could help shorten the sample preparation time significantly. In the detection step, the mono-segmentation technique provided a precise control of reaction mixture volume in a similar way to batch-wise practice, and thus the absolute concentration of the sample could be determined. This system could be applied for different analytical systems and further developed into a fully automated system for continuous multi-analyte analysis.

2.4. Preliminary Work in Application to a Real Sample

Using the developed SI-LAV system, preliminary work in the application to a real sample was performed for a urine sample, as shown in Table 2.

Table 2. The determination of albumin, glucose, and creatinine in a urine sample *.

Biomarker	Concentration (mg/dL)
Albumin	1.8
Glucose	not detectable **
creatinine	129

* A urine sample taken from a healthy female aged 31 years old; ** less than LOD as described earlier.

3. Experimental

3.1. Reagents and Chemicals

All chemicals used were of laboratory reagent grade. Deionized (DI) water, obtained from an Aquarius GSH-210 apparatus (Advantec, Tokyo, Japan), was used for preparing the solutions throughout.

For the albumin assay, human serum albumin (HSA) (Serologicals Proteins Inc., Kankakee, IL, USA), tetrabromophenolphthalein ethyl ester potassium salt ($C_{22}H_{13}Br_4KO_4$, Wako Pure Chemical Co., Osaka, Japan), and Triton X-100 ($C_{14}H_{22}O(C_2H_4O)_n$ (n = 9–10), Fisher Scientific UK Ltd., Leicestershire, UK) were used.

For the glucose assay, D-(+)-glucose ($C_6H_{12}O_6$, Sigma-Aldrich, St. Louis, MO, USA), p-anisidine (C_7H_9NO, Alfa Aesar, Lancashire, UK), iron(II) sulphate ($FeSO_4$, Wako Pure Chemical Co., Japan), 98% sulfuric acid (H_2SO_4, QRëc, Auckland, New Zealand), glucose oxidase (E.C.1.1.3.4) from Aspergillus sp., 200U/mg, (Sigma-Aldrich, St. Louis, MO, USA), and 30% hydrogen peroxide (H_2O_2, BDH Prolabo, Lutterworth, UK) were used.

For creatinine assay, creatinine ($C_4H_7N_3O$, Wako Pure Chemical Co., Osaka, Japan), potassium dihydrogen phosphate (KH_2PO_4, Wako Pure Chemical Co., Osaka, Japan), sodium picrate monohydrate ($C_6H_4N_3NaO_8$, Wako Pure Chemical Co., Osaka, Japan), and sodium hydroxide (NaOH, NACALAI TESQUE, INC., Kyoto, Japan) were used.

3.2. The Instrument Setup

The SI-LAV system (Figure 5) was similar to that reported previously [22]. The system consisted of C: Mixing chamber and detection cell (micro pipette tip 1000 µL (Thermo Fisher, San Diego, CA, USA), light path length flow cell approximately 5 mm), R1: Albumin reagent (mixed solution of TBPE and Triton X-100 buffered at pH 3.2), R2: Glucose reagent I (mixed solution of p-anisidine and iron(II) buffered at pH 4.5), R3: Glucose reagent II (glucose oxidase), R4: Creatinine reagent (mixed solution of picric acid and sodium hydroxide), S: Sample, SD: Standard solution, W: DI Water, waste, syringe pump (2500 µL, FIA lab instruments, Seattle, WA, USA), 10-port selection valve (Valco Instruments Co. Inc., Houston, TX, USA), holding coil (PTFE Tubing, 0.5 mm i.d., 2.5 m long), spectrometer (USB4000, Ocean Optics, Largo, FL, USA), and fiber optics (P200-2UV/Vis Ocean Optics, Inc., Largo, FL, USA).

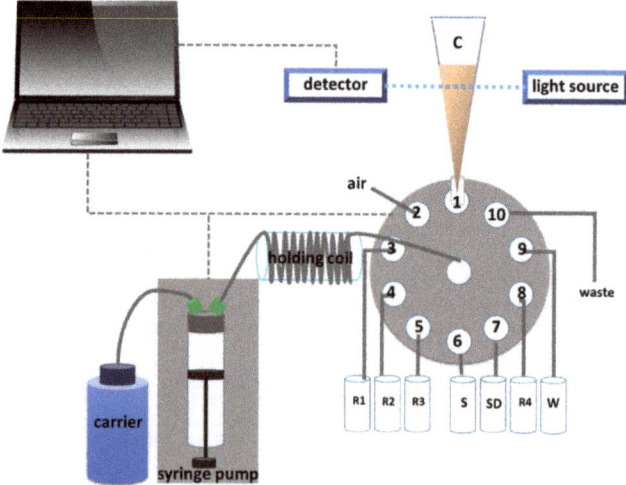

Figure 5. Schematic diagram of the sequential injection lab-at-valve (SI-LAV).

In this work, the SI-LAV system was used for the determination of glucose, albumin, and creatinine. Sample and reagents were sequentially aspirated into an automatic control system. In-house software was used to control the pump and multi-port selection valve with various operational control sequences, as shown in Figure 6.

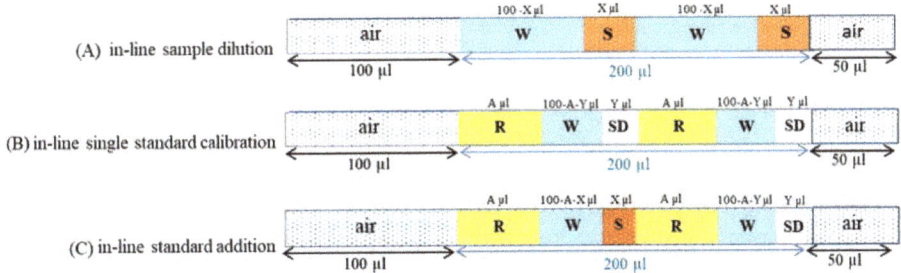

Figure 6. Operational sequences for (**A**) in-line sample dilution, (**B**) in-line single standard calibration, (**C**) in-line standard addition, W (DI water), S (sample), R (reagent), SD (standard solution).

Within each mono-segment, the sample/standard solution could be split into a few small portions and sequentially aspirated in between diluent portions to increase the contract area, resulting in better mixing within the fixed volume of the mono-segmented zone. The mono-segmentation was employed in three sample preparation and detection steps as follows:

(1) For in-line dilution, the variable ratio of either sample (X μL) or standard solution (Y μL) and diluent (either 100 − X μL or 100 − Y μL) was aspirated twice within the fixed total volume of a 200-μL mono-segment between two air segments to yield the desired concentration, as illustrated in Figure 6(A). For dilution factors greater than 20, repeated in-line dilution was done by aspirating the first dilution segment to the port number 2 prior to detection.

(2) For in-line single-standard calibration, the fixed volume of reagent (A μL) combined with the variable ratio of standard solution (Y μL), and diluent (100 − A − Y μL) was aspirated twice within the fixed total volume of 200 μL mono-segment, as illustrated in Figure 6B.

(3) For in-line standard addition, the fixed volume of reagent (A μL) and sample (X μL) combined with the variable ratio of standard solution (Y μL) and diluent (200 − 2A − X − Y μL) was aspirated within the fixed total volume of 200 μL mono-segment between two air segments, as illustrated in Figure 6C.

4. Conclusions

The determination of albumin, glucose, and creatinine, which are the key biomarkers for diabetic screening diagnosis and medical treatment, employing a single sequential injection lab-at-valve with a mono-segmented flow system enabling in-line dilution, in-line single-standard calibration, and in-line standard addition, was proposed. Without needing to change the system configuration, the system enabled downscaling and automation approaches to be performed via in-line dilution, in-line single-standard, and in-line standard addition throughout the analysis process, which is convenient. With the system, a shorter analysis time via flow methodology can be obtained from the enzymatic reaction (for glucose) and a slow reaction (for creatinine). Development of the assay of the three analytes in one urine sample using the proposed system is in progress, with the aim of application in small hospitals and clinics with a small number of samples. This would reduce the time taken to transfer a sample to a larger laboratory.

Supplementary Materials: The following are available online, Table S1: Operational steps of the SI-LAV system for albumin determination, Figure S1: Sequence profile for albumin determination, Table S2: Operational steps of the SI-LAV system for glucose determination, Figure S2: A sequence profile for glucose determination, Table S3:

Operational steps of the SI-LAV system for on-line sample pre-treatment for creatinine determination, Figure S3: Sequence profile for on-line sample pre-treatment creatinine solution.

Author Contributions: Conceptualization, K.G., T.S., N.T. and W.W.; experimental work, K.K. and W.W., data analysis and evaluation data; K.G., N.T., K.K. and W.W.; writing—original draft preparation; K.G. and K.K.; writing—review and editing, K.G., T.S., N.T., W.W. and K.K.; supervision, K.G., T.S. and N.T. All authors have read and agreed to the published version of the manuscript.

Funding: This work was supported by JSPS KAKENHI (19H02751), and The Thailand Research Fund (TRF) for a scholarship to Kanokwan Kiwfo (PHD/0349/2552) and The TRF Distinguished Research Professor Award Grant (DPG6080002 to Kate Grudpan).

Acknowledgments: This work was partly supported by Japan Society for the Promotion of Science, Grant-in-Aid for Scientific Research (B) (JSPS KAKENHI) Grant Number 19H02751 for Norio Teshima. The Royal Golden Jubilee PhD and The Thailand Research Fund (TRF) for a scholarship to Kanokwan Kiwfo (PHD/0349/2552) and The TRF Distinguished Research Professor Award Grant (DPG6080002 to Kate Grudpan) are acknowledged. The authors thank additional support from Chiang Mai University through the Center of Excellence on Innovation in Analytical Science and Technology (I-ANALY-S-T), and the Department of Applied Chemistry, Aichi Institute of Technology, and Pathinan Paengnakorn for some help in manuscript preparation.

Conflicts of Interest: The authors declare no conflict of interest.

References

1. Assali, M.; Raimundo, J.; Facchin, I. Simultaneous multiple injection to perform titration and standard addition in monosegmented flow analysis. *J. Autom. Methods Manag. Chem.* **2001**, *23*, 83–89. [CrossRef] [PubMed]
2. Silva, M.S.P.; Masini, J.C. Exploiting monosegmented flow analysis to perform in-line standard additions using a single stock standard solution in spectrophotometric sequential injection procedures. *Anal. Chim. Acta* **2002**, *466*, 345–352. [CrossRef]
3. Silva, M.S.P.; Galhardo, C.X.; Masini, J.C. Application of sequential injection-monosegmented flow analysis (SI-MSFA) to spectrophotometric determination of sulfide in simulated waters samples. *Talanta* **2003**, *60*, 45–52. [CrossRef]
4. Dos Santos, L.B.O.; Silva, M.S.P.; Masini, J.C. Developing a sequential injection-square wave voltammetry (SI-SWV) method for determination of atrazine using a hanging mercury drop electrode. *Anal. Chim. Acta* **2005**, *528*, 21–27. [CrossRef]
5. Dos Santos, L.B.O.; Masini, J.C. Determination of picloram in natural waters employing sequential injection square wave voltammetry using the hanging mercury drop electrode. *Talanta* **2007**, *72*, 1023–1029. [CrossRef] [PubMed]
6. Kozak, J.; Wójtowicz, M.; Wróbel, A.; Kościelniak, P. Novel approach to calibration by the complementary dilution method with the use of a monosegmented sequential injection system. *Talanta* **2008**, *77*, 587–592. [CrossRef]
7. Dos Santos, L.B.O.; Masini, J.C. Square wave adsorptive cathodic stripping voltammetry automated by sequential injection analysis. Potentialities and limitations exemplified by the determination of methyl parathion in water samples. *Anal. Chim. Acta* **2008**, *606*, 209–216. [CrossRef]
8. Siriangkhawut, W.; Grudpan, K.; Jakmunee, J. Sequential injection anodic stripping voltammetry with monosegmented flow and in-line UV digestion for determination of Zn(II), Cd(II), Pb(II) and Cu(II) in water samples. *Talanta* **2011**, *84*, 1366–1373. [CrossRef]
9. Kozak, J.; Wójtowicz, M.; Gawenda, N.; Kościelniak, P. An automatic system for acidity determination based on sequential injection titration and the monosegmented flow approach. *Talanta* **2011**, *84*, 1379–1383. [CrossRef]
10. Kochana, J.; Kozak, J.; Skrobisz, A.; Woźniakiewicz, M. Tyrosinase biosensor for benzoic acid inhibition-based determination with the use of a flow-batch monosegmented sequential injection system. *Talanta* **2012**, *96*, 147–152. [CrossRef]
11. Barreto, I.S.; Andrade, S.I.E.; Lima, M.B.; Silva, E.C.; Araújo, M.C.U.; Almeida, L.F. A monosegmented flow-batch system for slow reaction kinetics: Spectrophotometric determination of boron in plants. *Talanta* **2012**, *94*, 111–115. [CrossRef] [PubMed]

12. Khanhuathon, Y.; Siriangkhawut, W.; Chantiratikul, P.; Grudpan, K. Flow-Batch Method with a Sequential Injection System for Spectrophotometric Determination of Selenium(IV) in Selenium-Enriched Yeast Using o-Phenylenediamine. *Anal. Lett.* **2013**, *46*, 1779–1792. [CrossRef]
13. Khanhuathon, Y.; Siriangkhawut, W.; Chantiratikul, P.; Grudpan, K. Spectrophotometric method for determination of aluminium content in water and beverage samples employing flow-batch sequential injection system. *J. Food Compos. Anal.* **2015**, *41*, 45–53. [CrossRef]
14. Grudpan, K. Some recent developments on cost-effective flow-based analysis. *Talanta* **2004**, *64*, 1084–1090. [CrossRef]
15. Sakai, T.; Kito, Y.; Teshima, N.; Katoh, S.; Watla-iad; Grudpan, K. Spectrophotometric Flow Injection Analysis of Protein in Urine Using Tetrabromophenolphthalein Ethyl Ester and Triton X-100. *J. Flow Injection Anal.* **2007**, *24*, 23–26.
16. Watla-iad, K.; Sakai, T.; Teshima, N.; Katoh, S.; Grudpan, K. Successive determination of urinary protein and glucose using spectrophotometric sequential injection method. *Anal. Chim. Acta* **2007**, *604*, 139–146.
17. Sakai, T.; Ohta, H.; Ohno, N.; Imai, J. Routine assay of creatinine in newborn baby urine by spectrophotometric flow-injection analysis. *Anal. Chim. Acta* **1995**, *308*, 446–450. [CrossRef]
18. Siangproh, W.; Teshima, N.; Sakai, T.; Katoh, S.; Chailapakul, O. Alternative method for measurement of albumin/creatinine ratio using spectrophotometric sequential injection analysis. *Talanta* **2009**, *79*, 1111–1117. [CrossRef]
19. Songjaroen, T.; Maturos, T.; Sappat, A.; Tuantranont, A.; Laiwattanapaisal, W. Portable microfluidic system for determination of urinary creatinine. *Anal. Chim. Acta* **2009**, *647*, 78–83. [CrossRef]
20. Miller, J.N.; Miller, J.C. *Statistics and Chemometrics for Analytical Chemistry*; Pearson/Prentice Hall: Upper Saddle River, NJ, USA, 2005; ISBN 9780131291928.
21. Odebunmi, E.O.; Owalude, S.O. Kinetic and thermodynamic studies of glucose oxidase catalysed oxidation reaction of glucose. *J. Appl. Sci. Environ. Manag. December* **2007**, *11*, 95–100.
22. Paengnakorn, P.; Chanpaka, S.; Watla-iad, K.; Wongwilai, W.; Grudpan, K. Towards Green Titration: Downscaling the Sequential Injection Analysis Lab-at-valve Titration System with the Stepwise Addition of a Titrant. *Anal. Sci.* **2018**, *35*, 219–221. [CrossRef] [PubMed]

Sample Availability: The urine samples are not available from the authors.

© 2020 by the authors. Licensee MDPI, Basel, Switzerland. This article is an open access article distributed under the terms and conditions of the Creative Commons Attribution (CC BY) license (http://creativecommons.org/licenses/by/4.0/).

Article

Novel Approach to Automated Flow Titration for the Determination of Fe(III)

Joanna Kozak [1,*], Justyna Paluch [1], Marek Kozak [2], Marta Duracz [1], Marcin Wieczorek [1] and Paweł Kościelniak [1]

1 Faculty of Chemistry, Jagiellonian University, Gronostajowa 2, 30-387 Krakow, Poland; justyna.paluch@uj.edu.pl (J.P.); konieczna.marta11@wp.pl (M.D.); marcin.wieczorek@uj.edu.pl (M.W.); koscieln@chemia.uj.edu.pl (P.K.)
2 Oil and Gas Institute—National Research Institute, Lubicz 25A, 31-503 Krakow, Poland; kozak@inig.pl
* Correspondence: j.kozak@uj.edu.pl; Tel.: +48-126862416

Received: 14 March 2020; Accepted: 25 March 2020; Published: 27 March 2020

Abstract: A novel approach to automated flow titration with spectrophotometric detection for the determination of Fe(III) is presented. The approach is based on the possibility of strict and simultaneous control of the flow rates of sample and titrant streams over time. It consists of creating different but precisely defined concentration gradients of titrant and analyte in each successively formed monosegments, and is based on using the calculated titrant dilution factor. The procedure was verified by complexometric titration of Fe(III) in the form of a complex with sulfosalicylic acid, using EDTA as a titrant. Fe(III) and Fe(II) (after oxidation to Fe(III) with the use of H_2O_2) were determined with good precision (CV lower than 1.7%, $n = 6$) and accuracy (|RE| lower than 3.3%). The approach was applied to determine Fe(III) and Fe(II) in artesian water samples. Results of determinations were consistent with values obtained using the ICP–OES reference method. Using the procedure, it was possible to perform titration in 6 min for a wide range of analyte concentrations, using 2.4 mL of both sample and titrant.

Keywords: titration; flow analysis; Fe(III), Fe(II) determination; speciation analysis

1. Introduction

Techniques of flow analysis offer many possibilities of automation of analytical methods, including titration analysis. Using flow-based systems, titration has been performed based on the conventional procedure [1–7], and many novel approaches have also been developed for its implementation [8–35]. They differ in terms of the way of introducing the sample into the flow system, merging it with the titrant solution to create an analyte concentration gradient, and, consequently, the way of determining the analyte concentration. Generally, they can be divided [8] into titration based on the flow of a sample stream or segment, formation of a sample monosegment, or a combination of different ways of conducting titration in flow mode.

Titration based on continuous flow is generally performed in two modes. The first is based on changing the flow rate of only one stream (usually titrant [9–11]) and creating a single analyte concentration gradient using, e.g., a peristaltic pump. In this way, the analyte concentration was determined based on analytical calibration (a set of standard solutions was subjected to the titration procedure in the same instrumental conditions as the sample) [9,10] or on the relationship between the time necessary to achieve the endpoint of titration and the flow rate of the titrant (introduced using a double-plunger micropump) measured at the endpoint of titration [11]. The other approach is based on changing the flow rates of the sample and titrant streams while keeping the sum of the flow rates constant [12–22]. In this (latter) case, a single concentration gradient with analytical calibration [12] or a triangle-programmed concentration gradient of the analyte was formed [12–17]. To this end, e.g., two streams of titrant and analyte were introduced into the system using peristaltic pumps, first increasing the titrant flow rate from zero to a value corresponding to zero sample flow rate and then, at the same rate, reducing to zero again. As a result, two titration curves were obtained and the time corresponding to the difference between the endpoints of titration detected on the curves was calculated [12]. Triangle-programmed flow titration with linear concentration gradients was also performed using computer-controlled micropumps. Under defined conditions, it was possible to perform determinations without analytical calibration [13]. In flow titration, called continuous feedback-based titration [14–16], at a constant sum of the flow rates of the titrant and sample, the linear change of titrant flow rate was related to the controller output voltage dependency. During titration, the voltage increased until a certain value, U_H, was reached, at which the time corresponding to the endpoint of the titration was exceeded. Then, the voltage decreased in the same range until the U_L value was reached. The controller output voltage at the endpoint of titration was the arithmetic mean of the U_H and U_L values. The method enabled the determination of analyte concentration using an appropriate formula [14,16] or analytical calibration [15,16]. Coulometric titration with a continuous sample flow in which the concentration gradient of the analyte was associated with the magnitude of the rectangular [17] or triangular [18–20] current pulse generating the titrant in a flow-through electrolytic vessel was also proposed.

Monosegmented flow titration is characterized by the formation of segments of liquid—each segment consists of the sample, titrant, and possibly a diluent—separated on both sides from the carrier by a segment of air (or inert gas) [21–28]. The components of the created monosegment are mixed in a reaction coil. In order to facilitate homogenization, several smaller segments may be introduced alternately instead of a single segment of the sample and the titrant [21,23]. For a monosegment prepared in this way, usually, a peak with a plateau is registered. While maintaining the size of the entire monosegment, the described procedure is repeated, changing the volume of the titrant segment introduced into the monosegment [21,24] or selecting the volume of both the titrant and the sample [23,25]. In this method of titration, the endpoint is determined on the basis of titration curves representing the relation between signal and titrant volume [23] or time [24] or using an appropriate numerical algorithm [21,25]. Finally, the analyte concentration can be calculated. In the system described in [26], titration was performed in a monosegment containing the sample to which the titrant portion was dosed by a burette. The monosegment content was mixed in a reaction coil and directed to the detector. After the peak was recorded, the direction of the flow was changed in order to introduce another portion of the titrant into the monosegment and continue the procedure. Monosegmented titration was also carried out using Lab-on-Valve [27] or sequential injection [28] systems in which segments of air, sample, indicator, titrant, and then air were introduced into the holding coil in turn. The content of the monosegment was homogenized by changing the direction of flow, after which the second of the introduced air segments was removed to waste, and solutions were directed to the detection channel to measure the signal. Calibration was performed using a set of standard solutions.

There is a group of flow titration methods that can be considered as a combination of various approaches. Among them is multicommutated continuous flow titration [29–34], which can be classified as a method in which the sample is introduced both as a stream and as a segment. This titration, provided that the sum of the volumes of two successive segments remains constant, can be implemented, e.g., by introducing alternately larger segments of the titrant and smaller segments of the sample into the stream of the sample [29–31], larger segments of sample, and smaller segments of titrant into the titrant stream [32,33], or by using binary search concepts [34]. Based on the titration curve representing the relationship between the registered signal and the time of analysis [29–32] or the number of delivery cycles of the titrant and the sample [33], the titration time or the cycle number corresponding to the achievement of the endpoint of titration was determined. The concentration of the analyte in the sample was determined using the time corresponding to the endpoint of titration and the theoretical model describing the change in the analyte concentration during the delivery of the titrant and sample to the mixing chamber. The concentration of the analyte in the sample was also determined based on calibration using a set of standard methods.

A tracer-monitored flow titration with spectrophotometric detection was developed to omit the stage of analytical calibration [35]. The approach requires simultaneous monitoring of two absorbing species, the titration indicator and the dye tracer [36]. The tracer is applied to estimate the instant sample and titrant volumetric fractions without the need for volume, mass, or peak width measurements [36]. The method was implemented using a single flow system and applied to perform triangle-programmed titration, and to establish concentration gradients along the sample zone in the flow injection technique.

In the present work, a simple approach to the automation of titration with spectrophotometric detection based on merging (in a strictly controlled way) sample and titrant streams introduced continuously into the detection system in the form of monosegments is presented. It consists of creating in the flow system different, but precisely defined concentration gradients of titrant and analyte in each of successively formed monosegments and is based on using the calculated titrant dilution factor. The procedure was verified and applied to complexometric Fe(III) titration.

2. Results and Discussion

2.1. Flow System Developed for Titration

The flow system proposed for titration is presented in Figure 1. It consists of three syringe pumps equipped with nine-position selection valves. Pump I and Pump II were used for sample and titrant propelling, respectively, whereas the third pump was used for air introduction to reduce the dispersion of solutions. Streams of sample and titrant solutions were introduced simultaneously into the system with established flow rates, joined at the confluence point, and merged in the mixing coil to complete the reaction. The product of the reaction was directed in a continuous way to the flow cell and appropriate signals were measured by the detector. In this way, the change in the titration curve was monitored continuously. Segments of air were introduced at strictly defined, subsequent stages of titration. Using the pumps, the flow rate of streams could be strictly defined and controllably changed in a discrete or continuous, and reproducible way. This became the basis for developing titration procedures based on using the calculated titrant dilution factor.

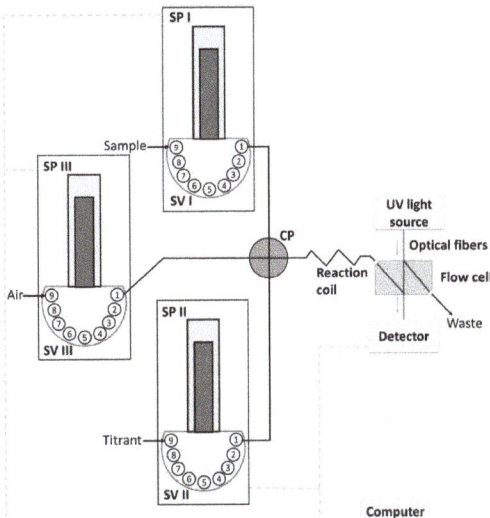

Figure 1. Flow system designed for the proposed titration procedure. SP—syringe pump, SV—selection valve, CP—confluence point.

2.2. Procedure of Titration

Titration procedure consisted of creating different but precisely defined concentration gradients of titrant and analyte in each successively formed monosegments in order to facilitate the mixing of titrant and sample solutions, and in order to obtain monosegments containing precisely defined titrant (V_T) and sample (V_S) volumes. The procedure was carried out by simultaneously introducing into the monosegment the sample and titrant, each with a known, properly selected, flow rate. The first monosegment contained only the sample, and in the subsequent monosegments, the volume of sample was gradually decreased, whereas the volume of titrant increased by the same value (e.g., V_S/V_T (μL): 300/0; 280/20; 260/40; 240/60; ... 40/260; 20/280; 0/300) until the formation of a monosegment that contained only the titrant. In practice, the titration procedure was ended after obtaining the endpoint of titration. During the whole titration procedure, the total volume of the mixture (V_M) in successive monosegments remained constant, whereas the titrant and analyte concentration increased and decreased linearly, respectively. Consequently, based on the volumes of appropriate solutions introduced with the use of syringe pumps, for each monosegment it was possible to calculate titrant dilution factor (f_T) (Equation (1)):

$$f_T = V_T/(V_T + V_S) \tag{1}$$

and sample dilution factor f_S (Equation (2)):

$$f_S = V_S/(V_T + V_S) \tag{2}$$

It should be noted that, therefore, $f_T = 1 - f_S$. The absorbance in the form of a steady signal was measured for each of the monosegments (Figure 2) and a titration curve showing the relationship between absorbance and the titrant dilution factor was prepared (Figure 3). Using this titration curve, the titrant dilution factor at the endpoint of titration f_{TEP} was determined. Analyte concentration (C_A, g L^{-1}) was calculated based on titrant concentration (C_T, mol L^{-1}) and titrant dilution factor determined at the endpoint of titration according to Equation (3):

$$C_A = (Q \cdot C_T \cdot M_A \cdot f_{TEP})/(1 - f_{TEP}) \tag{3}$$

where Q is the coefficient resulting from the stoichiometry of the titration reaction (Q = m/n, m = number of moles of analyte, n = number of moles of titrant) and M_A is the analyte molar mass.

Procedure Verification

Instrumental conditions were selected for the developed titration procedure. The procedure was verified by determination of Fe(III) using complexometric titration, in terms of precision, accuracy, and the range of application using synthetic samples and certified reference material. Determination of Fe(III) relied on titration of the sample with EDTA solution in the presence of sulfosalicylic acid used as an indicator [37,38]. Titration was performed in an acidic medium at a pH of about two. In these conditions, a red-purple complex between Fe(III) and sulfosalicylic acid (1:1) was formed in a flask. Then, the sample was introduced into Syringe Pump I and EDTA into Syringe II (Figure 1). The solutions and air met at the confluence point (Figure 1). During the titration process, Syringe Pump I dispensed known ever-decreasing sample volumes (e.g., from 300 to 0 μL, with a 20 μL step) to successive monosegments, while Pump II dispensed ever-increasing titrant volumes (from 0 to 300 μL, with a 20 μL step). Solutions were introduced at the same time (about 13 s) with different flow rates calculated to enable entirely overlapping introduced zones and to facilitate their mixing. Before and after introducing the defined volumes of sample and titrant, a segment of air was inserted to form a monosegment. EDTA mixed with the sample (in the monosegment) replaced sulfosalicylic acid to form a more stable colorless complex with Fe(III) (1:1). Polytetraflouroethylene (PTFE) tubing (ID 0.8 mm) of capacity 1 mL was selected to complete the reaction on the basis of previous experiments [39].

At the moment of the flow-through the flow cell of successive monosegments, analytical signals were recorded and then the solution was directed to waste. After recording all signals, water was used to wash the syringe and the flow cell. The details of the procedure are described in Table 1. The composition of monosegments formed during the titration procedure and signals registered during titration of Fe(III) of concentration 2.00 mg L^{-1} using EDTA of concentration 0.02 mmol L^{-1} as titrant are shown in Figure 2. During the titration the absorbance decreased to a value close to zero, showing that the whole amount of Fe(III) was bound to EDTA (the endpoint of titration). The Fe(III) concentration in the sample was determined on the basis of the EDTA concentration and the dilution factor of EDTA at the endpoint of titration. The way of determining the endpoint of titration and the titrant dilution factor is shown in Figure 3. Fe(II) was determined by applying the same chromogenic reagent, after oxidizing the analyte with the use of H_2O_2. The thus prepared sample was subjected to the titration procedure and the sum of Fe(II) and Fe(III) was determined. The Fe(II) concentration was determined from the difference between the sum and the Fe(III) concentration.

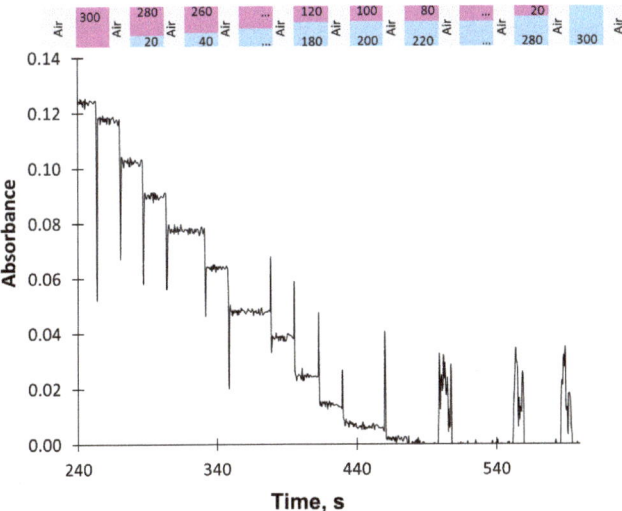

Figure 2. Composition of monosegments formed during the titration procedure (in µL) and signals registered during titration of Fe(III) (2.00 mg L^{-1}) using EDTA (0.02 mmol L^{-1}) as titrant.

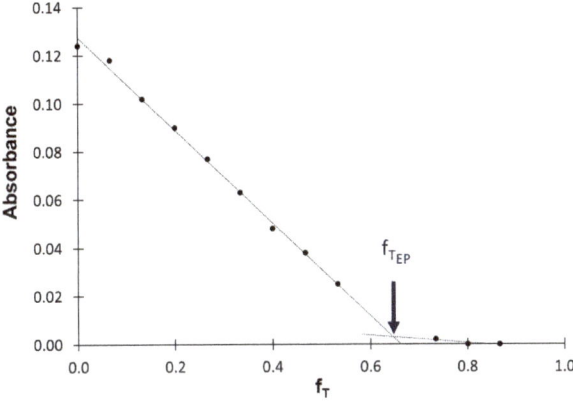

Figure 3. The way of determination of the endpoint of titration (EP) using titrant dilution factor f_T (details in the text).

Table 1. Titration procedure using the system presented in Figure 1. SV—selection valve, SP—syringe pump.

Step	SV Position			SP Flow Rate, μL s⁻¹			Volume, μL			Action
	I	II	III	I	II	III	I	II	III	
1	9	9	9	100	100	100	1000	1000	1000	Aspiration of sample, titrant, and air into syringes
2	1	1	1	0	0	100	0	0	100	Introduction of air into reaction coil
3	1	1	1	100	0	0	300	0	0	Formation of zone I in reaction coil
4	1	1	1	0	0	100	0	0	100	Introduction of air into reaction coil
5	1	1	1	94	10	0	280	20	0	Formation of zone II in reaction coil
6	1	1	1	0	0	100	0	0	100	Introduction of air into reaction coil
7	1	1	1	86	20	0	260	40	0	Formation of zone III in reaction coil
8	1	1	1	0	0	100	0	0	100	Introduction of air into reaction coil
9	9	9	9	100	0	100	840	0	400	Aspiration of sample and air into syringes
10	1	1	1	80	30	0	240	60	0	Formation of zone IV in reaction coil
11	1	1	1	0	0	100	0	0	100	Introduction of air into reaction coil
12	1	1	1	74	40	0	220	80	0	Formation of zone V in reaction coil
13	1	1	1	0	0	100	0	0	100	Introduction of air into reaction coil
14	1	1	1	67	50	0	200	100	0	Formation of zone VI in reaction coil
15	1	1	1	0	0	100	0	0	100	Introduction of air into reaction coil
16	1	1	1	60	60	0	180	120	0	Formation of zone VII in reaction coil
17	1	1	1	0	0	100	0	0	100	Introduction of air into reaction coil
18	1	1	1	54	70	0	160	140	0	Formation of zone VIII in reaction coil
19	1	1	1	0	0	100	0	0	100	Introduction of air into reaction coil
20	9	9	9	100	0	100	560	0	500	Aspiration of sample and air into syringes
21	1	1	1	47	80	0	140	160	0	Formation of zone IX in reaction coil
22	1	1	1	0	0	100	0	0	100	Introduction of air into reaction coil
23	1	1	1	40	90	0	120	180	0	Formation of zone X in reaction coil
24	1	1	1	0	0	100	0	0	100	Introduction of air into reaction coil
25	9	9	9	0	100	100	0	900	200	Aspiration of titrant and air into syringes
26	1	1	1	34	100	0	100	200	0	Formation of zone XI in reaction coil
27	1	1	1	0	0	100	0	0	100	Introduction of air into reaction coil
28	1	1	1	27	100	0	80	220	0	Formation of zone XII in reaction coil
29	1	1	1	0	0	100	0	0	100	Introduction of air into reaction coil
30	1	1	1	20	100	0	60	240	0	Formation of zone XIII in reaction coil
31	1	1	1	0	0	100	0	0	100	Introduction of air into reaction coil
32	1	1	1	14	100	0	40	260	0	Formation of zone XIV in reaction coil
33	1	1	1	0	0	100	0	0	100	Introduction of air into reaction coil
34	9	9	9	0	100	100	0	900	400	Aspiration of titrant and air into syringes
35	1	1	1	7	100	0	20	280	0	Formation of zone XV in reaction coil
36	1	1	1	0	0	100	0	0	100	Introduction of air into reaction coil
37	1	1	1	0	100	0	0	300	0	Formation of zone XVI in reaction coil
38	1	1	1	0	0	100	0	0	300	Introduction of air into mixing coil and transport of zones to detector

Two different total volumes of monosegment (100 and 300 µL) and volume change steps in a monosegment (10 and 20 µL, respectively) were tested, using a titrant of concentration 0.02 mmol L^{-1}. A volume of 300 µL and a step of 20 µL were selected because of the possibility of determination of Fe(III) in a wider range of concentrations. Using EDTA at a concentration of 0.02 mmol L^{-1}, it was possible to determine Fe(III) in the range 0.35–4.50 mg L^{-1}. In selected instrumental conditions, the titration curve was plotted over a period of about 6 min, because of the necessity of refilling the syringes during the titration process. A 2.4 mL volume of sample and the same volume of titrant were consumed for a single titration procedure.

The titration procedure was verified by titration of solutions containing Fe(III) at concentrations (mg L^{-1}): 0.50, 1.00, 1.50, 2.00, 3.00, and 4.00 using EDTA solution at a concentration of 0.02 mmol L^{-1}. Concentrations of analytes were selected on the basis of similarity to content expected in artesian water samples. Samples were analyzed three times. Calculated values of relative error (RE) and coefficient of variance (CE) for the results were lower than 1.6% and 2.1%, respectively. Therefore, the possibility of determination of Fe(III) in the presence of Fe(II) was studied. To this end, Fe(III) at a concentration of 1.50 mg L^{-1} was determined in the presence of Fe(II) at a concentration (mg L^{-1}) of: 0.50, 1.00, 1.50, and 2.00. As the RE for the results was lower than 2.2%, it was assumed that Fe(II) did not influence Fe(III) determination. Finally, the procedure was verified by the determination of Fe(III), the sum of Fe(II) and Fe(III), and, consequently, the content of Fe(II) in the sample. The results of titration are presented in Table 2.

Table 2. Verification of the developed titration procedure: results of determination of Fe(III), the sum of Fe(III) and Fe(II) and Fe(II) as the difference in synthetic samples, EDTA—0.02 mmol L^{-1}, CV—coefficient of variation (n = 3), RE—relative error.

No.	Fe(III), mg L^{-1}		CV, %	|RE|, %	Fe(III) + Fe(II), mg L^{-1}	Fe(II), mg L^{-1}		CV, %	|RE|, %
	Expected	Determined			Determined	Expected	Determined		
1	0.50	0.50	1.8	0.9	1.01	0.50	0.51	1.8	2.3
2	0.50	0.51	2.4	1.4	2.53	2.00	2.02	2.3	1.0
3	1.00	0.99	2.9	1.4	1.98	1.00	0.99	3.4	0.5
4	1.00	1.02	0.1	2.3	3.54	2.50	2.51	3.7	0.6
5	1.00	1.03	1.5	3.3	4.10	3.00	3.07	0.7	2.3
6	2.50	2.49	1.1	0.3	3.49	1.00	1.00	2.8	0.2

Using the developed procedure, Fe(III) was determined with RE lower than 3.3%. The accuracy of determination of Fe(II) (by difference) was also very good (|RE| < 2.3%). The precision of the procedure was also studied by determination of Fe(III) and Fe(II) in synthetic samples containing Fe(III)/Fe(II) at concentrations (mg L^{-1}): 0.50/0.50, 1.00/2.50, and 2.50/1.00. Each sample was analyzed six times. In all samples, Fe(III) and Fe(II) were determined with precision (CV) better than 1.7% and 3.1%, respectively. Figures of merit of the developed procedure are summarized in Table 3.

Table 3. Figures of merit of the developed titration procedure for Fe(III) determination.

Parameter	Value
Accuracy, |RE|, %	3.3
Precision, CV, % (n = 6)	1.7
Sample consumption, mL	2.4
Titrant consumption, mL	2.4
Time of single titration, min	6

2.3. Analysis of Real Samples

The developed titration procedure was applied to the determination of Fe(II)/Fe(III) in samples of artesian water. Samples were analyzed three times. The results of the determinations using the developed method and the ICP method with appropriate values of confidence intervals ($\alpha = 0.05$, $n = 3$) are presented in Table 4. The results of determination of the analytes in wastewater certified reference material (directly and with the addition of Fe(III) at a concentration of 0.50 mg L^{-1}) are also shown in the table.

Table 4. Results of determination of Fe(II) and Fe(III) in samples of artesian water and wastewater certified reference material (WWater, *WWater—wastewater and wastewater with addition of Fe(III) at a concentration of 0.50 mg L^{-1}, respectively, ICP–OES—Inductively Coupled Plasma—Optical Emission Spectrometry method; EDTA: 0.02 mmol L^{-1}, confidence interval, $n = 3$, $\alpha = 0.05$.

Sample	Fe(III), mg L^{-1}	Fe(II), mg L^{-1}	Fe (Total), mg L^{-1}	
			Developed Procedure	[1] ICP–OES/ [2] Certified Value
Water 1	0.63 ± 0.01	0.94 ± 0.04	1.56 ± 0.04	[1] 1.56 ± 0.01
Water 2	0.35 ± 0.01	2.40 ± 0.04	2.75 ± 0.04	[1] 2.73 ± 0.04
Water 3	0.57 ± 0.03	0.70 ± 0.03	1.27 ± 0.01	[1] 1.26 ± 0.02
WWater	0.36 ± 0.04	0.13 ± 0.05	0.49 ± 0.03	[2] 0.49 ± 0.02
*WWater	0.85 ± 0.04	0.13 ± 0.05	0.98 ± 0.04	[2] 0.99 ± 0.02

It can be assumed that the results of the determination of iron (presented as the sum of Fe(II) and Fe(III)) are consistent with those obtained with the use of the inductively coupled plasma–optical emission spectrometry (ICP–OES) technique and the certified value. It can be noticed that Fe(II) was determined in the wastewater with higher uncertainty, but it should be borne in mind that Fe(II) was determined at a low concentration by difference; however, the sum of Fe(II) and Fe(III) concentrations agreed with the certified value. In conclusion, the procedure was developed for Fe(III) determination, but it can also be applied to the determination of the sum of Fe(III) and Fe(II), and, consequently, Fe(II) determination.

3. Methodology

3.1. Reagents and Solutions

A stock standard solution of Fe(III) was prepared by dissolving 7.725 g of NH$_4$Fe(SO$_4$)$_2$ 12 H$_2$O (POCH S.A., Gliwice, Poland) in HNO$_3$ (0.1 mol L^{-1}) and making the solution up to 100.0 mL. A stock standard solution of Fe(II) was prepared daily by dissolving 0.176 g of (NH$_4$)$_2$Fe(SO$_4$)$_2$ 6 H$_2$O (Chempur, Piekary Śląskie, Poland) in 16 mL of H$_2$SO$_4$ (1 mol L^{-1}) and making it up to 25.0 mL with water. The stock solutions were diluted with water to form solutions of appropriate analyte concentration. A solution of sulfosalicylic acid was prepared by dissolving 0.2 g of C$_7$H$_6$O$_6$ 2H$_2$O (Chempur, Poland) in HCl (0.1mol L^{-1}) in a 100.0 mL volumetric flask. EDTA solution of concentration 0.01 mol L^{-1} was prepared by dissolving 0.373 g of C$_{10}$H$_{14}$N$_2$Na$_2$O$_8$ 2H$_2$O (Chempur, Piekary Śląskie, Poland) in water and making the solution up to 100.0 mL. A more diluted titrant solution was prepared by dilution of the above solution with water. A solution of hydrogen peroxide of concentration 0.0348 mol L^{-1} was prepared by appropriate dilution of H$_2$O$_2$ solution (35%) (Merck, Darmstadt, Germany) with water. Samples for iron determination were prepared by adding sulfosalicylic acid (4 mL, 0.2%) to the sample in the case of determination of Fe(III) or H$_2$O$_2$ (2 mL, 0.0348 mol L^{-1}) and sulfosalicylic acid (4 mL, 0.2%) in the case of determination of the sum of Fe(II) and Fe(III)), and making the solution up to 100 mL with water. Wastewater certified reference material EnviroMAT Waste Water, High (EU-H-3) Lot Number: SC8301825 (SCP SCIENCE, Quebec Canada) was diluted with water in accordance with the manufacturer's instructions. Samples of water from local artesian wells were collected and analyzed the same day. Samples were acidified [40] and nitrogen was passed through them for about 40 min in order to remove gas interferents (e.g., O_2, H_2S). Moreover, water samples were degassed for 15 min using an ultrasonic bath (Sonic 3, Warsaw, Poland). Reagents of analytical grade were used. Substances for preparation of stock standard solutions were weighed to the nearest 0.0001 g. Deionized water (0.05 µS cm^{-1}) obtained from the HLP5sp system (Hydrolab, Straszyn, Poland) was used throughout the study.

3.2. Instrumentation

A system (FIAlab® Instruments, Seattle, WA, USA) consisting of three syringe pumps (for introducing the sample, titrant, and air, respectively), each equipped with a 9-position selection valve and 1.0 mL Cavro glass barrel syringe was used for the studies. PTFE tubing (0.8 mm i.d.) was used as tubes. A PTFE reaction coil (ID 0.8 mm) of capacity 1 mL was used for iron determination, and a PTFE reaction coil (ID 0.5 mm) of capacity 160 µL was selected for acids and acidity determination. Signals were measured with the use of the USB 4000 Ocean Optics spectrophotometer (Ocean Optics, Dunedin, FL, USA) equipped with fiber optics cables, a halogen light source HL-2000 (Ocean Optics, Dunedin, FL, USA), and a flow cell of light path length 200 mm. Measurements were performed at wavelength 530 nm for determination of iron in the form of a complex with sulfosalicylic acid with a reference scan at wavelength 700 nm.

The ICP–OES method [41] was used as a reference method. Analyses were performed using a SPECTRO ARCOS SOP spectrometer with radial plasma observation (Spectro Analytical Instruments, Kleve, Germany) at operation conditions recommended by the instrument manufacturer. The emission intensities for iron were measured at 259.562 nm.

4. Conclusions

The developed approach to titration using a set of syringe pumps is automatic, simple, and convenient in use. From the analytical point of view, it allows you to obtain results of Fe(III) determination with good precision and accuracy. Titration procedure enables determination of the analyte in a wide concentration range, does not need calibration, lasts about 6 min, and consumes 2.4 mL of sample and titrant solutions. The procedure can fulfill the requirements of green analytical chemistry.

The procedure was developed based on a number of ingenious approaches developed earlier. Compared to other methods, it differs in terms of the kind of flow system used, the simplicity of implementation, calculation of titrant dilution factor, and the system automation. It is also characterized by the facility of application to various samples in a wide range of concentrations. This provides a real opportunity to apply it in routine analyses. Compared to tracer-monitored flow titration, the proposed approach uses a simple, reliable, but more expensive flow system which can be applied to tracer-monitored titration. The cost of the flow system can be reduced if, e.g., three-way solenoid valves are used instead of 9-position selection valves. The system also has the potential to be applied to titrations for which it is difficult to find an appropriate tracer. The procedure was designed for flow systems with spectrophotometric detection. The possibility of adapting it to other types of titration and kinds of detection is going to be studied.

Author Contributions: Conceptualization, J.K.; Data curation, M.D., J.K., J.P., M.K.; Formal analysis, J.K.; Funding acquisition, P.K., M.W.; Investigation, M.D., J.P., M.K., M.W.; Methodology, J.K.; Project administration, P.K.; Supervision, J.K.; Validation, M.D., J.P.; Visualization, J.K., J.P., M.D.; Writing—original draft, J.K.; Writing—review & editing, J.K., J.P. All authors have read and agreed to the published version of the manuscript.

Funding: This research was funded by National Science Centre, Poland (Opus, 2017-2020) grant number 2016/23/B/ST4/00789.

Conflicts of Interest: The authors declare no conflicts of interest.

References

1. Pasquini, C.; de Aquino, E.V.; das Virgens Reboucas, M.; Gonzaga, F.B. Robust flow-batch coulometric/biamperometric titration system: Determination of bromine index and bromine number of petrochemicals. *Anal. Chim. Acta* **2007**, *600*, 84–89. [CrossRef]
2. Tan, A.; Zhang, L.; Xiao, C. Simultaneous and automatic determination of hydroxide and carbonate in aluminate solutions by a micro-titration method. *Anal. Chim. Acta* **1999**, *388*, 219–223. [CrossRef]
3. Alerm, L.; Bartroli, J. Development of a sequential microtitration system. *Anal. Chem.* **1996**, *68*, 1394–1400. [CrossRef]
4. Hoogendijk, R. A compact titration configuration for process analytical applications. *Anal. Chim. Acta* **1999**, *378*, 211–217. [CrossRef]
5. Tan, A.; Xiao, C. An automatic back titration method for microchemical analysis. *Talanta* **1997**, *44*, 967–972. [CrossRef]
6. Zhang, H.; Zhu, J.; Chen, Q.; Jiang, S.; Zhang, Y.; Fu, T. New intelligent photometric titration system and its method for constructing chemical oxygen demand based on micro-flow injection. *Microchem. J.* **2018**, *143*, 292–304. [CrossRef]
7. Crispino, C.C.; Reis, B.F. Development of an automatic photometric titration procedure to determine olive oil acidity employing a miniaturized multicommuted flow-batch setup. *Anal. Methods* **2014**, *6*, 302–307. [CrossRef]
8. Wójtowicz, M.; Kozak, J.; Kościelniak, P. Novel approaches to analysis by flow injection gradient titration. *Anal. Chim. Acta* **2007**, *600*, 78–83. [CrossRef]
9. Marcos, J.; Ríos, A.; Valcárcel, M. Automatic titrations in unsegmented flow systems based on variable flow-rate patterns. Part 1. Principles and applications to acid-base titrations. *Anal. Chim. Acta* **1992**, *261*, 489–494. [CrossRef]
10. Marcos, J.; Ríos, A.; Valcárcel, M. Automatic titrations in unsegmented flow systems based on variable flow-rate patterns. Part 2. Complexometric and redox titrations. *Anal. Chim. Acta* **1992**, *261*, 495–503. [CrossRef]
11. Katsumata, H.; Teshima, N.; Kurihara, M.; Kawashima, T. Potentiometric flow titration of iron(II) and chromium(VI) based on flow rate ratio of a titrant to a sample. *Talanta* **1999**, *48*, 135–141. [CrossRef]
12. Lopez García, I.; Viñas, P.; Campillo, N.; Hernandez Córdoba, M. Linear flow gradients for automatic titrations. *Anal. Chim. Acta* **1995**, *308*, 67–76. [CrossRef]
13. Fuhrmann, B.; Spohn, U. Volumetric triangle-programmed flow titrations based on precisely generated concentration gradients. *Anal. Chim. Acta* **1993**, *282*, 397–406. [CrossRef]
14. Dasgupta, P.K.; Tanaka, H.; Jo, K.D. Continuous on-line true titrations by feedback based flow ratiometry: Application to potentiometric acid-base titrations. *Anal. Chim. Acta* **2001**, *435*, 289–297. [CrossRef]
15. Jo, K.D.; Dasgupta, P.K. Continuous on-line feedback based flow titrations. Complexometric titrations of calcium and magnesium. *Talanta* **2003**, *60*, 131–137. [CrossRef]
16. Tanaka, H.; Baba, T. High throughput continuous titration based on a flow ratiometry controlled with feedback-based variable triangular waves and subsequent fixed triangular waves. *Talanta* **2005**, *67*, 848–853. [CrossRef]
17. Dakashev, A.D.; Dimitrova, V.T. Pulse coulometric titration in continuous flow. *Analyst* **1994**, *119*, 1835–1838. [CrossRef]
18. He, Z.K.; Fuhrmann, B.; Spohn, U. Coulometric microflow titrations with chemiluminescent and amperometric equivalence point detection. Bromimetric titration of low concentrations of hydrazine and ammonium. *Anal. Chim. Acta* **2000**, *409*, 83–91. [CrossRef]
19. Guenat, O.T.; van der Schoolt, B.H.; Morf, W.E.; de Rooij, N.F. Triangle-programmed coulometric nanotitrations completed by continuous flow with potentiometric detection. *Anal. Chem.* **2000**, *72*, 1585–1590. [CrossRef]
20. Fuhrmann, B.; Spohn, U. A PC-based titrator for flow gradient titrations. *J. Autom. Chem.* **1993**, *15*, 209–216. [CrossRef]

21. Martelli, P.B.; Reis, B.F.; Korn, M.; Lima, J.L.F.C. Automatic potentiometric titration in monosegmented flow system exploiting binary search. *Anal. Chim. Acta* **1999**, *387*, 165–173. [CrossRef]
22. Aquino, E.V.; Rohwedder, J.J.R.; Pasquini, C. Monosegmented flow titrator. *Anal. Chim. Acta* **2001**, *438*, 67–74. [CrossRef]
23. Borges, E.P.; Martelli, P.B.; Reis, B.F. Automatic stepwise potentiometric titration in a monosegmented flow system. *Mikrochim. Acta* **2000**, *135*, 179–184. [CrossRef]
24. Assali, M.; Raimundo, I.M., Jr.; Facchin, I. Simultaneous multiple injection to perform titration and standard addition in monosegmented flow analysis. *J. Autom. Methods Manag. Chem.* **2001**, *23*, 83–89. [CrossRef]
25. Honorato, R.S.; Araújo, M.C.U.; Veras, G.; Zagatto, E.A.G.; Lapa, R.A.S.; Lima, J.L.F.C. A monosegmented flow titration for the spectrophotometric determination of total acidity in vinegars. *Anal. Sci.* **1999**, *15*, 665–668. [CrossRef]
26. Vidal de Aquino, E.; Rohwedder, J.J.R.; Pasquini, C. A new approach to flow-batch titration. A monosegmented flow titrator with coulometric reagent generation and potentiometric or biamperometric detection. *Anal. Bioanal. Chem.* **2006**, *386*, 1921–1930. [CrossRef]
27. Jakmunee, J.; Pathimapornlert, L.; Hartwell, S.K.; Grudpan, K. Novel approach for mono-segmented flow micro-titration with sequential injection using a lab-on-valve system: A model study for the assay of acidity in fruit juices. *Analyst* **2005**, *130*, 299–303. [CrossRef]
28. Kozak, J.; Wójtowicz, M.; Gawenda, N.; Kościelniak, P. An automatic system for acidity determination based on sequential injection titration and the monosegmented flow approach. *Talanta* **2011**, *84*, 1379–1383. [CrossRef]
29. Almeida, C.M.N.V.; Lapa, R.A.S.; Lima, J.L.F.C.; Zagatto, E.A.G.; Araújo, M.C.U. An automatic titrator based on a multicommutated unsegmented flow system. Its application to acid-base titrations. *Anal. Chim. Acta* **2000**, *407*, 213–223. [CrossRef]
30. Almeida, C.M.N.V.; Araújo, M.C.U.; Lapa, R.A.S.; Lima, J.L.F.C.; Reis, B.F.; Zagatto, E.A.G. Precipitation titrations using an automatic titrator based on a multicommutated unsegmented flow system. *Analyst* **2000**, *125*, 333–340. [CrossRef]
31. Almeida, C.M.N.V.; Lapa, R.A.S.; Lima, J.L.F.C. Automatic flow titrator based on a multicommutated unsegmented flow system for alkalinity monitoring in wastewater. *Anal. Chim. Acta* **2001**, *438*, 291–298. [CrossRef]
32. Paim, A.P.S.; Almeida, C.M.N.V.; Reis, B.F.; Lapa, R.A.S.; Zagatto, E.A.G.; Lima, J.L.F.C. Automatic potentiometric flow titration procedure for ascorbic acid determination in pharmaceutical formulations. *J. Pharm. Biomed. Anal.* **2002**, *28*, 1221–1225. [CrossRef]
33. Wang, X.D.; Cardwell, T.J.; Cattrall, R.W.; Dyson, R.P.; Jenkins, G.E. Time-division multiplex technique for producing concentration profiles in flow analysis. *Anal. Chim. Acta* **1998**, *368*, 105–110. [CrossRef]
34. Lima, M.J.A.; Reis, B.F. Fully automated photometric titration procedure employing a multicommuted flow analysis setup for acidity determination in fruit juice, vinegar, and wine. *Microchem. J.* **2017**, *135*, 207–212. [CrossRef]
35. Sasaki, M.K.; Rocha, D.L.; Rocha, F.R.P.; Zagatto, E.A.G. Tracer-monitored flow titrations. *Anal. Chim. Acta* **2016**, *902*, 123–128. [CrossRef]
36. Martz, T.R.; Dickson, A.G.; DeGrandpre, M.D. Tracer monitored titrations: Measurement of total alkalinity. *Anal. Chem.* **2006**, *78*, 1817–1826. [CrossRef]
37. Kochana, J.; Parczewski, A. New method of simultaneous determination of Fe (II) and Fe (III) by two-component photometric titration. *Chem. Anal.* **1997**, *42*, 411–416.
38. Kozak, J.; Gutowski, J.; Kozak, M.; Wieczorek, M.; Kościelniak, P. New method for simultaneous determination of Fe(II) and Fe(III) in water using flow injection technique. *Anal. Chim. Acta* **2010**, *668*, 8–12. [CrossRef]
39. Paluch, J.; Kozak, J.; Wieczorek, M.; Kozak, M.; Kochana, J.; Widurek, K.; Konieczna, M.; Kościelniak, P. Novel approach to two-component speciation analysis. Spectrophotometric flow-based determinations of Fe(II)/Fe(III) and Cr(III)/Cr(VI). *Talanta* **2017**, *171*, 275–282. [CrossRef]
40. *Water Quality—Sampling—Part 3: Guidance on the Preservation and Handling of Water Samples*; PN-EN ISO 5667-3:2005; Polish Committee for Standardization: Warsaw, Poland, 2005.

41. *Water Quality—Determination of Selected Elements by Inductively Coupled Optical Emission Spectrometry (ICP-OES)*; ISO 11885:2007(E); International Organization for Standardization: Geneva, Switzerland, 2007.

Sample Availability: Samples of the compounds are not available from the authors.

 © 2020 by the authors. Licensee MDPI, Basel, Switzerland. This article is an open access article distributed under the terms and conditions of the Creative Commons Attribution (CC BY) license (http://creativecommons.org/licenses/by/4.0/).

Article

Novel Approach to Sample Preconcentration by Solvent Evaporation in Flow Analysis

Justyna Paluch [1], Joanna Kozak [1,*], Marcin Wieczorek [1], Michał Woźniakiewicz [1], Małgorzata Gołąb [1], Ewelina Półtorak [1], Sławomir Kalinowski [2] and Paweł Kościelniak [1]

[1] Faculty of Chemistry, Jagiellonian University, Gronostajowa 2, 30-387 Krakow, Poland; justyna.paluch@uj.edu.pl (J.P.); marcin.wieczorek@uj.edu.pl (M.W.); michal.wozniakiewicz@uj.edu.pl (M.W.); mgolab@doctoral.uj.edu.pl (M.G.); ewelina.poltorak@o2.pl (E.P.); pawel.koscielniak@uj.edu.pl (P.K.)

[2] Department of Chemistry, University of Warmia and Mazury, Plac Łódzki 4, 10-957 Olsztyn, Poland; kalinow@uwm.edu.pl

* Correspondence: j.kozak@uj.edu.pl; Tel.: +48-1268-62416

Academic Editor: Paraskevas D. Tzanavaras
Received: 11 March 2020; Accepted: 16 April 2020; Published: 18 April 2020

Abstract: A preconcentration module operated in flow mode and integrated with a sequential injection system with spectrophotometric detection was developed. Using the system, preconcentration was performed in continuous mode and was based on a membraneless evaporation process under diminished pressure. The parameters of the proposed system were optimized and the system was tested on the example of the spectrophotometric determination of Cr(III). The preconcentration effectiveness was determined using the signal enhancement factor. In the optimized conditions for Cr(III), it was possible to obtain the signal enhancement factors of around 10 (SD: 0.9, $n = 4$) and determine Cr(III) with precision and intermediate precision of 8.4 and 5.1% (CV), respectively. Depending on the initial sample volume, signal enhancement factor values of about 20 were achieved. Applicability of the developed preconcentration system was verified in combination with the capillary electrophoresis method with spectrophotometric detection on the example of determination of Zn in certified reference materials of drinking water and wastewater. Taking into account the enhancement factor of 10, a detection limit of 0.025 mg L^{-1} was obtained for Zn determination. Zn was determined with precision less than 6% (CV) and the results were consistent with the certified values.

Keywords: preconcentration; evaporation; flow analysis; sequential injection analysis

1. Introduction

Sample pretreatment is an integral and indispensable stage of many analytical methods. Performed in a traditional mode, it is often laborious, multi-stage, and time-consuming, especially when analyte preconcentration is necessary before the determination. On the other hand, it is necessary to achieve the concentration of the analyte within the limits of quantification of the method, and the way of its implementation is important because of the need to obtain satisfactory precision and accuracy of analytical results.

In recent years, special attention has been paid to the development of environmentally friendly analytical methods. The rules of green analytical chemistry (GAC) emphasize, among others, the importance of developing direct methods, integrating analytical operations, developing automated and/or miniaturized systems, using multi-analyte methods, reducing the use of toxic reagents and waste production, and ensuring work safety [1]. These principles should also be met at the stage of sample pretreatment [2]. Therefore, improving existing and developing new sustainable analytical procedures is becoming a significant trend in analytical chemistry [1–5].

In this field, the significance of flow techniques should be emphasized, which by enabling the automation (also miniaturization) of analytical procedures, integration of analytical operations, limiting the use of reagents, acceleration of analyses or improving work safety, meet the important principles of green analytical chemistry. The achievements in sample pretreatment performed using flow-based methods were presented in a large numbers of research articles, review papers [6–13], as well as in monographs [14–19].

The sample preparation step is considered as the crucial one in the light of GAC principles [2]. One of the ways used very often for sample preparation is sample preconcentration coupled with the analyte isolation from the sample matrix. It is usually performed by well-known extraction techniques (liquid–liquid, solid-phase) in various macro- and micro systems. In the literature one can find many examples of modern extraction approaches used for sample preparation [3,4], a number of them have been successfully adapted to flow analysis [6–10,13]. However, independently of the specificity of micro-extraction techniques, the necessity of using reagents, including organic solvents can be regarded as a drawback from the green chemistry point of view.

One of few non-reagent-based sample preparation ways is the evaporation technique. The approach based on evaporation has been quite often implemented in flow systems equipped with pervaporation modules [20–26]. In the developed systems, temperature assisted sample preconcentration, analyte (analytes) separation, or analyte postconcentration was coupled to process of diffusion using properly selected permeable membranes. In these systems, it is important to select a membrane with the right properties to ensure good contact between the solution and the membrane, and adequate pervaporation efficiency. Various approaches to the sample preconcentration stage based on the pervaporation/evaporation have been also developed using microfluidic and paper-based devices [27–29]. Many of these approaches are characterized by high preconcentration factors. They are usually designed for specific determinations and cannot be used directly for other analytical determinations.

Although the membraneless evaporation approach is still quite popular in batch analysis, only few proposals of flow-based membraneless systems have been reported in the literature. One of the reasons may be difficulties in designing appropriate modules ensuring obtaining controlled preconcentration factor values and sufficiently repeatable, and accurate analytical results. Membraneless evaporation systems have been developed for both preconcentrating the analyte [30] and separating the analyte from the sample matrix [31–33]. Regarding the flow injection (FI) system employing a membraneless evaporation for an analyte preconcentration, a method for the determination of anthocyanins in wine based on continuous liquid–solid extraction, evaporation, HPLC separation and photometric detection was developed [30]. In the system, the analytes were removed from the wine in a continuous way using a C_{18} minicolumn and the eluted fraction was preconcentrated by solvent evaporation assisted by heat and removing off the vapor using a flow of N_2. Using the developed evaporation system, the extract was subjected to evaporation (5 min) at a temperature of 140 °C, with N_2 flow equal to 600 mL min^{-1} and transported to the chromatographic system. Values of the coefficient of variance (CV) of the results of determination of anthocyanins were between 10% and 15% and the values of enrichment factor were not provided in the article. The second group of methods based on the membraneless evaporation, was developed to separate the (volatile) analyte from the sample matrix. To this aim, FI systems for the determination of ethanol in alcoholic drinks based on the reduction of dichromate by ethanol vapor [31,32] were developed. The system [31] was equipped with a module containing two parallel channels of appropriate configuration, applied as a donor and acceptor channel. Using the system, the detection limit about three times lower than using conventional gas diffusion unit was achieved. In [32], the diffusion of the volatile analyte took place in the space between specially designed two chambers. The chambers were additionally aerated to increase the diffusion. Detection limit was found to be 2.7% (v/v) of ethanol for diffusion time of 60 s and a precision of 3.7% CV (coefficient of variance). The approach based on the use of the evaporation unit described in [31] was also adapted for direct quantitation of calcium carbonate in cement with contactless conductivity detection [33]. The results of the determinations were obtained with the precision (CV) better than 5.3%.

In the present paper, an instrumental system with an original preconcentration module operated in a flow mode and integrated with a sequential injection (SI) system with spectrophotometric detection has been proposed. The novelty of the proposed system consists in the possibility of preconcentrating the sample based on a membraneless evaporation under diminished pressure in a continuous mode, using a mechanized flow system. The system has the potential to be used for various samples and to perform the preconcentration from different initial sample volumes. Parameters of the operation of the evaporation module and the flow system were optimized. The operation of the developed system was tested on the example of preconcentration and determination of Cr(III) and verified by the determination of Zn in certified reference materials of drinking water and wastewater with the use of the capillary electrophoresis method.

2. Results and Discussion

Research included: Design of a flow system containing a module for preconcentration, optimization of evaporation conditions in order to achieve effective preconcentration factor with acceptable precision, testing, and verification of the system on the example of the determination of selected analytes.

2.1. Evaporation Module

The main element of the constructed flow system was the module for preconcentration of a sample by evaporation under diminished pressure that is schematically shown in Figure 1 (and presented in Figure A1). The evaporation module (EM) consists of a glass tube with a capacity of about 15 mL (1.4 (i.d.) × 10 (height) cm), which is located inside an aluminum block. The block is responsible for thermostating the system in the temperature range from ambient to 150 °C. At the top the cylinder is covered with a PTFE block with an inlet (i.d. 0.8 mm) for dispensing solution of a sample. In the upper part there are also connections with a vacuum pump and a pressure gauge. The sample is introduced to the vessel through an inverted cone-shaped drop-forming element. Due to the selected height of the glass tube and the specific shape of this element the sample loss caused by its suction into the vacuum pump has been reduced. The lower part of the cylinder is closed by another PTFE block. The inner part of the bottom block is of the shape of an inverted cone with a narrow outlet (i.d. 0.8 mm) allowing even a small amount of preconcentrated sample to be collected and introduced into the flow system. During the evaporation process, the dosed drops of solution fell freely to the bottom of the vessel. Evaporation conditions (flow, pressure, and temperature) were selected so that the concentrated solution accumulated at the bottom of the vessel without filling it, so that as little solution as possible remained or splashed on the walls of the vessel.

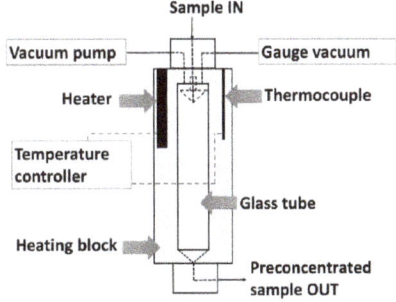

Figure 1. Scheme of the developed evaporation module.

2.2. Flow System

A sequential injection analysis system was used to mechanize the sample evaporation process in the developed EM and to enable repeatable sampling from the module for measurement after the evaporation process. The scheme of the system is presented in Figure 2. The system was equipped with a pressure syringe pump (SP, with a capacity of syringe of 4 mL) enabling the collection of precisely defined portions of liquid from the EM to a flow cell with the selected flow rate. The use of a ten-position selection valve (SV) allowed control of the direction of the sample flow between individual elements of the system. The SV was connected to the upper part of the EM (Figure 2; tubing T1, 3 cm) and simultaneously with the lower part of the EM (tubing T2, 3 cm), and to the SP with PTFE tubing (i.d. 0.5 mm). A flow cell (1 cm wide) has been installed in the path between the SV (tubing T3, 3 cm) and the pump (holding coil, HC, 5 cm).

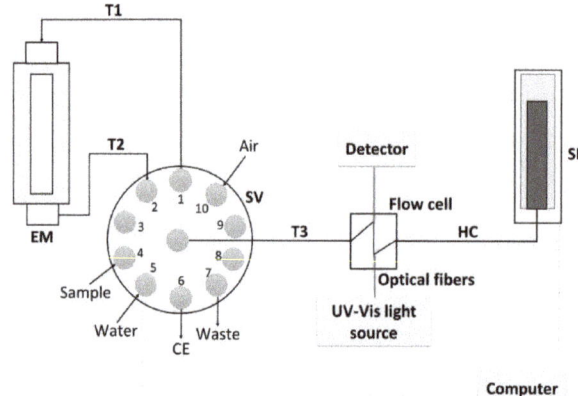

Figure 2. Scheme of the developed flow system with the evaporation module (EM); SV—selection valve, SP—syringe pump, HC—holding coil, T (1–3)—tubing, CE—capillary electrophoresis system.

2.3. Preconcentration Procedure

The measurement procedure developed for the flow system with the EM for the evaporation of a 6 mL of sample is presented in Table 1. At the beginning, T3, HC, and SP were washed with the sample (Table 1, Steps 1,2). Next the defined volume of sample was introduced into the SP, and after reversing the flow direction, sample was dosed to EM (through the HC, T3, and T1) with a selected flow rate (Steps 3–6). The evaporation process was carried out under diminished pressure and at a selected temperature. Evaporation conditions (sample dosing flow rate, pressure and temperature) were selected so that the evaporation process took place during the introduction of the sample into the vessel.

After completing the sample dispensing, the vacuum pump was turned off (the system was opened to equalize the pressure, Step 7, 30 s). Then a segment of air (1000 µL) was introduced into the tube T3 and HC (to prevent the dilution of the preconcentrated sample, Table 1, Step 8) and the sample was taken from the bottom of the vessel (through T2 and T3 tubing) to the flow cell and HC using the SP (Step 9)—to measure the signal for the sample after the preconcentration procedure. The segment of air was used as the carrier, its volume was selected so that after introducing the preconcentrated sample to the HC (after the evaporation process), air was moved to the SP and later (after reversing the flow direction) it allowed the whole sample to be introduced into a vial, in case of collecting the sample for further analysis (Figure 2, port 6, Table 2, Steps 10, 11) or to waste (in case of Cr(III) determination, during testing the system). The SP, HC, tubing and the EM were rinsed carefully with water between analyses (Table 1, Steps 12–18) and filled with air. To prevent drops of solution from remaining on walls, the whole system was washed daily with ethanol solution (50%, v/v) and again filled with air.

Using the designed flow system, the evaporation process was carried out for the initial sample volume equal the syringe pump capacity (4 mL) and was continued in continuous way (without interrupting the preconcentration procedure) by introducing another (subsequent) portion of the sample to the syringe pump (Table 1, Steps 5, 6). In this way, the preconcentration process could be carried out continuously for different initial sample volumes.

Table 1. Procedure developed for the preconcentration of 6 mL of sample in the proposed flow system; S_{pre}—preconcentrated sample signal, SV—selection valve, SP—syringe pump, HC—holding coil, EM—evaporation module, T—tubing, CE—capillary electrophoresis method; *—for Zn determination, **—for Cr(III) determination.

Step	Vacuum Pump	SV Position	SP Flow Rate, $\mu L\ s^{-1}$	Substance	Volume, μL	Action
1	Off	4	200	Sample	1000	Introducing the sample into the SP (washing T3, HC and SP with sample)
2	Off	7	200	Sample	1000	Transport of the sample to waste
				Two repetitions of stages 1–2		
				Preconcentration		
				Turning on the vacuum pump		
3	On	4	200	Sample	4000	Aspiration of the sample to the SP
4	On	1	3	Sample	4000	Introducing the sample into the EM Sample signal measurement
5	On	4	200	Sample	2000	Aspiration of the sample to the SP
6	On	1	3	Sample	2000	Introducing the sample into the EM; Sample signal measurement
				Turning off the vacuum pump		
7	Off	1	0	-	-	Delay 30 s
				Signal S_{pre} Measurement		
8	Off	10	200	Air	1000	Aspiration of air to the HC
9	Off	2	10	S_{pre}	500	Aspiration of the S_{pre} to the flow cell; S_{pre} signal measurement
				Collecting the sample, Washing SP, HC, Tubing and EM		
10	Off	6	50	S_{pre}	600 */0 **	Transport of the S_{pre} to a vial for further analysis
11	Off	7	200	Solutions and Air	900 */ 1500 **	Transport of the solutions and air to waste
12	Off	5	200	Carrier (H_2O)	1000	Introducing water into the SP
13	Off	7	200	Carrier (H_2O)	1000	Transport of water to waste
				Two repetitions of stages 12–13		
14	Off	5	200	Carrier (H_2O)	4000	Introducing water into the SP
15	On	1	200	Carrier (H_2O)	4000	Introducing water into the EM
16	On	1	0	-	-	Delay 30 s
17	Off	2	200	Carrier (H_2O)	4000	Introducing water into the SP
18	Off	7	200	Carrier (H_2O)	4000	Transport of water to waste
				Two repetitions of stages 14–18		
				Repetition of stages 14-18 using air instead of water (SV position 10, to remove water from T1, EM, and T2)		

Table 2. Results of the determination of Cr(III) using the developed flow system ($n = 3$).

No.	Cr(III) Concentration, mmol L^{-1}		|RE|, %	CV, %
	Expected	Found		
1		0.11	6.9	2.0
2	0.10	0.11	11.8	2.1
3		0.11	5.5	8.4
4		0.21	3.4	1.0
5	0.20	0.20	0.6	1.6
6		0.21	3.3	2.9
7		0.30	1.6	1.3
8	0.30	0.29	3.8	1.4
9		0.30	0.2	2.5
10		0.38	4.1	1.4
11	0.40	0.38	5.6	1.3
12		0.40	0.8	4.9
13		0.47	5.2	1.4
14	0.50	0.52	4.4	1.2
15		0.49	1.2	4.6

2.4. Preconcentration Factor

Signal enhancement factor (EF) was proposed to be determined as a measure of the effectiveness of the preconcentration process. EF was determined based on the signal values measured before and after the evaporation procedure. To this aim, a substance (tracer) was proposed to be added to the sample solution. Signal (absorbance) A_0 was measured by introducing the sample directly into the flow cell, then the evaporation process was started. After the evaporation process, a selected volume of the preconcentrated sample was taken from the module and signal A_{pre} was measured. EF was calculated as:

$$EF = \frac{A_{pre}}{A_0} \quad (1)$$

EF was selected as reliable for determining the degree of the sample preconcentration due to the possibility of sample loss in the membraneless evaporation process. This factor can be applied for samples whose absorbance, measured at the selected wavelength, does not change after the sample preconcentration (i.e., the absorbance of the sample itself does not affect the absorbance value measured for the tracer). As a tracer, one can select a substance for which the signal does not coincide with the analyte signal. Since the accuracy of the results depends on the accuracy of the EF determination, it is important that signal A_0 is stable and repeatable.

2.5. Preliminary Studies

The preliminary tests included the selection of an appropriate flow rate for dispensing a sample into the EM. Cr(III) solution of concentration of 2.5 mmol L^{-1} was used as a sample because for this solution the analytical signal was stable and reproducible. The initial sample volume (V_s) of 6 mL was subjected to the evaporation process. To make the preconcentration process as short as possible, the research was carried out at the thermostatic block heated up to 100 °C, under a vacuum of about 80 kPa. The tests were performed at four flow rates in the range from 2 to 4 µL s^{-1}. For a given flow rate, the evaporation procedure was repeated four times. The determined mean EF values with the intervals of standard deviation (SD) are shown in Figure 3. At the same time, the volume of the sample remained after the evaporation process (V_{pre}), was determined by taking the sample at a selected flow rate (10 µL s^{-1}) from the EM, through the flow cell to the SP and measuring the duration of the sample signal. The obtained V_{pre} values together with the standard deviation intervals are also shown in Figure 3.

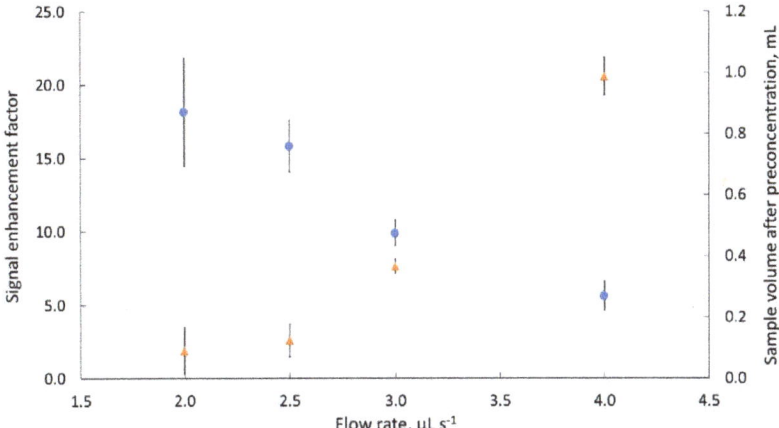

Figure 3. Dependences of the signal enhancement factor (●) and of the sample volume collected after the evaporation process (▲) (with SD intervals; $n = 4$) depending on the sample dispensing flow rate.

Figure 3 shows that under selected conditions, as the flow rate increased, increased amount of the preconcentrated sample (from about 0.1 to 1.0 mL) remained in the vessel after evaporation and decreased EF values (from 18 to 5, respectively) were obtained. In addition, for smaller volumes, the EF values were characterized by worse precision. The best and acceptable precision ($n = 4$) was obtained at a flow rate of 3 µL s^{-1}, namely SD = 0.9 and 0.024 mL for EF ≈ 10 and V_{pre} ≈ 0.4 mL, respectively. Thus, this flow rate value was decided to be used for further research.

In the next step, it was decided to examine whether the V_s volume affects the EF factor and the volume V_{pre}. To this end, 4 samples with a volume of 4 to 12 mL were evaporated under the same instrumental conditions. In this case evaporation was carried out once for the sample. The results are presented in Figure 4. It can be noticed that in the V_s range from 6 to 12 mL, the EF values ranged from 12 to 10, and the V_{pre} volume was about 0.20 mL. In further studies, volume $V_s = 6$ mL was taken for the evaporation process. Under these conditions, the evaporation process lasted about 34 min.

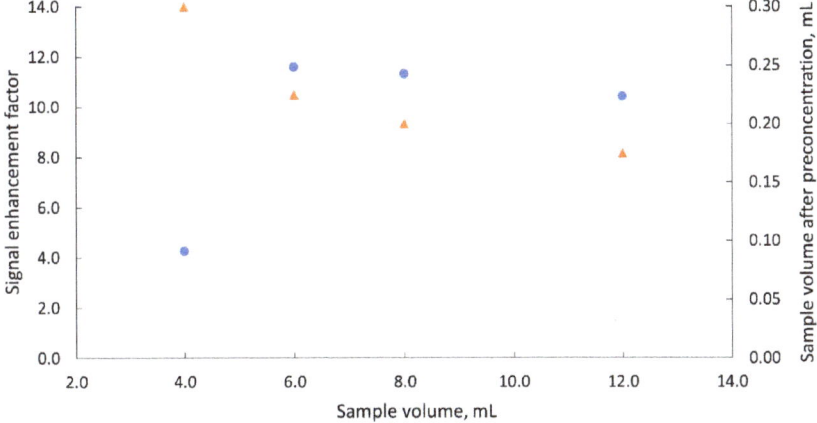

Figure 4. Dependences of the signal enhancement factor (●) and of the sample volume collected after the evaporation process (▲) on the initial sample volume subjected to evaporation; dispensing flow rate 3 µL s^{-1}.

In order to check whether using the evaporation system, it is possible to achieve higher concentration coefficient values, tests were also carried out for a wider range of sample volume V_s, from 8 to 44 mL, subjected to evaporation. To reduce the time needed to conduct the experiment, samples were dosed at a flow rate of 4 µL s^{-1}. In this case, evaporation was carried out once for a sample. The results are shown in Figure 5. It can be stated that when V_s was increased more, even greater EF value, approaching 23, was possible to be achieved. However, in this case the preconcentration time increased significantly to several hours.

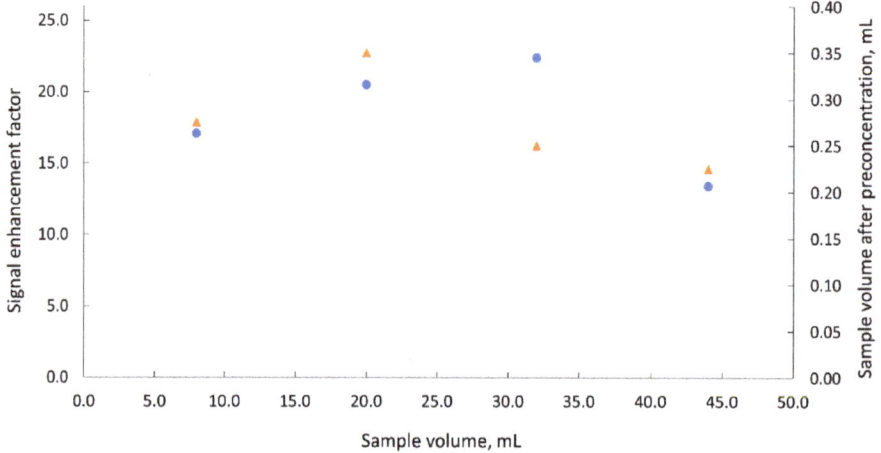

Figure 5. Dependences of the signal enhancement factor (●) and of the sample volume collected after the evaporation process (▲) on the initial sample volume subjected to evaporation; dispensing flow rate 4 µL s^{-1}.

2.6. Testing of the Flow-Based Preconcentration System

The operation of the flow system with the EM module was tested on the example of Cr(III) determination in synthetic samples. To determine the concentration of analyte in the sample, C_S, the calibration based on the standard addition method was applied. The standard with analyte of concentrations, $\Delta C_1, \ldots, \Delta C_4$ was added to the sample in accordance with the standard addition method. All solutions were subjected to measurements separately after introducing them one after the other to the flow cell and the calibration graph was prepared. The analyte concentration in the first standard (ΔC_1) was high enough to produce well repeatable signal A_0. Then, the sample with the first standard addition (ΔC_1) was introduced into the evaporation vessel, subjected to the evaporation process and transferred to the flow cell to measure the signal A_{pre}. EF value was calculated from Equation (1) and the analyte concentration in the sample, C_s, was determined using the equation:

$$C_s = \frac{C_{pre}}{EF} - \Delta C_1 \qquad (2)$$

Samples with Cr(III) in concentrations between 0.1 and 0.5 mmol L^{-1} were subjected to the preconcentration process in different days. Each sample was evaporated three times in the same conditions. Evaporation was carried out at 100 °C, vacuum of 80 kPa and at a sample flow rate of 3 µL s^{-1} into the evaporation vessel. The signals were measured directly at 530 nm. Linear calibration graphs were obtained for Cr(III) in the concentration range 1.0–50.0 mmol L^{-1} (e.g., y = 0.0143x + 0.0006; R = 0.999; y—absorbance, x—concentration). The Table 2 presents the mean analytical results (with the values of coefficient of variance, CV and relative error, RE).

Analyzing Table 2, it can be concluded that Cr(III) was determined with satisfactory accuracy: Relative error (RE) values exceeded 7% (11.8%) only in one case, for the lowest concentration of Cr(III). The results are also characterized by good precision, only in one case, CV value (also for the lowest Cr(III) concentration) exceeded 5% (8.4%), and intermediate precision (CV < 5.1%). With regard to accuracy, the highest RE values (5.5%–11.8%) were obtained for samples containing the lowest amounts of Cr(III). For samples containing more Cr(III), consistence of found and expected values was fully satisfactory. However, the method is dedicated to samples in which the analyte concentration is below the limit of quantification, therefore it can be assumed that for these concentration levels relative errors close to 10% may be acceptable.

2.7. Verification of the Flow Preconcentration System

The operation of the system was verified on the example of determination of zinc by capillary electrophoresis with spectrophotometric detection. Zn was determined in certified reference materials of drinking and waste water. To collect the entire sample volume after evaporation, the HC (Figure 2) was extended to 90 cm (i.d. 0.8 mm). First, the segment of air was introduced into the HC (to separate the sample from the solution in the pump and prevent the sample dilution, Table 1), then the sample (S_{pre}) was introduced into the HC through the flow cell (to measure A_{pre}) and finally (after changing the flow direction), the sample was directed to CE vial. To determine the EF value, Cr(III) at a concentration of 4.0 mmol L^{-1} was added to the sample. Limit of detection (LOD) and limit of quantification (LOQ) for the CE method were evaluated by calculating the signal-to-noise (S/N) ratio, considering the general rule that for LOD and LOQ the S/N ratio should be 3 and 10, respectively, which provided the concentration levels of 0.25 and 0.90 mg L^{-1}, respectively. One should note that taking into account the EF delivered by the EM (EF = 10), the appropriate LOD and LOQ levels for Zn determination were 0.025 and 0.09 mg L^{-1}, respectively.

The mean values of the determined Zn concentration and the corresponding values of the confidence interval ($n = 3$, $\alpha = 0.05$) were calculated for drinking water (2.4 ± 0.5) mg L^{-1} and wastewater (0.80 ± 0.15) mg L^{-1}. To evaluate the agreement between obtained and certified concentrations, the values were compared with each other using the Student's t-test ($\alpha = 0.05$). Certified values were treated as true values. It was confirmed that the determined concentrations were consistent with the certified values (2.49 ± 0.06) mg L^{-1} and (0.862 ± 0.010) mg L^{-1}, respectively. Generally, in the certified reference material samples Zn was determined with precision better than 6% (CV). The results showed the possibility of using the developed approach (for sample preconcentration and determination of the preconcentration degree) for the analyte determination using another method. The total analysis time included time necessary for the sample preconcentration and time of CE analysis. In the case of Zn determination, it was about 40 and 15 min, respectively.

To sum up, it can be stated that the developed mechanized flow system with the module for evaporation has a chance to be used to preconcentrate samples with the ability to determine the degree of sample preconcentration. The preconcentrated sample can be introduced into another detection system to determine the concentration of the specific analyte/analytes. For this purpose, e.g., capillary electrophoresis technique can be used, which allows the separation of analytes, while at the same time very small sample volumes can be used for analysis. This technique with spectrophotometric detection is characterized by relatively high values of the detection limit. Therefore, the use of the developed flow system with the preconcentration module creates greater possibilities for determining various analytes. Compared to other sample preconcentration systems based on the membraneless evaporation, its advantage may be the ability to easily adapt to preconcentrate different samples of various volumes. In the developed system, preconcentration time depends on the initial sample volume, the flow rate of sample introduced to the EM and the applied conditions of pressure and temperature. These conditions should be selected to a sample individually. The use of a tracer makes it possible to determine the signal enhancement factor for a particular determination; however, attention should be paid to the selection of the tracer and its concentration, so that it is possible to determine the analyte with acceptable accuracy. Further research can include the possibility of combining the developed flow-based evaporation system with CE instrument and selecting the CE analysis conditions to determine more analytes.

Compared to the membraneless evaporation system described in the literature [30], using the developed approach better precision of the results was obtained (8.4%, CV) but the evaporation time was longer. For the reported system [30] the values of enrichment factors were not given. Regarding other literature examples of the membraneless evaporation [31–33], they were characterized by better precision and shorter time of analysis; however, they were designed to determine (volatile) analyte in the evaporated fraction.

3. Materials and Methods

3.1. Instruments and Materials

The flow system consisted of bidirectional syringe pump SIChrom (FIAlab, Seattle, WA, USA) with 10-position selection valve (VICI Valco Instruments, Houston, USA), light source (deuterium and halogen lamps) DH-2000 (Ocean Optics, Dunedin, FL, USA), light fibers, spectrometer USB 4000 (Ocean Optics, USA), Ultem flow cell SMA-Z—10 mm optical path, and PTFE tubing, i.d. 0.8 or 0.5 mm (IDEX Health & Science, Rohnert Park, California, CA, USA). FIAlab software (FIAlab, Seattle, WA, USA) served in data acquisition, signals visualization and measuring. The preconcentration module included aluminum heating block (KPS Elektronika Laboratoryjna, Olsztyn, Poland), vacuum pump ME 2C NT + 2AK (Vacuubrand, Wertheim, Germany), temperature controller Transmit PID G6 (Termipol, Lubliniec, Poland), prsssure gauge ZSE4/ISE4 (SMC, Chiyoda, Japan), and electronic controller 60/16 (KPS Elektronika Laboratoryjna, Olsztyn, Poland).

The capillary electrophoresis analyses were performed using the PA/800 Plus system (Beckman-Coulter, Miami, Florida, FL, USA), equipped with the UV detector operating at 200 nm in an indirect detection mode. Separations were carried out in the bare fused silica capillary (75 μm i.d., total length of 60.2 cm, effective length of 50.2 cm), thermostated to 30 °C. Every working day, before the first analysis the capillary was rinsed (20 psi = 137.9 kPa) in a sequence with MeOH (20 min), 1 mol L^{-1} HCl (5 min), water (1 min), 1 mol L^{-1} NaOH (5 min), 0.1 mol L^{-1} NaOH (20 min), water (2 min), and background electrolyte (2 min). Between runs, the capillary was flushed only with BGE for 3 min (20 psi, 137.9 kPa). Samples and calibration solutions were stored in the thermostated sample garage (25 °C) and they were hydrodynamically injected (0.5 psi = 3.45 kPa, 5 s) into the capillary before applying the separation voltage of 20 kV (anode at inlet; 0.5 min ramping time) for 10 min.

3.2. Reagents and Solutions

Standard solutions and synthetic samples containing Cr(III) were prepared using chromium(III) nitrate 9-hydrate, Cr(NO$_3$)$_3$ · 9H$_2$O, (POCh SA, Gliwice, Poland). The Cr(III) stock solution (0.05 g L^{-1}) was prepared by dissolving appropriate amount of the Cr(NO$_3$)$_3$ · 9H$_2$O in water. Standard solutions and synthetic samples were prepared by appropriately diluting the stock solution with water. Certified reference materials of drinking water (EnviroMAT Drinking water, high (EP-H) and wastewater (EnviroMAT Wastewater, high (EU-H) (SCP SCIENCE, Quebec, ON, Canada) were used for the system verification. The materials were diluted with water according to the producer instructions. Zn calibration solutions were prepared by appropriate dilution of Zn standard solution (1000 mg, Merck, Darmstadt, Germany) with water. Reagents of analytical grade were used. Deionized water (0.05 μS cm^{-1}) obtained from HLP5sp system (Hydrolab, Straszyn, Poland) was used throughout the study. For the determination using capillary electrophoresis method, the background electrolyte (BGE) composition was optimized for the purpose and it consisted of 15 mmol L^{-1} imidazole (Bioshop, Burlington, Canada), 2 mmol L^{-1} 18-crown-6 ether (Sigma Aldrich, Saint Louis, MO, USA), 5% (v/v) methanol (Sigma Aldrich, USA), 5% (v/v) acetonitrile (Sigma Aldrich, Saint Louis, MO, USA), and it was adjusted to pH = 3.7 with acetic acid (Sigma Aldrich, Saint Louis, MO, USA). BGE was then filtered through the regenerated cellulose syringe filter (0.45 µm) and—before analyses—degassed by centrifugation. Fifty microliters of each of the preconcentrated sample was diluted with 40 µL of BGE and 10 µL of deionized water. The Zn calibration solutions were prepared the same way to keep the BGE dilution constant. The calibration plot was constructed by taking the time-corrected peak area as the analytical signal.

4. Conclusions

In summary, a mechanized flow system with an original module was developed in which the sample preconcentration based on a membraneless evaporation takes place in a continuous mode. Using the system and the proposed mode of determining the signal enhancement factor, samples of various initial volumes can be preconcentrated and the degree of sample preconcentration can be determined. Using different evaporation conditions, various values of the signal enhancement factor (from several to 20) were obtained. In the optimized conditions for Cr(III) determination, it was possible to obtain signal enhancement factors of around 10 with a precision (CV) of less than 10%. Preconcentrated samples can be taken for further analyses. The operation of the flow system with the preconcentration module was positively verified on the example of the determination of Zn in certified reference materials of drinking water and wastewater using capillary electrophoresis method.

Author Contributions: Conceptualization P.K., S.K., J.K.; Data curation, J.P., J.K., M.W. (Marcin Wieczorek), M.W. (Michał Woźniakiewicz), M.G.; Formal analysis, J.P., J.K.; Funding acquisition, P.K.; Investigation, J.P., J.K., M.W. (Marcin Wieczorek), M.W. (Michał Woźniakiewicz), M.G., E.P. S.K.; Methodology, J.P., J.K., P.K.; Project administration, P.K.; Supervision, J.K.; Validation, J.P., J.K., M.W. (Marcin Wieczorek), M.W. (Michał Woźniakiewicz); Visualization, J.P., J.K.; Writing—original draft, J.K.; Writing—review & editing, J.K., P.K. All authors have read and agreed to the published version of the manuscript.

Funding: Authors thank for financial support received from the National Science Centre, Poland (Opus, 2017-2020, grant no. 2016/23/B/ST4/00789).

Conflicts of Interest: The authors declare no conflict of interest.

Appendix A

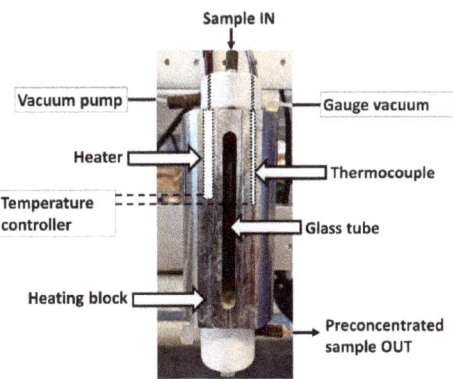

Figure A1. Evaporation module.

References

1. Armenta, S.; Garrigues, S.; De La Guardia, M. Green Analytical Chemistry. *TrAC Trends Anal. Chem.* **2008**, *27*, 497–511. [CrossRef]
2. Tobiszewski, M.; Mechlinska, A.; Zygmunt, B.; Namiesnik, J. Green analytical chemistry in sample preparation for determination of trace organic pollutants. *TrAC Trends Anal. Chem.* **2009**, *28*, 943–951. [CrossRef]

3. Pena-Pereira, F.; Lavilla, I.; Bendicho, C. Liquid-phase microextraction techniques within the framework of green chemistry. *TrAC Trends Anal. Chem.* **2010**, *29*, 617–628. [CrossRef]
4. Xie, H.-Y.; He, Y.-Z. Green analytical methodologies combining liquid-phase microextraction with capillary electrophoresis. *TrAC Trends Anal. Chem.* **2010**, *29*, 629–635. [CrossRef]
5. Płotka, J.; Tobiszewski, M.; Sulej, A.; Kupska, M.; Górecki, T.; Namieśnik, J. Green chromatography. *J. Chromatogr. A* **2013**, *1307*, 1–20.
6. Miro, M.; Hansen, E.H. Solid reactors in sequential injection analysis: Recent trends in the environmental field. *TrAC Trends Anal. Chem.* **2006**, *25*, 267–281. [CrossRef]
7. Anthemidis, A.; Miro, M. Recent Developments in Flow Injection/Sequential Injection Liquid-Liquid Extraction for Atomic Spectrometric Determination of Metals and Metalloids. *Appl. Spectrosc. Rev.* **2009**, *44*, 140–167. [CrossRef]
8. Theodoridis, G.A.; Zacharis, C.K.; Voulgaropoulos, A.N. Automated sample treatment by flow techniques prior to liquid-phase separations. *J. Biochem. Biophys. Methods* **2007**, *70*, 243–252. [CrossRef]
9. Kocúrová, L.; Balogh, I.S.; Andruch, V. Solvent microextraction: A review of recent efforts at automation. *Microchem. J.* **2013**, *110*, 599–607. [CrossRef]
10. Silvestre, C.I.; Santos, J.L.; Lima, J.; Zagatto, E.A. Liquid–liquid extraction in flow analysis: A critical review. *Anal. Chim. Acta* **2009**, *652*, 54–65. [CrossRef]
11. De Castro, M.L.; Gámiz-Gracia, L. Analytical pervaporation: An advantageous alternative to headspace and purge-and-trap techniques. *Chromatographia* **2000**, *52*, 265–272. [CrossRef]
12. Sae-Khow, O.; Mitra, S. Pervaporation in chemical analysis. *J. Chromatogr. A* **2010**, *1217*, 2736–2746. [CrossRef] [PubMed]
13. Fang, Z.-L. Trends of flow injection sample pretreatment approaching the new millennium. *Anal. Chim. Acta* **1999**, *400*, 233–247. [CrossRef]
14. Fang, Z.; Tao, G. New developments in flow injection separation and preconcentration techniques for electrothermal atomic absorption spectrometry. *Anal. Bioanal. Chem.* **1996**, *355*, 576–580. [CrossRef] [PubMed]
15. Ruzicka, J. Flow Injection Analysis. *Anal. Chem.* **1983**, *55*, 1040A–1053A. [CrossRef]
16. Karlberg, B.; Pacey, G.E. *Flow Injection Analysis—A Practical Guide*; Elsevier: Amsterdam, The Netherlands, 1989; pp. 102–123.
17. Valcárcel, M.; De Castro, M.D.L. Continuous separation techniques in flow injection analysis. *J. Chromatogr. A* **1987**, *393*, 3–23. [CrossRef]
18. Miro, M.; Hansen, E.H. On-Line Sample Processing Methods in Flow Analysis. In *Advances in Flow, Analysis*; Trojanowicz, M., Ed.; Wiley–VCH: Weinheim, Germany, 2008; pp. 291–320.
19. Trojanowicz, M. Flow Injection Analysis—Instrumentation and Applications. *Flow Inject. Anal.* **2000**, 154–196. [CrossRef]
20. Shishov, A.; Penkova, A.; Zabrodin, A.; Nikolaev, K.; Dmitrenko, M.; Ermakov, S.; Bulatov, A. Vapor permeation—stepwise injection simultaneous determination of methanol and ethanol in biodiesel with voltammetric detection. *Talanta.* **2016**, *148*, 666–672. [CrossRef]
21. Mataix, E.; De Castro, M.L. Sequential determination of total and volatile acidity in wines based on a flow injection-pervaporation approach. *Anal. Chim. Acta* **1999**, *381*, 23–28. [CrossRef]
22. Cuadrado, M.U.; De Castro, M.L.; Gómez-Nieto, M. Ángel Fully Automated Flow Injection Analyser for the Determination of Volatile Acidity in Wines. *J. Wine Res.* **2006**, *17*, 127–134. [CrossRef]
23. Vallejo, B.; Richter, P.; Toral, I.; Tapia, C.; De Castro, M.L. Determination of sulphide in liquid and solid samples by integrated pervaporation–potentiometric detection. *Anal. Chim. Acta* **2001**, *436*, 301–307. [CrossRef]
24. Gonzalez-Rodriguez, J. Determination of ethanol in beverages by flow injection, pervaporation and density measurements. *Talanta* **2003**, *59*, 691–696. [CrossRef]
25. Delgado-Reyes, F.; Romero, J.M.F.; De Castro, M.L. Selective determination of pectinesterase activity in foodstuffs using a pervaporator coupled to an open-closed dynamic biosensing system. *Anal. Chim. Acta* **2001**, *434*, 95–104. [CrossRef]
26. Takeuchi, M.; Dasgupta, P.K.; Dyke, J.V.; Srinivasan, K. Postcolumn Concentration in Liquid Chromatography. On-Line Eluent Evaporation and Analyte Postconcentration in Ion Chromatography. *Anal. Chem.* **2007**, *79*, 5690–5697. [CrossRef]

27. Kachel, S.; Zhou, Y.; Scharfer, P.; Vrančić, C.; Petrich, W.; Schabel, W. Evaporation from open microchannel grooves. *Lab Chip* **2014**, *14*, 771–778. [CrossRef]
28. Fornells, E.; Barnett, B.; Bailey, M.; Shellie, R.; Hilder, E.; Breadmore, M. Membrane assisted and temperature controlled on-line evaporative concentration for microfluidics. *J. Chromatogr. A* **2017**, *1486*, 110–116. [CrossRef]
29. Fornells, E.; Hilder, E.; Breadmore, M.C. Preconcentration by solvent removal: Techniques and applications. *Anal. Bioanal. Chem.* **2019**, *411*, 1715–1727. [CrossRef]
30. Mataix, E.; De Castro, M.L. Determination of anthocyanins in wine based on flow-injection, liquid–solid extraction, continuous evaporation and high-performance liquid chromatography–photometric detection. *J. Chromatogr. A* **2001**, *910*, 255–263. [CrossRef]
31. Choengchan, N.; Mantim, T.; Wilairat, P.; Dasgupta, P.K.; Motomizu, S.; Nacapricha, D. A membraneless gas diffusion unit: Design and its application to determination of ethanol in liquors by spectrophotometric flow injection. *Anal. Chim. Acta* **2006**, *579*, 33–37. [CrossRef]
32. Ratanawimarnwong, N.; Pluangklang, T.; Chysiri, T.; Nacapricha, D. New membraneless vaporization unit coupled with flow systems for analysis of ethanol. *Anal. Chim. Acta* **2013**, *796*, 61–67. [CrossRef]
33. Sereenonchai, K.; Teerasong, S.; Chaneam, S.; Saetear, P.; Choengchan, N.; Uraisin, K.; Amornthammarong, N.; Motomizu, S.; Nacapricha, D. A low-cost method for determination of calcium carbonate in cement by membraneless vaporization with capacitively coupled contactless conductivity detection. *Talanta* **2010**, *81*, 1040–1044. [CrossRef] [PubMed]

Sample Availability: Samples of the compounds are not available from the authors.

© 2020 by the authors. Licensee MDPI, Basel, Switzerland. This article is an open access article distributed under the terms and conditions of the Creative Commons Attribution (CC BY) license (http://creativecommons.org/licenses/by/4.0/).

Article

Automated Photochemically Induced Method for the Quantitation of the Neonicotinoid Thiacloprid in Lettuce

J. Jiménez-López, E.J. Llorent-Martínez, S. Martínez-Soliño and A. Ruiz-Medina *

Department of Physical and Analytical Chemistry, Faculty of Experimental Sciences, University of Jaén, Campus Las Lagunillas, E-23071 Jaén, Spain; jujimene@ujaen.es (J.J.-L.); ellorent@ujaen.es (E.J.-L.M.); saramsolinho@gmail.com (S.M.-S.)
* Correspondence: anruiz@ujaen.es; Tel.: +34-953-212759; Fax: +34-953-212940

Received: 11 September 2019; Accepted: 6 November 2019; Published: 12 November 2019

Abstract: In this work, we present an automated luminescence sensor for the quantitation of the insecticide thiacloprid, one of the main neonicotinoids, in lettuce samples. A simple and automated manifold was constructed, using multicommutated solenoid valves to handle all solutions. The analyte was online irradiated with UV light to produce a highly fluorescent photoproduct ($\lambda_{exc}/\lambda_{em}$ = 305/370 nm/nm) that was then retained on a solid support placed in the flow cell. In this way, the pre-concentration of the photoproduct was achieved in the detection area, increasing the sensitivity of the analytical method. A method-detection limit of 0.24 mg kg^{-1} was achieved in real samples, fulfilling the Maximum Residue Limit (MRL) of The European Union for thiacloprid in lettuce (1 mg kg^{-1}). A sample throughput of eight samples per hour was obtained. Recovery experiments were carried out at values close to the MRL, obtaining recovery yields close to 100% and relative standard deviations lower than 5%. Hence, this method would be suitable for routine analyses in quality control, as an alternative to other existing methods.

Keywords: neonicotinoid; thiacloprid; solid-phase spectroscopy; optosensor; luminescence

1. Introduction

Neonicotinoid pesticides are the most widely used class of insecticides worldwide, representing a 25% share of the insecticides market in 2014 [1]. They have a wide range of applications: plant protection (crops, vegetables, and fruits), veterinary products, and biocides to invertebrate pest control in fish farming. However, their use is a controversial subject, as several toxicological studies proved that some neonicotinoids (imidacloprid, clothianidin, and thiamethoxam) produce the collapse of honey-producing bee colonies [2]. In 2018, the European Union decided to ban the outdoor use of these three pesticides [3], and the Environmental Protection Agency announced on May 2019 that the registration for 12 neonicotinoid-based products would be canceled. However, the mentioned ban does not affect thiacloprid (TCP) and other neonicotinoids, which makes it important to develop accurate and quick analytical methods for their reliable quantitation in a wide variety of food samples, in order to ensure their safe consumption.

Among neonicotinoids, TCP is one of the most commonly used, and it belongs to the so-called "first generation" neonicotinoids. The usual analytical methods for TCP quantitation in food samples use liquid chromatography [4–8]. In particular, HPLC-MS/MS [9,10] and UHPLC-MS/MS [11,12] have been reported for their determination in lettuce. Moreover, electrochemistry [13,14], micellar electrokinetic chromatography [15], immunoassays [16,17], and luminescence [18–20] have been also proposed for TCP quantitation. The main goal of this work was to develop an alternative luminescence analytical method for TCP routine analysis in lettuce, one of the most widely consumed vegetables, paying special attention to the simplicity, economy, and sample throughput of the system developed.

The use of luminescence sensors has increased in the last decade, minimizing reagents consumption and increasing the degree of automation. In this sense, the use of automated methodologies, such as multicommutated devices, provide advantages, such as increased precision, robustness, and high automation. The combination of flow methodologies and solid-phase spectroscopy (SPS) is a successful approach that maintains the key advantages of automated flow systems, increasing the sensitivity and selectivity of the analytical methods due to the retention and pre-concentration of the target compounds on a solid support placed in the detection area [21]. For instance, a previous method was reported that used sequential injection analysis for the fluorometric determination of hydroxytyrosol (phenolic phytochemical with antioxidant properties in vitro) in food samples, measuring its native fluorescence [22]. The use of multicommutation has proved successful, too, for the quantitation of clothianidin by photochemically induced fluorescence (PIF) in drinking water, rice, and honey [23]. As a follow-up to previous works, we report a multicommutated flow-injection analysis (MCFIA)-based method, using PIF detection to overcome the handicap of the absence of native fluorescence of TCP. The main difference from the previous paper is the selected food sample, lettuce, which made it necessary to carry out a different extraction procedure due to the different matrix. In addition, the novel instrumental and chemical conditions made it possible to discriminate between TCP and other neonicotinoids. The analyte is UV-irradiated to produce a fluorescence photoproduct which is retained and detected on a solid support placed in the flow cell. By means of the MCFIA manifold, this irradiation takes place online, simplifying the procedure and increasing sample throughput. The proposed method allows for the fulfillment of the Maximum Residue Limit (MRL) of the European Union [24] for TCP in lettuce.

2. Experimental

2.1. Reagents and Solutions

TCP (Sigma-Aldrich, Madrid, Spain) stock solution of 100 mg L^{-1} was prepared in Milli-Q water (Millipore); it was kept in the dark at 4 °C, and working solutions were prepared daily. Acetonitrile, graphitized carbon black (GBC), primary–secondary amine (PSA), hydrochloric acid (HCl), sodium hydroxide (NaOH), sodium acetate, acetic acid, ammonium chloride (NH$_4$Cl), ammonia (NH$_3$), and magnesium sulphate (MgSO$_4$) were purchased from Sigma (Sigma-Aldrich). Isolute QuEChERS extraction kit was acquired from Biotage (Sweden). Sephadex QAE A-25 and Sephadex SP C-25 in sodium form, both of them 40–120 µm average particle size (Sigma-Aldrich, Buchs, Switzerland), and C$_{18}$ bonded phase silica gel beads (Waters, Milford, MA, USA) of 55–105 µm average particle size, were tested as solid supports.

Ultrapure water (Milli-Q Waters purification system, Millipore, Milford, MA, USA) was used for all analyses.

2.2. Instrumentation and Apparatus

A Cary-Eclipse Luminescence Spectrometer (Varian Inc., Mulgrave, Australia) with Cary-Eclipse (Varian) software and a Hellma flow cell 176.752-QS (Hellma, Mülheim, Germany) (25 µL of inner volume, and a light path length of 1.5 mm) were used. The cell was filled with the solid support and was blocked at the outlet with glass wool, to prevent displacement of the particles.

A four-channel Gilson Minipuls-3 (Villiers Le Bel, France) peristaltic pump with rate selector and methanol-resistant pump tubes type Solvflex (Elkay Products, Shrewsbury, MA, USA) were used. An electronic interface based on ULN 2803 integrated circuit (Motorola, Phoenix, AZ, USA) was employed to generate the electric potential (12 V) and current (100 mA) required to control the three 161T031 NResearch three-way solenoid valves (Neptune Research, West Caldwell, NJ, USA).

The software for controlling the system was written in Java. Flow lines of 0.8 mm internal diameter PTFE tubing and methacrylate connections were used.

For UV-irradiation, a homemade continuous photochemical reactor was constructed by coiling PTFE tubing (180 cm, 0.8 mm i.d.) around a low-pressure mercury lamp (30 W, 254 nm). A Sonorex Digital 10P (Bandelin Electronic, Berlin, Germany) ultrasonic bath, a pH-meter Crison GLP21 (Crison Instruments, Barcelona, Spain), a centrifuge Mixtasel-BL (Selecta, Barcelona, Spain), and a rotary evaporator (Heidolf, Schawabach, Germany) were also used.

2.3. Sample Preparation

All samples (iceberg lettuce, baby Romaine lettuce, and green oak leaf lettuce) were purchased at local markets. Approximately 200 g of each sample was ground and homogenized with a high-speed laboratory homogenizer. TCP was extracted, following a modified QuEChERS method [25]. An extraction kit (Isolute QuEChERS) containing GCB was used for all samples. This nonpolar sorbent allowed for the removal of hydrophobic interaction-based compounds, such as chlorophyll and carotenoids. The method used was as follows: 10 g of sample was weighed in a 50 mL PTFE centrifuge tube, and acetonitrile (10 mL) was added. Then, the content of a 15 mL tube extraction kit (4 g of $MgSO_4$, 1 g of sodium citrate, 0.5 g of sodium citrate sesquihydrate, and 1 g of NaCl) was added, and the samples were vortexed for 1 min. After centrifugation (5 min, 4000 rpm), 6 mL of the supernatant was transferred into a 15 mL dispersive SPE tube containing 150 mg of PSA, 900 mg of $MgSO_4$, and 15 mg of GCB. Samples were vortexed for 1 min and centrifuged for 5 min, at 4000 rpm. In this way, the acetonitrile (supernatant) contained the analyte. Prior to analysis, an appropriate volume of the acetonitrile extract was diluted with acetate buffer (0.05 mol L^{-1}, pH 4.6).

2.4. General Procedure

The flow manifold is shown in Figure 1. In the initial status, all valves are switched off, and the carrier (0.05 mol L^{-1} acetate buffer, pH 4.6) flows through the flow-through cell, while all other solutions are recycled to their vessels. The sample (20–250 µg L^{-1} prepared in 0.05 mol L^{-1} of acetate buffer, pH 4.6) is introduced by simultaneously switching valves V_1 and V_2 on for 200 s. In this way, the analyte is carried toward the photochemical reactor, where it is UV-irradiated for 150 s, obtaining its fluorescent photoproduct. Then, the photoproduct is carried toward the flow cell, which is filled with Sephadex SP C-25 microbeads. TCP photoproduct is strongly retained on the solid microbeads, and the fluorescence signal is recorded ($\lambda_{exc}/\lambda_{em}$ = 305/370 nm/nm). Then, an eluting solution (0.05 mol L^{-1} of NH_3/NH_4Cl buffer, pH 9.0) is inserted into the system by activating valves V_1 and V_3 for 50 s, desorbing the photoproduct from the solid support. Finally, the carrier solution flows again through the system until the next sample insertion. All calibration standards and samples were analyzed in triplicate.

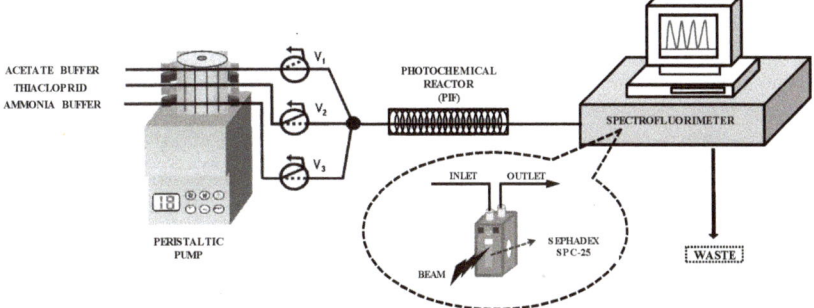

Figure 1. Flow manifold. Vi = three-way solenoid valves.

3. Results and Discussion

Neonicotinoids do not present native fluorescence (or very low luminescence in some cases). Therefore, different strategies are required to develop luminescent analytical methods for their determination. In this case, we tested the possibility of irradiating TCP with a UV-lamp, in order to generate possible fluorescent photoproducts. The absorption spectrum of thiacloprid (200–380 nm; maximum at 242 nm) makes this compound an interesting candidate to perform PIF with the low-pressure mercury lamp (emission of 200–300 nm; maximum at 254 nm). The different parameters of the system were optimized to obtain the highest sensitivity.

3.1. Instrumental Variables and Selection of Solid Support

We tested different solid supports (Sephadex QAE A-25, Sephadex SP C-25, and C_{18} silica gel) in the flow cell, to select the optimum one for the retention of TCP photoproduct. The optimum sample pH was obtained for pH values of 4–6 (see Section 3.2); as expected, TCP photoproduct was not retained on the anion-exchange QAE A-25, which is suitable for anionic species at basic pH values. On the other hand, although both the cation-exchange SP C-25 and nonionic C_{18} silica gel beads could retain the photoproduct, the signal obtained in C_{18} was very low, observing the highest signal with SP C-25, which was the selected solid support. However, it is important to consider that, when the signal is recorded on a solid support, there is a considerable background signal. Therefore, instrumental parameters have to be carefully studied to achieve the maximum sensitivity without compromising the linear dynamic range due to a high background signal. Excitation and emission slit widths were optimized between 5 and 20 nm, whereas the voltage of the photomultiplier tube (PMT) was studied in the range of 400–800 V. Wide slit widths and high PMT voltages increased the sensitivity, as well as the background signal produced by the solid support. Overall, the best results were obtained for excitation/emission slit widths of 5/10 nm/nm, respectively, and a PMT voltage of 780 V.

3.2. Chemical Variables

The chemical variables can affect the performance of the analytical methods not only from the point of view of the generation of the fluorescent photoproduct but also in terms of its retention/elution kinetics on the solid support. We thus optimized the pH value of the sample solution in the first place, adjusting the pH with HCl and NaOH solutions.

TCP generated fluorescence photoproducts in a wide range of pH values, obtaining the highest sensitivity in the range of 4–6. For acidic pH values, low luminescence was obtained. Likewise, the luminescence signal decreased drastically as the pH value increased (see Figure 2). Different buffer solutions were tested at this range (acetate, citrate, and succinate), observing an enhancement of approximately 20% when using an acetate buffer solution respect to adjust only with HCl solution. The concentration of acetate buffer was tested in the range 0.01–0.1 mol L^{-1}, selecting as optimum 0.05 mol L^{-1} and a pH value of 4.6. However, when using this buffer in both carrier and sample solutions, the photoproduct was not completely eluted from the Sephadex SP C-25 solid support, due to its high retention. An increase in carrier ionic strength (higher buffer concentrations) resulted in a decrease of the analytical signal, as a consequence of a lower retention of TCP photoproduct on the solid support in these conditions. We thus introduced an additional eluting solution to regenerate the solid support after the photoproduct had developed its analytical signal. Due to the nature of the solid support (cation-exchanger), changes in pH values resulted in different retention/elution kinetics of TCP photoproduct. Eluting solutions with pH higher than 8 provided the desorption of the photoproduct, hence selecting a solution of ammonia/chloride ammonium of 0.05 mol L^{-1}, at pH 9 (tested in the range 0.01–0.1 mol L^{-1}). In this way, TCP photoproduct provided the highest sensitivity when retained on the solid support at pH 4.6, and, after the signal was recorded, the solid support was regenerated by the eluting solution at pH 9.

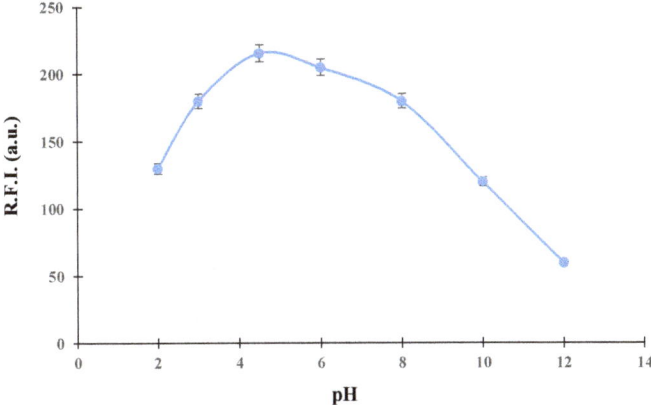

Figure 2. Effect of pH values on the analytical signal.

3.3. Irradiation Time

The irradiation time is an essential variable for the generation of fluorescent photoproducts. To optimize this parameter, different UV lamps (8, 15, and 30 W) and irradiation times (30–230 s) were tested for a TCP solution of 100 µg L^{-1}. First of all, the 30 W UV lamp was selected, as a higher analytical signal was obtained compared to the other lamps. Second, the irradiation time was studied with this lamp, inserting the sample solution in the system and stopping the flow when the whole plug of the sample was within the photoreactor. Then, the sample was irradiated for increasing periods of time; the results are shown in Figure 3. The analytical signal increased up to an irradiation time of 150 s, decreasing for higher values. The shape of the irradiation time curve suggests a two-step photolysis mechanism, in which the photoproduct observed at 150 s suffered a posterior photodegradation into nonfluorescent product(s) or different photoproduct(s) with lower fluorescence emission. This kind of behavior was previously reported for imidacloprid [26]; hence, the irradiation time was fixed at 150 s. To obtain this irradiation time without the need to stop the flow, flow parameters were optimized. Although the exact structure of TCP photoproduct could not be elucidated, a previous work reported the formula of the photoproduct as $C_{10}H_{11}N_4OS$ [27]. This means that TCP ($C_{10}H_9ClN_4S$) suffered a C–Cl bond cleavage to produce the photoproduct; the loss of Cl results in an enhancement of the fluorescence.

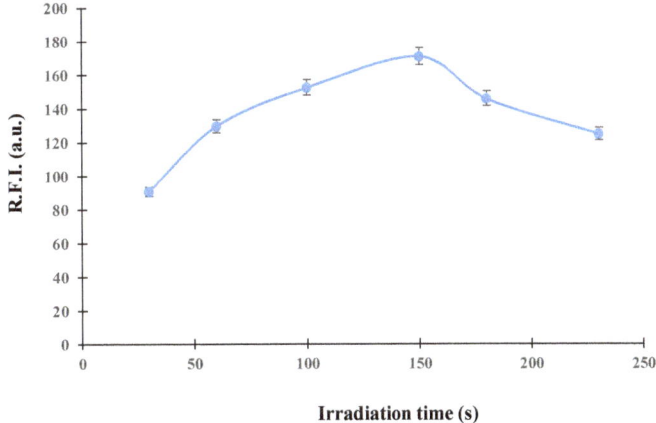

Figure 3. Effect of irradiation time on the analytical signal.

3.4. Flow Parameters

The flow rate selected for the manifold is critical to improving the sample throughput of the analytical method. However, it can also affect it in other ways: (a) a high flow rate may produce overpressures due to the solid support placed in the flow cell; (b) the flow rate and photoreactor length are critical to keeping the optimum irradiation time. As a result, a flow rate of 1.3 mL min^{-1} was selected. Using this flow rate, the length of the photoreactor was adjusted so that the sample plug required 150 s to go through the whole photoreactor.

In flow-through optosensors, the sensitivity of the method improves by increasing the sample volume inserted. The higher the sample volume (keeping the same concentration), the higher the amount of analyte inserted in the system and retained on the solid support. In this way, a pre-concentration of the analyte takes place on the solid microbeads. However, increases in sample volumes also imply lower sample throughput, so a compromise solution is usually needed. When multicommutation is used, sample insertion time, instead of sample volume, is used (when the time and flow rate are known, the volume can be calculated). We thus checked the influence of sample-insertion times between 20 and 300 s. The analytical signal increased up to 200 s, being constant for higher insertion times; hence, 200 s was selected as the optimum insertion time, achieving the required sensitivity for the applications.

3.5. Analytical Parameters

The analytical parameters of the system were studied under the optimized conditions previously discussed. They are shown in Table 1, and all of them correspond to a sample insertion time of 200 s.

Table 1. Analytical parameters.

Parameter	
Linear dynamic range/µg L^{-1}	20–250
Calibration graph	
Intercept	3.0769
Slope/L µg^{-1}	1.8754
Correlation coefficient	0.9997
Detection limit/µg L^{-1}	6
Quantification limit/µg L^{-1}	20
Repeatability (%)	4.5
Intermediate precision (%)	7.8
Sample throughput (samples h^{-1})	8

The calibration graph was constructed, fitting the data by standard least-squares treatment. Detection and quantitation limits, calculated by the 3σ and 10σ criterion, were 6 and 20 µg L^{-1}, respectively. Considering sample preparation, these values corresponded to method detection and quantitation limits of 0.24 and 0.8 mg kg^{-1}, respectively, in real samples. Precision was studied by analyzing lettuce extracts spiked with 50–200 µg L^{-1} of TCP. Repeatability ($n = 10$, within the same day) and intermediate precision ($n = 9$, 3 consecutive days) were lower than 5% and 8%, respectively. A sample throughput of approximately eight samples per hour was obtained. We also evaluated the robustness of the method by modifying some parameters from the optimum values: excitation/emission wavelengths (±2 nm), photomultiplier tube voltage (±10 V), and flow rate (±0.1 mL min^{-1}). The differences observed were always lower than 5% compared to the optimum values. When the proposed method is compared with other methods for TCP determination (Table 2), it can be observed that the precision is similar or better. When comparing the sensitivity, chromatographic methods usually present lower detection limits. However, the detection limit of the present method compares favorably with other non-chromatographic methods.

Table 2. An overview on reported determination of TCP.

Technique	Sample	Sample Treatment	DL	RSD (%)	Ref.
LC–MS/MS	Cucumber, soil	QuEChERS	0.71 µg kg^{-1}	<13.2	[4]
LC–MS/MS	Tea	QuEChERS	50 µg kg^{-1} *	≤7.2	[5]
UHPLC–MS/MS	Edible fungi	QuEChERS	0.08 µg kg^{-1}	≤4.3	[7]
LC–MS/MS	Lettuce, orange	SLE	10 µg kg^{-1} *	≤19	[11]
UHPLC–MS/MS	Lettuce	QuEChERS	2 µg L^{-1}	<6	[12]
SMEKC	Cucumber	DLLME	0.8 µg kg^{-1}	≤6.5	[15]
SWV	River water		270 µg L^{-1}	<5	[14]
ELISA	Water, soil, pear, tomato	SLE	0.47 µg L^{-1}	≤10	[16]
TRFIA	Water, tomato, pear, soil	SLE	0.0019 µg L^{-1}	≤11.3	[17]
TSL	Tea	SLE, SPE	60 µg L^{-1}	<5	[18]
Fluorescence	Waters		30 µg L^{-1}	<5	[19]
PICL	Waters	SPE	0.8 µg L^{-1}	<10	[20]
TSL	Waters		60 µg L^{-1}	<4	[28]
Proposed	Lettuces	QuEChERS	6 µg L^{-1}	≤4	

* Limit of quantification; DL: detection limit; RSD: relative standard deviation; LC–MS/MS: liquid chromatography–tandem mass spectrometry; UHPLC–MS/MS: ultra-high performance liquid chromatography–tandem mass spectrometry; SMEKC: sweeping micellar electrokinetic chromatography; SWV: square-wave voltammetry; ELISA: enzyme-linked immunosorbent assay; TRFIA: time-resolved fluoroimmunoassay; TSL: terbium-sensitized luminescence; PICL: photo-induced chemiluminescence; QuEChERS: quick, easy, cheap, effective, rugged, and safe; SLE: solid–liquid extraction; DLLME: dispersive liquid–liquid microextraction; SPE: solid-phase extraction.

3.6. Interference Study

We studied the potential interference caused by different pesticides that may be present in the analyzed samples (Table 3). Among them were some fluorescent common pesticides (carbendazim and o-phenylphenol), neonicotinoids (clothianidin, imidacloprid, nitenpyram, and thiamethoxam), and other pesticides that have been found in lettuce samples (cypermethrin, chlorpyrifos, and λ-cyhalothrin) [29]. This study was performed by analyzing a solution of 50 µg L^{-1} TCP, adding increasing concentrations of each individual pesticide. Tolerance level was the level of interferent that caused an error of ±3% compared to the analytical signal obtained in the absence of the potential interference. In all cases, interferent levels similar or higher than TCP concentration did not produce any deviation of the analytical signal, therefore allowing the selective quantitation of TCP in the presence of other pesticides that may be present in lettuce samples.

Table 3. Interference study carried out for 50 µg L^{-1} TCP.

Foreign Species	Tolerance Interferent/Analyte (w/w) Ratio
Clothianidin	75
Carbendazim, thiamethoxam	20
o-phenylphenol, cypermethrin, λ-cyhalothrin	6
Acetamiprid, chlorpyrifos, imidacloprid, nitenpyram	1

3.7. Analytical Applications

The proposed method was applied to the determination of TCP in lettuce samples, in which the MRL established by the European Union is 1 mg kg^{-1} [22]. First of all, extracts of all samples (iceberg lettuce, baby Romaine lettuce, and green oak leaf lettuce) were analyzed to check the absence of TCP (or other potential pesticides that may cause any interference). Neither of the extracts produced an analytical signal. Then, two calibration graphs were prepared for each sample, using external calibration and standard-addition methodology. Some of the samples presented a matrix effect; in this case, standard addition was used for the quantitation of TCP. As none of the extracts had TCP, recovery experiments were carried out to assess the accuracy of the analytical method. Taking into account TCP

MRL in lettuce samples, this study was performed by spiking the samples at analyte levels between 0.8 and 10 mg kg^{-1} (spiking was done previously to QuEChERS extraction).

Recovery yields varied between 91% and 108% in all cases (Table 4), with relative standard deviations (RSD; n = 3) lower than 5%, confirming the accuracy and precision of the proposed method. We also checked the accuracy by the method of the average recovery [30], obtaining an experimental t value of 0.093, lower than the tabulated t value at a 95% confidence level.

Table 4. Recovery study of TCP in lettuce samples.

Sample	Spiked (mg kg^{-1})	Found (mg kg^{-1})	Recovery ± RSD (%)
Iceberg lettuce-1	1	0.99 ± 0.02	99 ± 2
	2	2.12 ± 0.06	106 ± 3
	4	4.32 ± 0.08	108 ± 2
Iceberg lettuce-2	0.8	0.83 ± 0.03	104 ± 4
	5	5.25 ± 0.1	105 ± 3
	8	7.8 ± 0.3	97 ± 4
Baby Romaine lettuce-1	1	0.99 ± 0.03	99 ± 3
	2	1.82 ± 0.07	91 ± 4
	4	4.1 ± 0.1	102 ± 3
Baby Romaine lettuce-2	1	0.93 ± 0.03	93 ± 3
	4	3.9 ± 0.1	98 ± 4
	6	6.1 ± 0.2	102 ± 3
Green oak leaf lettuce-1	0.8	0.74 ± 0.03	93 ± 4
	3	3.15 ± 0.09	105 ± 3
	6	5.8 ± 0.2	96 ± 3
Green oak leaf lettuce-2	1	0.98 ± 0.02	98 ± 2
	4	3.8 ± 0.1	96 ± 3
	7	7.4 ± 0.3	106 ± 4

4. Conclusions

MCFIA-SPS-PIF implementation is a very attractive and fruitful research field. In this work, a novel and sensitive optosensor for the quantitation of TCP in vegetables was developed by using QuEChERS for sample pretreatment. The online photodegradation of TCP takes place, followed by the preconcentration and detection of its fluorescent photoproduct on a solid support placed in the flow cell (detection area). The high selectivity obtained is the result of the use of this solid support, as well as the derivatization of the analyte, allowing the fulfillment of the MRL of the European Union for TCP in lettuce.

Chromatographic methods are usually employed for the determination of TCP, but they require many cleanup steps and expensive instruments. However, this study's results show that flow-through optosensors are a good option for the analysis of neonicotinoids in food. Recoveries close to 100% are obtained in all cases. The simplicity, robustness and high sample frequency of the method developed makes it an interesting prescreening alternative.

Author Contributions: Conceptualization, E.J.L.-M. and A.R.-M.; methodology, J.J.-L. and A.R.-M.; software, J.J.-L. and S.M.-S.; validation, J.J.-L. and S.M.-S.; formal analysis, J.J.-L. and E.J.L.-M.; investigation, J.J.-L. and A.R.-M.; resources, J.J.-L. and E.J.L.-M.; data curation, E.J.L.-M.; writing—original draft preparation, E.J.L.-M. and A.R.-M.; writing—review and editing, A.R.-M.; visualization, A.R.-M.; supervision, E.J.L-M. and A.R.-M.; project administration, A.R.-M.; funding acquisition, A.R.-M.

Funding: This study was funded by the Ministerio de Economía y Competitividad (grant number CTQ2016-7511-R).

Conflicts of Interest: The authors declare no conflict of interest.

References

1. Bass, C.; Denholm, I.; Williamson, M.S.; Nauen, R. The global status of insect resistance to neonicotinoid insecticides. *Pestic. Biochem. Physiol.* **2015**, *121*, 78–87. [CrossRef] [PubMed]
2. Blacquière, T.; Smagghe, G.; van Gestel, C.A.M.; Mommaerts, V. Neonicotinoids in bees: A review on concentrations, side-effects and risk assessment. *Ecotoxicology* **2012**, *21*, 973–992. [CrossRef] [PubMed]
3. Butler, D. EU expected to vote on pesticide ban after major scientific review. *Nature* **2018**, *55*, 150–151. [CrossRef] [PubMed]
4. Abdel-Ghany, M.F.; Hussein, L.A.; El Azab, N.F.; El-Khatib, A.H.; Linscheid, M.W. Simultaneous determination of eight neonicotinoid insecticide residues and two primary metabolites in cucumbers and soil by liquid chromatography–tandem mass spectrometry coupled with QuEChERS. *J. Chromatogr. B* **2016**, *1031*, 15–28. [CrossRef] [PubMed]
5. Jiao, W.; Xiao, Y.; Qian, X.; Tong, M.; Hu, Y.; Hou, R.; Hua, R. Optimized combination of dilution and refined QuEChERS to overcome matrix effects of six types of tea for determination eight neonicotinoid insecticides by ultra performance liquid chromatography-electrospray tandem mass spectrometry. *Food Chem.* **2016**, *210*, 26–34. [CrossRef] [PubMed]
6. Wang, F.; Li, S.; Feng, H.; Yang, Y.; Xiao, B.; Chen, D. An enhanced sensitivity and cleanup strategy for the nontargeted screening and targeted determination of pesticides in tea using modified dispersive solid-phase extraction and cold-induced acetonitrile aqueous two-phase systems coupled with liquid chromatography-high resolution mass spectrometry. *Food Chem.* **2019**, *275*, 530–538. [PubMed]
7. Lu, Z.; Fang, N.; Zhang, Z.; Wang, B.; Hou, Z.; Li, Y. Simultaneous determination of five neonicotinoid insecticides in edible fungi using ultrahigh-performance liquid chromatography-tandem mass spectrometry (UHPLC-MS/MS). *Food Anal. Methods* **2018**, *11*, 1086–1094. [CrossRef]
8. Martínez-Domínguez, G.; Nieto-García, A.J.; Romero-González, R.; Frenich, A.G. Application of QuEChERS based method for the determination of pesticides in nutraceutical products (Camellia sinensis) by liquid chromatography coupled to triple quadrupole tandem mass spectrometry. *Food Chem.* **2015**, *177*, 182–190. [CrossRef] [PubMed]
9. Han, Y.; Zou, N.; Song, L.; Li, Y.; Qin, Y.; Liu, S.; Li, X.; Pan, C. Simultaneous determination of 70 pesticide residues in leek, leaf lettuce and garland chrysanthemum using modified QuEChERS method with multi-walled carbon nanotubes as reversed-dispersive solid-phase extraction materials. *J. Chromatogr. B* **2015**, *1005*, 56–64. [CrossRef] [PubMed]
10. Konatu, F.R.B.; Breitkreitz, M.C.; Jardim, I.C.S.F. Revisiting quick, easy, cheap, effective, rugged, and safe parameters for sample preparation in pesticide residue analysis of lettuce by liquid chromatography–tandem mass spectrometry. *J. Chromatogr. A* **2017**, *1482*, 11–22. [CrossRef] [PubMed]
11. Hanot, V.; Goxcinny, S.; Deridder, M. A simple multi-residue method for the determination of pesticides in fruits and vegetables using a methanolic extraction and ultra-high-performance liquid chromatography-tandem mass spectrometry: Optimization and extension of scope. *J. Chromatogr. A* **2015**, *1384*, 53–66. [CrossRef] [PubMed]
12. Konatu, F.R.B.; Jardim, I.C.S.F. Development and validation of an analytical method for multiresidue determination of pesticides in lettuce using QuEChERS-UHPLC-MS/MS. *J. Sep. Sci.* **2018**, *41*, 1726–1733. [CrossRef] [PubMed]
13. Li, Z.; Yu, Y.; Li, Z. A review of biosensing techniques for detection of trace carcinogen contamination in food products. *Anal. Bioanal. Chem.* **2015**, *407*, 2711–2726. [CrossRef] [PubMed]
14. Brycht, M.; Vajdle, O.; Papp, Z.; Guzsvány, V.; Obradovića, T.D. Renewable silver-amalgam film electrode for direct cathodic SWV determination of clothianidin, nitenpyram and thiacloprid neonicotinoid insecticides reducible in a fairly negative potential range. *Int. J. Electrochem. Sci.* **2012**, *7*, 10652–10665.
15. Zhang, S.; Yang, X.; Yin, X.; Wang, C.; Wang, Z. Dispersive liquid-liquid microextraction combined with sweeping micellar electrokinetic chromatography for the determination of some neonicotinoid insecticides in cucumber samples. *Food Chem.* **2012**, *133*, 544–550. [CrossRef] [PubMed]
16. Liu, Z.; Li, M.; Shi, H. Development and evaluation of an enzyme-linked immunosorbent assay for the determination of thiacloprid in agricultural samples. *Food Anal. Methods* **2013**, *6*, 691. [CrossRef]
17. Liu, Z.; Yan, X.; Hua, X.; Wang, M. Time-resolved fluoroimmunoassay for quantitative determination of thiacloprid in agricultural samples. *Anal. Methods* **2013**, *5*, 3572–3576. [CrossRef]

18. Llorent-Martínez, E.J.; Soler-Gallardo, M.I.; Ruiz-Medina, A. Determination of thiacloprid, thiamethoxam and imidacloprid in tea samples by quenching terbium luminescence. *Luminescence* **2019**, *34*, 460–464. [CrossRef] [PubMed]
19. Liu, Y.; Cao, N.; Gui, W.; Ma, Q. Nitrogen-doped graphene quantum dots-based fluorescence molecularly imprinted sensor for thiacloprid detection. *Talanta* **2018**, *183*, 339–344. [CrossRef] [PubMed]
20. Catalá-Icardo, M.; López-Paz, J.L.; Pérez-Plancha, L.M. Fast determination of thiacloprid by photoinduced chemiluminescence. *Appl. Spectrosc.* **2014**, *68*, 642–648. [CrossRef] [PubMed]
21. Llorent-Martínez, E.J.; Ortega-Barrales, P.; Fernández-de Córdova, M.L.; Ruiz-Medina, A. Contribution to automation for determination of drugs based on flow-through optosensors. *App. Spectrosc. Rev.* **2011**, *46*, 339–367. [CrossRef]
22. Llorent-Martínez, E.J.; Jiménez-López, J.; Fernández-de Córdova, M.L.; Ortega-Barrales, P.; Fernández-de Córdova, M.L.; Ruiz-Medina, A. Quantitation of hydroxytirosol in food prodcuts using a sequential injection analysis fluorescence oprtosensor. *J. Food Comp. Anal.* **2013**, *32*, 99–104. [CrossRef]
23. Jiménez-López, J.; Ortega-Barrales, P.; Ruiz-Medina, A. Determination of clothianidin in food products by using an automated system with photochemically induced fluorescence detection. *J. Food Comp. Anal.* **2016**, *49*, 49–56. [CrossRef]
24. The European Comission, EU Pesticide Database. Available online: http://ec.europa.eu/food/plant/pesticides/eu-pesticides-database/public/?event=homepage&language=EN (accessed on 1 September 2019).
25. Jiménez-López, J.; Ortega-Barrales, P.; Ruiz-Medina, A. A photochemically induced fluorescence based flow-through optosensor for screening of nitenpyram residues in cruciferous vegetables. *Food Addit. Contam. Part. A* **2018**, *35*, 941–949. [CrossRef] [PubMed]
26. Jeria, Y.; Bazaes, A.; Báez, M.E.; Espinoza, J.; Martínez, J.; Fuentes, E. Photochemically induced fluorescence coupled to second-order multivariate calibration as analytical tool for determining imidacloprid in honeybees. *Chemom. Intell. Lab. Syst.* **2017**, *160*, 1–7. [CrossRef]
27. Lu, Z.; Challis, J.K.; Wong, C.S. Quantum yields for direct photolysis of neonicotinoid insecticides in water: Implications for exposure to nontarget aquatic organisms. *Environ. Sci. Technol. Lett.* **2015**, *2*, 188–192. [CrossRef]
28. Ruiz-Medina, A.; Soler-Gallardo, M.I.; Llorent-Martínez, E.J. Enhanced quenching effect of neonicotinoid pesticides on time-resolved terbium luminescence in presence of surfactants. *J. Chem.* **2018**. [CrossRef]
29. Skovgaard, M.; Encinas, S.R.; Jensen, O.C.; Andersen, J.H.; Condarco, G.; Jørs, E. Pesticide residues in commercial lettuce, onion, and potato samples from Bolivia–A threat to public health? *Environ. Health Insights* **2017**, *11*, 1–8. [CrossRef] [PubMed]
30. González, A.G.; Herrador, M.A.; Asuero, A.G. Intra-laboratory testing of method acccuracy from recovery assays. *Talanta* **1999**, *48*, 729–736. [CrossRef]

© 2019 by the authors. Licensee MDPI, Basel, Switzerland. This article is an open access article distributed under the terms and conditions of the Creative Commons Attribution (CC BY) license (http://creativecommons.org/licenses/by/4.0/).

Article

Dual-Purpose Photometric-Conductivity Detector for Simultaneous and Sequential Measurements in Flow Analysis

Thitirat Mantim [1,2,3,*], Korbua Chaisiwamongkhol [1,4,5], Kanchana Uraisin [1,3], Peter C. Hauser [6], Prapin Wilairat [1,7] and Duangjai Nacapricha [1,3,*]

1. Flow Innovation-Research for Science and Technology Laboratories (FIRST Labs), Bangkok 10400, Thailand; korbua.cha@mfu.ac.th (K.C.); kanchana.ura@mahidol.ac.th (K.U.); prapin.wil@mahidol.ac.th (P.W.)
2. Department of Chemistry, Faculty of Science, Srinakharinwirot University, Sukhumwit 23 Road, Bangkok 10110, Thailand
3. Center of Excellence for Innovation in Chemistry and Department of Chemistry, Faculty of Science, Mahidol University, Rama 6 Road, Bangkok 10400, Thailand
4. School of Science, Mae Fah Luang University, Chiang Rai 57100, Thailand
5. Center of Chemical Innovation for Sustainability (CIS), Mae Fah Luang University, Chiang Rai 57100, Thailand
6. Department of Chemistry, University of Basel, Klingelbergstrasse 80, 4056 Basel, Switzerland; peter.hauser@unibas.ch
7. National Doping Control Centre, Mahidol University, Rama 6 Road, Bangkok 10400, Thailand
* Correspondence: thitiratm@g.swu.ac.th (T.M.); duangjai.nac@mahidol.ac.th (D.N.)

Academic Editors: Pawel Koscielniak and Dimosthenis Giokas
Received: 3 April 2020; Accepted: 7 May 2020; Published: 13 May 2020

Abstract: This work presents a new dual-purpose detector for photometric and conductivity measurements in flow-based analysis. The photometric detector is a paired emitter–detector diode (PEDD) device, whilst the conductivity detection employs a capacitively coupled contactless conductivity detector (C4D). The flow-through detection cell is a rectangular acrylic block (ca. 2 × 2 × 1.5 cm) with cylindrical channels in Z-configuration. For the PEDD detector, the LED light source and detector are installed inside the acrylic block. The two electrodes of the C4D are silver conducting ink painted on the PEEK inlet and outlet tubing of the Z-flow cell. The dual-purpose detector is coupled with a sequential injection analysis (SIA) system for simultaneous detection of the absorbance of the orange dye and conductivity of the dissolved oral rehydration salt powder. The detector was also used for sequential measurements of creatinine and the conductivity of human urine samples. The creatinine analysis is based on colorimetric detection of the Jaffé reaction using the PEDD detector, and the conductivity of the urine, as measured by the C4D detector, is expressed in millisiemens (mS cm^{-1}).

Keywords: paired emitter–detector diode detector; contactless conductivity detector; flow-based analysis; simultaneous detection; sequential detection

1. Introduction

Flow injection analysis (FIA) [1,2] and its later generation techniques, including sequential injection analysis (SIA) [3], lab-on-valve (LOV) [4], lab-at-valve (LAV) [5], all injection analysis (AIA) [6], simultaneous injection effective mixing flow analysis (SIEMA) [7,8] and cross injection analysis (CIA) [9,10], are effective techniques that have been used as tools for liquid handling in automated analysis. Sample introduction, reagent and sample mixing, detection and rinsing of the flow-path in every cycle are operated by computer technology. Thus, sample throughputs are up to 60 samples per hour [1,11–13]. Almost all kinds of detection can be coupled to the above flow-based techniques, including UV–Vis absorbance [14–22], fluorescence [22–24], chemiluminescence [25,26], electrochemical detections (amperometry) [27–30], conductivity [31] and contactless conductivity [32–37].

Some flow-based systems are equipped with two or more different types of detector in the same system, depending upon the purpose of analysis [38]. Detectors are arranged either in series [39–43] or in parallel [21,34,37,44]. The aim for integrating more than one detector is for multi-component analysis. For the "in-series" configuration, detectors are located along the same flow-path or in the same flow cell to accomplish sequential detections. An example of the "in-series detection" is multi-detection of different ionic species using several ion-selective electrodes but with a single reference electrode [40]. Another example of the "in-series" detection is a flow-based system that consists of an optical detection (UV–Vis absorbance) together with an amperometric detection for the analysis of total polyphenol content and antioxidant activity, respectively [39].

Unlike the "in-series" configuration, detectors with "in-parallel" configuration are placed in different flow paths. Usually, the sample is injected from the same port and the zone is split into fractions. Each of the zone-fractions is then transported into separate flowing streams for reaction (if needed), mixing (if needed) and detection. An example of the "in-parallel" configuration is the simultaneous determination of urea and creatinine in human urine using an FIA system equipped with a contactless conductivity detector (C4D) and a colorimeter [21]. Salinity, carbonate and ammoniacal nitrogen in water samples were simultaneously measured by FIA using a dual C4D detection cell arranged in an "in-parallel" configuration [34]. Salinity and carbonate were sequentially detected in C4D channel 1, and ammonical ammonia was detected after gas diffusion in the second C4D. Liquid handling by an automated SIA system is also suitable for "in-parallel" detection. Parallel analyses of sugar, colour and dissolved CO_2 contents in soft drinks can be carried out by connecting a near infrared LED photometer and a contactless conductivity detector using an "in- parallel" configuration to completely automate flow control by SIA [44].

In this work we developed a flow cell that has dual detection capability by combining optical detection with a contactless conductivity detection in a single flow cell. The photometric detection employs paired emitter–detector diodes (PEDD), as reported by Tymecki et al. [45]. LEDs are utilized as both light sources and light detectors. The voltage of the detector LED varies with the logarithm of the intensity of the incident light. The measured voltage therefore decreases linearly with the absorbance of the solution. The liquid flow cell was fabricated from an acrylic block with a Z-shape fluidic channel. The PEEK (poly ether ether ketone) tubing attached at the inlet and outlet of the Z-shape channel is painted with silver conducting paint to form two electrode bands on the exterior wall of the tube. These two silver bands are the electrodes of the contactless conductivity detector [46,47]. An alternating voltage is applied to the electrode at the input end of the flow cell. When the channel between the two silver bands is filled with a conducting aqueous solution an AC current is generated. This current is monitored at the outlet electrode band. The amplitude of the alternating current monitored corresponds to the sample conductivity for fixed applied AC voltage and frequency. By this arrangement, it is possible to simultaneously measure the absorbance and conductivity of a solution in the flow cell.

To the best of our knowledge, this work presents the first flow cell that provides dual photometric-conductivity detection for an FIA-based system. The dual detection approach is different from the "in-series" and "in-parallel" configurations. In dual detection the detectors are combined within the same flow cell (2-in-1). C4D has previously been combined with optical detection methods, such as with UV–Vis [48,49], fluorescence [50–52] or UV/VIS/fluorescence [53]. However, these dual detections with C4D were all developed for and applied to capillary electrophoresis (CE) [48–50,53] or to microchip CE [51,52]. C4D has also been combined with amperometric detection to constitute dual detection for CE [54]. However, PEDD has never been employed in combination with C4D for dual detection in either CE or FIA-based analysis. This work presents a combination of PEDD and C4D for a simple and low-cost flow-through cell for simultaneous photometric-conductometric detection. The dual-purpose detection cell was employed with a SIA system. The first application was the simultaneous analysis of conductivity of dissolved oral rehydration salts (ORS) and the absorbance of the orange dye in the product. These data are useful for checking the quantity of electrolytes and colorant in samples of a manufacturer for consumer protection. The same flow cell was also employed for the measurements of conductivity and determination of creatinine in human urine. Measurement of urine conductivity is a fast and convenient way to ascertain the hydration status of an athlete. Urine creatinine concentration is one of the commonly required parameters for medical diagnosis. Additionally, the amount of creatinine in urine is used to correct for variation in the volume of spot urine collected for measurements of excreted drugs.

2. Results and Discussion

2.1. Design of the Detector

The dual-purpose detector was designed in a "2-in-1" arrangement (see Figure 1a). The electrodes E1 and E2 of the C4D are positioned at the inlet and at the outlet of the flow cell, respectively. The PEDD is used as the photometric detector for monitoring absorbance of the solution and the C4D detects the change in conductivity of the same solution passing through the flow cell. The detector is connected to a SIA system as shown in Figure 2. Upon injecting a sample solution of NaCl and orange food colorant into the SIA system, the signals of both the C4D and PEDD appear at the same time (see Figure 3). Figure 1b shows the photograph of the dual-purpose detector used throughout the work.

2.2. Optimization of Frequency and Voltage for the C4D

Since the configuration of the flow cell for PEDD is sufficient for this work using matching LEDs and path length of 10 mm, optimization experiments for the dual C4D-PEDD flow cell focused on the parameters affecting C4D detection. The frequency of a sinusoidal AC voltage applied to electrode E1 was varied from 100 to 500 kHz. As expected, raising the frequency from 100 to 500 kHz increased the sensitivity (slope of calibration line) of the C4D detection. At a constant voltage of 2 V_{pp} (V_{pp}: peak-to-peak voltage), it was observed that while increasing the frequency from 100 to 300 kHz, the linearity range widened from 1–25 mM to 1–50 mM NaCl, respectively. The linearity ranges for 400 and 500 kHz were narrower than for 300 kHz (10–50 mM NaCl compared to 1–50 mM NaCl). However, we chose the frequency of 500 kHz, since it provided the highest sensitivity. The input voltage was then changed to 10 V_{pp}. It was observed that using this condition (500 kHz and 10 V_{pp}), a linear range was obtained that was suitable for the samples, and it had sufficient sensitivity. The calibration is linear with coefficient of determination (r^2) of 0.999. Thus, for this study, the frequency and voltage at 500 kHz and 10 V_{pp} were employed.

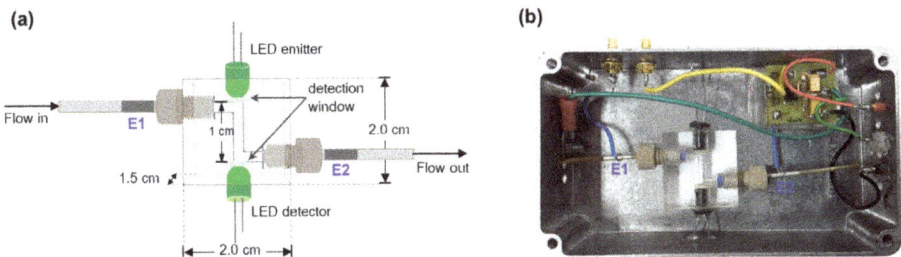

Figure 1. (a) Schematics of C4D-PEDD flow-through detection cell for the "2-in-1" detector for simultaneous detection. (b) Photograph of the "2-in-1" C4D-PEDD flow-through detection cell and the pre-amplifier board.

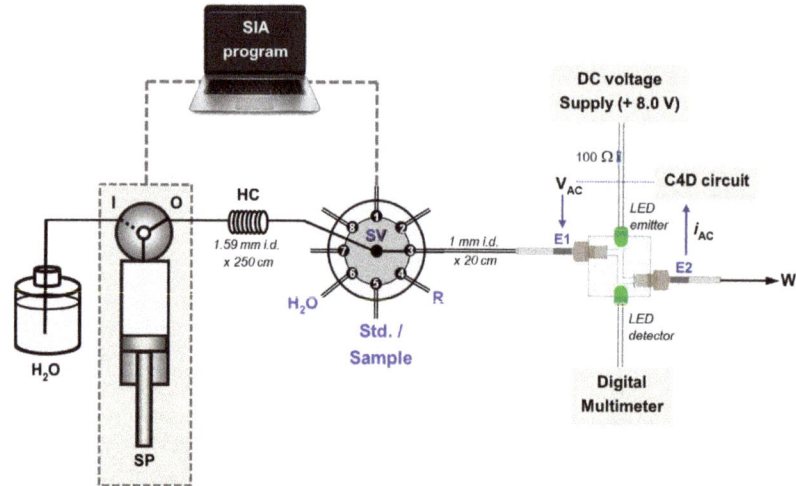

Figure 2. Schematic diagram of the SIA system coupled with the "2-in-1" flow cell for dual C4D-PEDD detection. Note: For the analysis of oral rehydration salt (ORS) samples, port 4 (labelled R) is not used. Port 4 is used as the inlet for the alkaline picrate reagent (R) in the sequential measurements of creatinine and conductivity of urine samples. SV—switching valve, SP—syringe pump, HC—holding coil, W—waste.

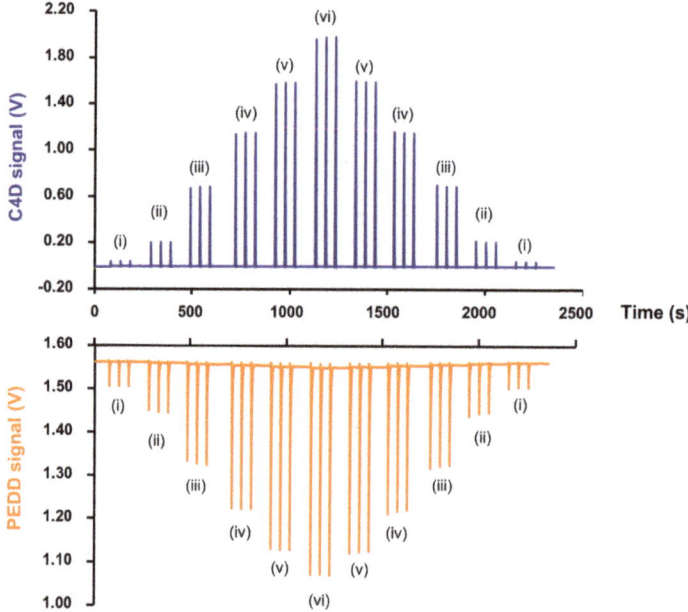

Figure 3. The series of signals recorded simultaneously by the C4D (top) and PEDD (bottom) detectors for triplicate injections of aqueous standard solutions in increasing order of concentrations and then in decreasing order, as carried out using SIA system (see Figure 2). The concentration of NaCl and the absorbance at 525 nm of orange dye in each standard solution: (i) 5 mM and 0.1124, (ii) 10 mM and 0.2228, (iii) 20 mM and 0.4466, (iv) 30 mM and 0.6657, (v) 40 mM and 0.8844 and (vi) 50 mM and 1.0332.

2.3. Quantification of Salts and Dye in ORS Samples

As shown in Figure 2, the dual detection cell is connected to a sequential injection analysis (SIA) system for simultaneous measurements of conductivity and absorbance at 525 nm. The sample/standard solution (100 µL) is aspirated into the holding coil (HC) and then propelled to the flow cell by the water carrier. There is simultaneous recording of the C4D and PEDD signals (see Figure 3).

Calibration was performed by triplicate injections of a series of six aqueous solutions of NaCl and pure orange dye. The concentrations of NaCl and the colored dye were first increased from the lowest to highest concentrations (see details of the concentrations in the caption of Figure 3) and then in decreasing order. Figure 3 shows the peak profiles for the 33 consecutive injections. It should be noted that the C4D signal increases with increasing concentration of NaCl solution, whereas the PEDD signal decreases with increasing absorbance of the dye (as described in [45]). The profiles did not show any signs of carryover for both C4D and PEDD detectors, when the order of injections was from high to low concentrations, respectively.

In order to convert the measured C4D signal into conductivity unit, i.e., mS cm^{-1}, the conductivity of a set of saline solutions (with concentrations covering the range employed in the C4D calibration) was measured using a commercial conductivity probe. The calibrating equation is y(mS cm^{-1}) = $(0.094 \pm 0.006)x + (0.18 \pm 0.06)$, $r^2 = 0.9996$, using 10.0–50.0 mM standard NaCl solutions. From this equation the C4D signal can be converted to conductivity unit, mS cm^{-1}. The plot of the C4D signal (V) against conductivity is shown in Figure S1a. The linear calibration equation is y(V) = $(0.47 \pm 0.01)x - (0.27 \pm 0.02)$, $r^2 = 0.9993$. The slopes and intercepts of the calibration lines obtained with the data for increasing concentrations and the data for decreasing concentrations are not statistically different, as shown in Figure S2a,b, respectively.

The linear calibration of the PEDD is the plot of the negative of the difference ($-\Delta V$) of the signal minimum (in volts, V) from the baseline value (see Figure 3) against the absorbance (au, at 525 nm) of the standard solutions of the orange dye (measured previously with a spectrophotometer). A typical calibration equation is: $-\Delta V = ((4.86 \pm 0.12) \times 10^{-1}) \cdot au + ((0.92 \pm 0.66) \times 10^{-2})$, $r^2 = 0.998$. Similarly to the C4D measurements, the slopes and intercepts of the calibration lines obtained with the data for increasing concentrations and that for decreasing concentrations are not statistically different, as shown in Figure S2c,d, respectively.

Sixteen sachets of ORS powder products with orange flavoring were employed in the analysis: eight sachets of "Brand A" for adults (A1–A8), and eight sachets of "Brand B" for children under 6 years old (B1–B8). The amount of electrolyte salts in a sachet for children is approximately one half of the amount provided for adults (17% (w/w) compared to 34% (w/w)). The powder in each sachet of "Brand A" was totally dissolved in 150.0 mL DI water, whereas the content of "Brand B" was dissolved in 120.0 mL DI water. These volumes were the recommended volumes of water, as given on the sachets. A 100 µL aliquot of each sample was aspirated into the SIA system (see Figure 2). The measured C4D signals were converted into mS cm^{-1}, using the calibration line in Figure S1a. The PEDD signals were similarly converted to absorbance units, and indicated the contents of the orange colorants in the ORS powders. The results are shown in Figure 4a,b, respectively.

The bar graphs show the mean values of the C4D measurements, converted to mS cm^{-1} unit, for these two products with values in red text together with their standard deviations. The small variation of the conductivity and the absorbance about their mean values (within ± 2SD) is a measurement of the consistency of the contents of each sachet, ensuring consumer protection.

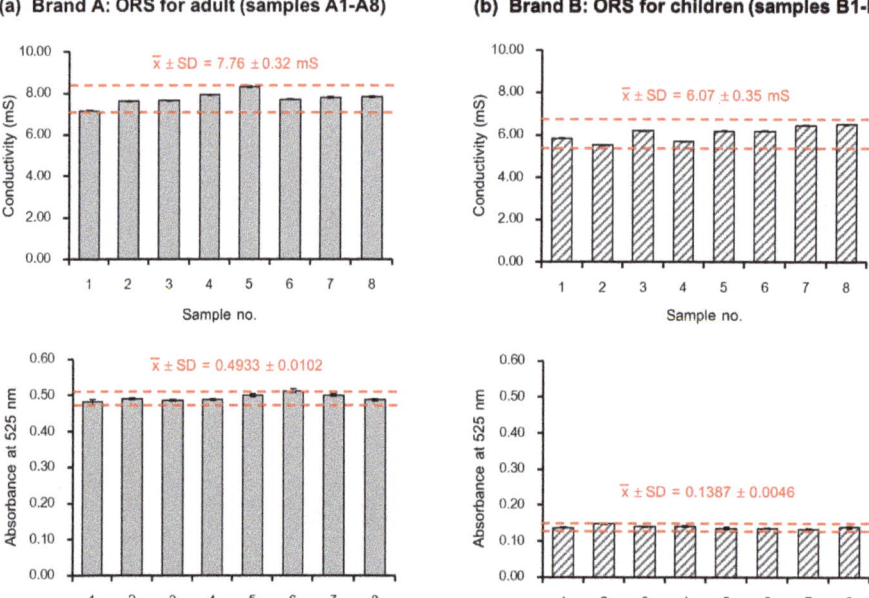

Figure 4. Bar graphs of the mean conductivity (mS cm^{-1}) of each sachet (dose) and the mean absorbances of 150.0 mL solutions for Brand A or 120.0 mL solutions of Brand B, with error bars ($n = 3$). The red text shows the means and standard deviations for samples A1–A8 and for samples B1–B8, respectively. The red dotted lines indicate the range for the mean ± 2SD of these two products. (**a**) Eight samples from "Brand A" for adults, labelled net weight of 3.645 g. (**b**) Eight samples from "Brand B" for children, labelled net weight of 3.3 g.

2.4. Determination of Creatinine and Measurement of Conductivity of Urine Samples

The electrical conductivity of urine is due to ionic species in urine [55]. Measurement of urine conductivity is a fast and convenient way to ascertain the hydration statuses of athletes [56] and to evaluate the diuretic effects of medicinal administration [57]. Determination of creatinine concentration in urine is one of the commonly required parameters for medical diagnosis. The amount of creatinine in urine is also used to correct for variation in the volume of spot urine collected for measurements of excreted drugs.

2.4.1. Zone Sequence and Signal Profiles

The SIA system in Figure 2 was employed for the sequential measurements of creatinine and the conductivity of human urine. For this application, port 4 of the selection valve was connected to the reagent reservoir containing the Jaffé reagent of alkaline picrate. This reagent is used for colorimetric detection of creatinine with PEDD. Similarly to the previous application of ORS samples, C4D was used to directly measure the conductivity of urine.

Preliminary experiments were carried out (data not shown) to optimize the zone sequence required for sequential detection of the urine conductivity followed by the colorimetric measurement of the complex formed between creatinine and the Jaffé reagent. The final selected zone sequence is shown in Figure 5a. The first sequence, designated Sequence 1, consists of three liquid segments; i.e., H$_2$O carrier (in the flow line)|urine (150 µL)|water (1000 µL). Sequence 1 contains the plug of urine which is propelled to the flow cell. Sequence 2, directly following Sequence 1, consists of four liquid segments; i.e., urine (100 µL)|Jaffe reagent R (350 µL)|urine (100 µL)|H$_2$O (3000 µL).

An example of the signal profile recorded from the C4D detector is shown in Figure 5b. The peak labelled "Con" corresponds to the passage of the urine segment (150 µL) of Sequence 1. The large signal following it is the large conductivity of the mixed three segments in "Sequence 2"—viz., urine (100 µL) |R (350 µL) | urine (100 µL), which is not used in the analysis of the C4D data. When all liquid zones comprising Sequence 1 have moved out of the flow cell, the signal (labelled as "Cre") for the product of the reaction of the two urine plugs with the plug of the Jaffé reagent is seen via the PEDD channel, as shown in Figure 5c. Note that there is no absorbance signal for the urine zone of the leading Sequence 1.

It was necessary to optimize the volume of the water segment (labelled as H_2O 1000 µL in Sequence 1) separating the first urine sample plug from the urine segments for the creatinine measurements. When using a segment of water less than 1000 µL, the two signal peaks were not baseline separated. For photometric measurements, the schlieren effect may occur at the interface of the sample zone and the water carrier due to differences in the refractive indices [19,44,58,59]. This was also observed with the PEDD signal for some urine samples. For these samples, a small sharp peak was observed in front of the large creatinine peak (see examples in Figure S3 in the Supplementary materials). Since the peak height of the PEDD signal was employed, the schlieren signal did not affect the analysis.

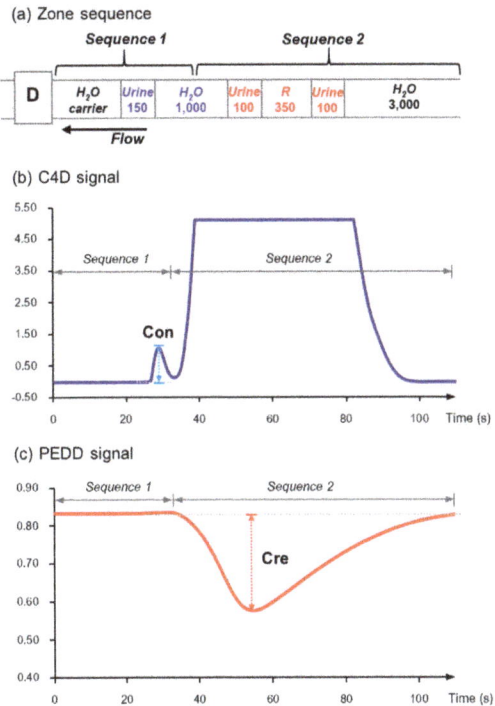

Figure 5. Measurements of urine conductivity and determination of creatinine using SIA equipped with the "2-in-1" C4D-PEDD detector. (**a**) Zone sequence of sample and reagent segments with aspirated volumes in microliters (D: detector). (**b**) Peak profile of urine conductivity (Con) from C4D. The large peak is due to the conductivity of the ions in the mixed zones of Sequence 2. (**c**) Peak profile of creatinine signal (Cre) from PEDD.

2.4.2. Analysis of Human Urine Samples

Employing the optimized sequence of sample and reagent introduction, standard solutions containing NaCl and creatinine (at concentrations covering the range of values expected for the diluted urine samples) were measured. It should be noted that the calibration equations for the C4D are not the same for the ORS and urine measurements (see Figure S1) due to the different volumes of sample aspirated into the flow-line (100 µL for ORS and 150 µL for urine) leading to different dispersion of the sample plug [60].

For PEDD, the negative of the difference ($-\Delta V$) values of the signal minimum (in volts, V) from the baseline value ("Cre" in Figure 5c) were plotted against the concentrations of standard creatinine. The calibration of PEDD was $y = ((2.06 \pm 0.01) \times 10^{-3})x + ((1.37 \pm 0.01) \times 10^{-1})$: $r^2 = 0.999$. The C4D and PEDD calibration equations were used to calculate urine conductivity and the concentration of creatinine in urine, respectively. The results are shown in Table 1.

The determination of creatinine using the SIA flow-cell was also compared with a batch method employing the Jaffé reaction [61–63]. Data in Table 1 show good agreement between the two methods, as confirmed by the statistical paired t-test (t_{stat} = 2.09, t_{crit} = 2.45).

Table 1. Analysis of conductivity and creatinine of human urine samples.

Sample	Conductivity (mS cm^{-1})	Creatinine (mg L^{-1}) *	
	C4D Detector (n = 3)	PEDD Detector (n = 3)	Spectrophometer (n = 3)
S1	19.18 ± 0.03	(1.06 ± 0.02) × 10^3	(1.08 ± 0.01) × 10^3
S2	27.10 ± 0.12	(1.40 ± 0.02) × 10^3	(1.41 ± 0.03) × 10^3
S3	26.57 ± 0.05	(1.54 ± 0.01) × 10^3	(1.56 ± 0.02) × 10^3
S4	31.60 ± 0.01	(1.17 ± 0.02) × 10^3	(1.14 ± 0.02) × 10^3
S5	12.42 ± 0.02	(7.36 ± 0.05) × 10^2	(7.13 ± 0.07) × 10^2
S6	10.07 ± 0.02	(5.60 ± 0.05) × 10^2	(5.06 ± 0.05) × 10^2
S7	4.99 ± 0.02	(1.52 ± 0.04) × 10^2	(1.22 ± 0.01) × 10^2
S8	11.96 ± 0.06	(4.39 ± 0.06) × 10^2	(4.09 ± 0.06) × 10^2

* Paired t-test: t_{stat} = 2.09, t_{crit} = 2.45.

2.4.3. Analytical Features

The analysis is simple with only dilution of urine with water (10-fold or 20-fold). The sample can be directly injected into the SIA system shown in Figure 2 and the data obtained within 2 min for a single injection. Linearity ranges are 0.37–3.93 mS cm^{-1} for the C4D and 5–100 mg L^{-1} for creatinine, respectively. Limits of quantitation (LOQs) for conductivity and creatinine were found to be 0.33 mS cm^{-1} (10 S/N) and 2.38 mg L^{-1} (10 SD intercept/slope), respectively. Compared to other Jaffé methods operated using SIA [64] or cross injection analysis (CIA) [10] flow platforms, this method gave a significantly lower LOQ for creatinine (2.38 mg L^{-1} compared with 11.7 mg L^{-1} and 20 mg L^{-1} for the other SIA and CIA flow platforms, respectively). Ten replicate analyses of a standard mixture of 10 mM NaCl and 10 mg L^{-1} creatinine were carried out to investigate the precision of the method. The flow method gave precisions for conductivity and creatinine of 1.2% RSD and 1.4% RSD, respectively. The precisions in analysis of human urine samples (n = 8) were %RSDs of 0.03–0.50% and 0.68–2.6%, for conductivity and creatinine, respectively. The precision for creatinine is improved from previously reported SIA [64] and CIA [10] methods. There are no other flow-based methods for direct analysis of urine conductivity. We therefore were not able to compare our method in the terms of LOQ and precision. Conductivity detection is integrated in flow cytometer (FC) instruments [65,66]. However, there is no information of LOQ and precision for conductivity, since the major role of FC instruments is for cell counting.

3. Materials and Methods

3.1. Chemicals, Reagents and Samples

All chemicals used in this work were analytical-reagent grade. Deionized (DI) Milli-Q® water was used throughout.

3.1.1. Preparation of Standard Solutions and Oral Rehydration Salt Samples

For the analysis of an oral rehydration salt (ORS) sample, a stock solution of 1.00 M NaCl was prepared using NaCl crystals (Fluka, Switzerland). A stock solution of Sunset Yellow was prepared by diluting a 220 µL aliquot of the liquid food colorant (Winner's orange color, Greathill Co., Ltd., Bangkok, Thailand) in 100.0 mL of water. The standard working solutions of 5–50 mM NaCl were prepared by adding an appropriate volume of the stock dye solution to obtain the desirable absorbance values of the saline solution (see caption of Figure 3 for the absorbance values). The absorbance was measured on a Lambda 25 UV-Vis spectrophotometer (Perkin Elmer, Waltham, MA, USA) and the conductivity measured using a commercial conductivity probe (Model 145, Thermo Orion, Beverly, MA, USA). These standard solutions were used to construct the calibration graphs for the C4D measurements and the amount of colorant in the ORS product, respectively.

Two brands of local ORS products ("Brand A": 8 samples, and "Brand B": 8 samples) produced by two different pharmaceutical companies were purchased and analyzed for the salt contents. "Brand A" products are for adults, whilst "Brand B" is labelled for children. For a single dose, a patient is recommended to empty the contents of the sachet (3.3 g net weight per sachet) and dissolve them in 150 mL drinking water. "Brand B" ORS product for children (3.645 g net weight per sachet) is recommended to be dissolved in 120 mL of drinking water. These samples were dissolved in DI water, as recommended, and injected into the SIA system shown in Figure 2.

3.1.2. Standard Solutions, Reagents and Samples for Urine Analysis

Stock creatinine solution at 5.00 g L^{-1} was prepared by dissolving 500 mg (to the nearest mg) of anhydrous creatinine (Sigma, St. Louis, MO, USA) in 100.0 mL of water. This stock creatinine solution was mixed with appropriate volumes of the stock standard 1.00 M NaCl solution (Section 3.1.1) to obtain a series of standard solutions for calibration; i.e., 2 mM NaCl+5 mg L^{-1} creatinine, 10 mM NaCl+20 mg L^{-1} creatinine, 30 mM NaCl+50 mg L^{-1} creatinine, 35 mM NaCl+75 mg L^{-1} creatinine and 40 mM NaCl+100 mg L^{-1} creatinine, respectively.

The Jaffé reagent was prepared by mixing 9.6 mL of aqueous saturated solution of picric acid (ca. 52 mM) with 4.0 mL of 2.5 M NaOH and 11.4 mL DI water, to give the alkaline picrate reagent solution (20 mM picrate in 0.4 M NaOH). Urine samples were anonymous samples from the National Doping Control Centre (NDCC), Mahidol University, Thailand. Before analysis, urine samples were diluted at least 10-fold with DI water. The absorbance of the creatinine was measured on a Lambda 25 UV–Vis spectrophotometer (Perkin Elmer, Waltham, MA, USA).

3.2. Flow Cell with Dual PEDD-C4D Detectors

The flow cell was made by drilling a square acrylic block (2 × 2 × 1.5 cm) to contain a cylindrical flow channel (0.3 cm i.d.) in a Z-configuration. The inlet and the outlet of a flow cell were fitted with screw fittings for inserting and connecting the PEEK tubes (0.75 mm i.d.). Silver conducting ink (SPI Supplies, West Chester, PA, USA) was painted on the exterior of these tubes to make electrodes E1 and E2 of the C4D (Figure 1a). The width of an electrode was 1 cm. The length of tubing between the inner edges of electrodes E1 and E2 was 4.5 cm. The C4D detector is an in-house, purpose-built electronic device. A sinusoidal AC voltage (V_{AC}) from the device (XR-2206 monolithic function generator, EXAR Corporation, Fremont, CA, USA) at 500 kHz and 10 V_{pp} was applied to electrode E1. The resulting AC current was measured at the E2 electrode using the pick-up amplifier (OPA655, Burr-Brown, Tucson, AZ, USA), which was connected to a rectifier (AD630, Analog Devices, Norwood, MA, USA) [67]. The final DC output voltage was recorded using an e-corder 210 (eDaq, Denistone East, NSW, Australia). The Z-flow cell and the pre-amplifier were installed inside an aluminum case, as shown in b, for electrical shielding of the C4D detection.

The paired emitter–detector diode (PEDD) detector [45] is integrated into the Z-flow cell for photometric detection. The PEDD detector consists of a pair of green light emitting diodes (InGaN

LED, λ_D 525 nm, ∅ 0.5 cm, Kingbright, New Taipei City, Taiwan). The vertical channel shown in Figure 1a,b is the light path for the optical detection by PEDD. An acrylic disc (∅ 0.5 cm and 0.1 cm thick) is glued at each end of the channel as optical windows and to seal the channel. One of the LEDs is used as the light source, whilst the second LED, placed at the opposite end of the light path, is the light detector. The optical path length of this cell is 1 cm. The LED light source is connected in series to a 100 ohm current limiting resistor and a DC voltage power supply (Model BK-1502D+, Baku, China), set at 8.00 V. A 6-digit multimeter (10 MΩ impedance, Model 8845A, Fluke, Everett, WA, USA) is used to directly measure the output signal from the LED detector to a computer via a RS232 to USB cable, since the input impedance of the e-recorder was too low. LabVIEW 8.0™ software is employed for recording and displaying the PEDD signal on a notebook computer.

3.3. Analysis by SIA with Dual-Detection Flow Cell

The SIA system in Figure 2 was employed for all work. A syringe pump (PSD/4 Hamilton, Reno, NV, USA) with an 8-port selection valve (MVP Hamilton, Reno, NV, USA) was used to control the carrier and reagent flows. Teflon tubes (0.1 cm i.d.) were used as flow lines. The schematics of the C4D and PEDD detectors are depicted in Figure 2. The inlet of flow-through detection cell (Figure 2) was connected to the SIA system via port 3 of the selection valve (SV) using 20-cm PTFE tubing. The MGC-MPV LMPro (version 5.2) software was used for controlling the syringe pump and the selection valve of the SIA system. The flow protocols for the analysis of ORS samples and for urine samples are given in Tables S1 and S2, respectively.

4. Conclusions

We have constructed a flow cell that incorporates a PEDD and C4D as dual detectors that can measure the absorbance and conductivity of a solution simultaneously. This was demonstrated by the measurements of the conductivity and absorbance (at 525 nm) of a solution of dissolved oral rehydration salts (ORS). Eight samples from two commercial brands were analyzed. The results show that there were no statistical differences between the measured conductivities and absorbances of the samples intra-brand. This is a useful method for consumer protection of ORS products.

When the dual-detector is linked to a SIA system, the conductivity of the sample can be directly measured in one flow sequence, whilst the ensuing flow sequence contains the sample that has reacted with a color-forming reagent, and the absorbance is measured in a sequential manner. Thus, two variables of a sample can be measured, such as conductivity of urine and its creatinine content after reacting with the Jaffé reagent added on-line. The conductivity values are useful for monitoring the hydration status of an athlete. Urine creatinine content is an index of various health problems and is employed as the normalization factor in the measurement of excreted drugs in spot urine. The creatinine concentrations were validated against a batch method using the Jaffé reagent and a spectrophotometer.

Supplementary Materials: The following are available online. Figure S1: (**a**) The calibration plot of C4D signal (V) against conductivity (mS cm^{-1}) for the ORS analysis. The equation of the regression line is y(V) = (0.47 ± 0.01)x − (0.27 ± 0.02), r^2 = 0.9993. (**b**) The calibration plot of C4D signal (V) against conductivity (mS cm^{-1}) for urine measurement. The equation of the regression line is y(V) = (1.47 ± 0.02)x − (0.47 ± 0.05), r^2 = 0.9996. It should be noted that the two calibration equations are not the same due to the different volumes of sample aspirated into the flow-line (100 µL for ORS and 150 µL for urine) leading to different dispersions of the sample plug. Figure S2: The slopes and intercepts of the calibration lines from the simultaneous C4D and PEDD measurements of a series of saline solutions containing orange dye. Column "a" is from the consecutive measurements using increasing concentrations of the calibration solutions. Column "b" is from the consecutive measurements using decreasing concentrations of the calibration solutions. Figure S3: Examples of triplicate PEDD signals showing small reproducible schlieren peaks appearing before the creatinine peaks of three urine samples. Table S1: Procedure of the SIA system for the analysis of ORS samples. Table S2: Procedure of the SIA system for the analysis of urine samples.

Author Contributions: Conceptualization, D.N. and T.M.; methodology, T.M. and K.U.; validation, T.M. and K.C.; formal analysis, P.W.; investigation, T.M. and D.N.; resources, D.N.; data curation, K.U.; writing—original draft preparation, T.M.; writing—review and editing, T.M., D.N., P.W. and P.C.H.; visualization: T.M. and D.N.; supervision, D.N.; project administration, D.N.; funding acquisition, D.N. All authors have read and agreed to the published version of the manuscript.

Funding: This research was funded by postdoctoral scholarship of Mahidol University, grant number MU-PD_2013_03.

Acknowledgments: Equipment was partially supported by the Center of Excellence for Innovation in Chemistry (PERCH-CIC), Ministry of Higher Education, Science, Research and Innovation. The authors also wish to thank the five reviewers for their constructive comments.

Conflicts of Interest: The authors declare no conflict of interest.

References

1. Růžička, J.; Hansen, E.H. Flow injection analysis Part I. A new concept of fast continuous flow analysis. *Anal. Chim. Acta* **1975**, *78*, 145–157.
2. Trojanowicz, M.; Kołacińska, K. Recent advances in flow injection analysis. *Analyst* **2016**, *141*, 2085–2139. [CrossRef] [PubMed]
3. Ruzicka, J.; Marshall, G.D. Sequential injection: A new concept for chemical sensors, process analysis and laboratory assays. *Anal. Chim. Acta* **1990**, *237*, 329–343. [CrossRef]
4. Wang, J.; Hansen, E.H. Sequential injection lab-on-valve: The third generation of flow injection analysis. *Trends Anal. Chem.* **2003**, *22*, 225–231. [CrossRef]
5. Grudpan, K. Some recent developments on cost-effective flow-based analysis. *Talanta* **2004**, *64*, 1084–1090. [CrossRef] [PubMed]
6. Itabashi, H.; Kawamoto, H.; Kawashima, T. A novel flow injection technique: All injection analysis. *Anal. Sci.* **2001**, *17*, 229–231. [CrossRef] [PubMed]
7. Teshima, N.; Noguchi, D.; Joichi, Y.; Lenghor, N.; Ohno, N.; Sakai, T.; Motomizu, S. Simultaneous injection-effective mixing analysis of palladium. *Anal. Sci.* **2010**, *26*, 143–144. [CrossRef]
8. Ratanawimarnwong, N.; Ponhong, K.; Teshima, N.; Nacapricha, D.; Grudpan, K.; Sakai, T.; Motomizu, S. Simultaneous injection effective mixing flow analysis of urinary albumin using dye-binding reaction. *Talanta* **2012**, *96*, 50–54. [CrossRef]
9. Nacapricha, D.; Sastranurak, P.; Mantim, T.; Amornthammarong, N.; Uraisin, K.; Boonpanaid, C.; Chuyprasartwattana, C.; Wilairat, P. Cross injection analysis: Concept and operation for simultaneous injection of sample and reagents in flow analysis. *Talanta* **2013**, *110*, 89–95. [CrossRef]
10. Choengchan, N.; Mantim, T.; Inpota, P.; Nacapricha, D.; Wilairat, P.; Jittangprasert, P.; Waiyawat, W.; Fucharoen, S.; Sirankpracha, P.; Phumala Morales, N. Tandem measurements of iron and creatinine by cross injection analysis with application to urine from thalassemic patients. *Talanta* **2015**, *133*, 52–58. [CrossRef]
11. Hansen, E.H.; Miró, M. How flow-injection analysis (FIA) over the past 25 years has changed our way of performing chemical analyses. *Trends Anal. Chem.* **2007**, *26*, 18–26. [CrossRef]
12. Melchert, W.R.; Reis, B.F.; Rocha, F.R.P. Green chemistry and the evolution of flow analysis. A review. *Anal. Chim. Acta* **2012**, *714*, 8–19. [CrossRef] [PubMed]
13. Worsfold, P.J.; Clough, R.; Lohan, M.C.; Monbet, P.; Ellis, P.S.; Quétel, C.R.; Floor, G.H.; McKelvie, I.D. Flow injection analysis as a tool for enhancing oceanographic nutrient measurements–A review. *Anal. Chim. Acta* **2013**, *803*, 15–40. [CrossRef] [PubMed]
14. Yaftian, M.R.; Almeida, M.I.G.S.; Cattrall, R.W.; Kolev, S.D. Flow injection spectrophotometric determination of V(V) involving on-line separation using a poly(vinylidene fluoride-co-hexafluoropropylene)-based polymer inclusion membrane. *Talanta* **2018**, *181*, 385–391. [CrossRef]
15. Ma, J.; Shu, H.; Yang, B.; Byrne, R.H.; Yuan, D. Spectrophotometric determination of pH and carbonate ion concentrations in seawater: Choices, constraints and consequences. *Anal. Chim. Acta* **2019**, *1081*, 18–31. [CrossRef]
16. Sández, N.; Calvo-López, A.; Vidigal, S.S.M.P.; Rangel, A.O.S.S.; Alonso-Chamarro, J. Automated analytical microsystem for the spectrophotometric monitoring of titratable acidity in white, rosé and red wines. *Anal. Chim. Acta* **2019**, *1091*, 50–58. [CrossRef]
17. Michalec, M.; Koncki, R.; Tymecki, Ł. Optoelectronic detectors for flow analysis systems manufactured by means of rapid prototyping technology. *Talanta* **2019**, *198*, 169–178. [CrossRef]

18. Cecil, F.; Guijt, R.M.; Henderson, A.D.; Macka, M.; Breadmore, M.C. One step multi-material 3D printing for the fabrication of a photometric detector flow cell. *Anal. Chim. Acta* **2020**, *1097*, 127–134. [CrossRef]
19. Saetear, P.; Khamtau, K.; Ratanawimarnwong, N.; Sereenonchai, K.; Nacapricha, D. Sequential injection system for simultaneous determination of sucrose and phosphate in cola drinks using paired emitter-detector diode sensor. *Talanta* **2013**, *115*, 361–366. [CrossRef]
20. Sitanurak, J.; Inpota, P.; Mantim, T.; Ratanawimarnwong, N.; Wilairat, P.; Nacapricha, D. Simultaneous determination of iodide and creatinine in human urine by flow analysis with an on-line sample treatment column. *Analyst* **2015**, *140*, 295–302. [CrossRef]
21. Chaneam, S.; Kaewyai, K.; Mantim, T.; Chaisuksant, R.; Wilairat, P.; Nacapricha, D. Simultaneous and direct determination of urea and creatinine in human urine using a cost-effective flow injection system equipped with in house contactless conductivity detector and LED colorimeter. *Anal. Chim. Acta* **2019**, *1073*, 54–61. [CrossRef] [PubMed]
22. Bzura, J.; Fiedoruk-Pogrebniak, M.; Koncki, R. Photometric and fluorometric alkaline phosphatase assays using the simplest enzyme substrates. *Talanta* **2018**, *190*, 193–198. [CrossRef] [PubMed]
23. Giakisikli, G.; Anthemidis, A.N. Automatic pressure-assisted dual-headspace gas-liquid microextraction. Lab-in-syringe platform for membraneless gas separation of ammonia coupled with fluorimetric sequential injection analysis. *Anal. Chim. Acta* **2018**, *1033*, 73–80. [CrossRef] [PubMed]
24. Inpota, P.; Strezelak, K.; Koncki, R.; Sripumkhai, W.; Jeamsaksri, W.; Ratanawimarnwong, N.; Wilairat, P.; Choengchan, N.; Chantiwas, R.; Nacapricha, D. Microfluidic analysis with front-face fluorometric detection for the determination of total inorganic iodine in drinking water. *Anal. Sci.* **2018**, *34*, 161–167. [CrossRef] [PubMed]
25. Miró, M.; Estela, J.M.; Cerdà, V. Potentials of multisyringe flow injection analysis for chemiluminescence detection. *Anal. Chim. Acta* **2005**, *541*, 57–68. [CrossRef]
26. Lara, F.J.; García-Campaña, A.M.; Aaron, J.-J. Analytical applications of photoinduced chemiluminescence in flow systems–A review. *Anal. Chim. Acta* **2010**, *679*, 17–30. [CrossRef]
27. Chailapakul, O.; Ngamukot, P.; Yoosamran, A.; Siangproh, W.; Wangfuengkanagul, N. Recent electrochemical and optical sensors in flow-based analysis. *Sensors* **2006**, *6*, 1383–1410. [CrossRef]
28. Upan, J.; Reanpang, P.; Chailapakul, O.; Jakmunee, J. Flow injection amperometric sensor with a carbon nanotube modified screen printed electrode for determination of hydroquinone. *Talanta* **2016**, *146*, 766–771. [CrossRef]
29. Chailapakul, O.; Amatatongchai, M.; Wilairat, P.; Grudpan, K.; Nacapricha, D. Flow-injection determination of iodide ion in nuclear emergency tablets, using boron-doped diamond thin film electrode. *Talanta* **2004**, 1253–1258. [CrossRef]
30. Nontawong, N.; Amatatongchai, M.; Thimoonnee, S.; Laosing, S.; Jarujamrus, P.; Karuwan, C.; Chairam, S. Novel amperometric flow-injection analysis of creatinine using amolecularly-imprinted polymer coated copper oxidenanoparticle-modified carbon-paste-electrode. *J. Pharm. Biomed. Anal.* **2019**, *175*, 112770. [CrossRef]
31. Danchana, K.; Clavijo, S.; Cerdà, V. Conductometric determination of sulfur dioxide in wine using a multipumping system coupled to a gas-diffusion cell. *Anal. Lett.* **2019**, *52*, 1363–1378. [CrossRef]
32. Sereenonchai, K.; Teerasong, S.; Chan-Eam, S.; Saetear, P.; Choengchan, N.; Uraisin, K.; Amornthammarong, N.; Motomizu, S.; Nacapricha, D. A low-cost method for determination of calcium carbonate in cement by membraneless vaporization with capacitively coupled contactless conductivity detection. *Talanta* **2010**, *81*, 1040–1044. [CrossRef] [PubMed]
33. Sereenonchai, K.; Saetear, P.; Amornthammarong, N.; Uraisin, K.; Wilairat, P.; Motomizu, S.; Nacapricha, D. Membraneless vaporization unit for direct analysis of solid sample. *Anal. Chim. Acta* **2007**, *597*, 157–162. [CrossRef] [PubMed]
34. Chaneam, S.; Inpota, P.; Saisarai, S.; Wilairat, P.; Ratanawimarnwong, N.; Uraisin, K.; Meesiri, W.; Nacapricha, D. Green analytical method for simultaneous determination of salinity, carbonate and ammoniacal nitrogen in waters using flow injection coupled dual-channel C4D. *Talanta* **2018**, *189*, 196–204. [CrossRef]
35. Chantipmanee, N.; Alahmad, W.; Sonsa-ard, T.; Uraisin, K.; Ratanawimarnwong, N.; Mantim, T.; Nacapricha, D. Green analytical flow method for the determination of total sulfite in wine using membraneless gas-liquid separation with contactless conductivity detection. *Anal. Methods* **2017**, 6107–6116. [CrossRef]

36. Alahmad, W.; Pluangklang, T.; Mantim, T.; Cerdà, V.; Wilairat, P.; Ratanawimarnwong, N.; Nacapricha, D. Development of flow systems incorporating membraneless vaporization units and flow-through contactless conductivity detector for determination of dissolved ammonium and sulfide in canal water. *Talanta* **2018**, *177*, 34–40. [CrossRef] [PubMed]
37. Kraikaew, P.; Pluangklang, T.; Ratanawimarnwong, N.; Uraisin, K.; Wilairat, P.; Mantim, T.; Nacapricha, D. Simultaneous determination of ethanol and total sulfite in white wine using on-line cone reservoirs membraneless gas-liquid separation flow system. *Microchem. J.* **2019**, *149*, 104007. [CrossRef]
38. Luque de Castro, M.D.; Valcàrcel Cases, M. Simultaneous determinations in flow injection analysis—A review. *Analyst* **1984**, *109*, 413–419. [CrossRef]
39. Ricci, A.; Teslic, N.; Petropolus, V.-I.; Parpinello, G.P.; Andrea, V. Fast analysis of total polyphenol content and antioxidant activity in wines and oenological tannins using a flow injection system with tandem diode array and electrochemical detections. *Food Anal. Methods* **2019**, *12*, 347–354. [CrossRef]
40. Najib, F.M.; Othman, S. Simultaneous determination of Cl^-, Br^-, I^- and F^- with flow-injection/ion-selective electrode systems. *Talanta* **1992**, *39*, 1259–1267. [CrossRef]
41. Hansen, E.H.; Růžička, J.; Ghose, A.K. Flow injection analysis for calcium in serum, water and waste waters by spectrophotometry and by ion-selective electrode. *Anal. Chim. Acta* **1978**, *100*, 151–165. [CrossRef]
42. Mascini, M.; Palleshi, G. A flow-through detector for simultaneous determination of glucose and urea in serum samples. *Anal. Chim. Acta* **1983**, *145*, 213–217. [CrossRef]
43. Ramsing, A.U.; Janata, J.; Růžička, J.; Levy, M. Miniaturization in analytical, chemistry—A combination of flow injection analysis and ion-sensitive field effect transistors for determination of pH, and potassium and calcium ions. *Anal. Chim. Acta* **1980**, *118*, 45–52. [CrossRef]
44. Teerasong, S.; Chan-Eam, S.; Sereenonchai, K.; Amornthammarong, N.; Ratanawimarnwong, N.; Nacapricha, D. A reagent-free SIA module for monitoring of sugar, color and dissolved CO_2 content in soft drinks. *Anal. Chim. Acta* **2010**, *668*, 47–53. [CrossRef] [PubMed]
45. Tymecki, Ł.; Pokrzywnicka, M.; Koncki, R. Paired emitter detector diode (PEDD)-based photometry—An alternative approach. *Analyst* **2008**, *133*, 1501–1504. [CrossRef] [PubMed]
46. Kubáň, P.; Hauser, P.C. Contactless conductivity detection for analytical techniques—Developments from 2014 to 2016. *Electrophoresis* **2017**, *38*, 95–114. [CrossRef]
47. Kubáň, P.; Hauser, P.C. Contactless conductivity detection for analytical techniques: Developments from 2016 to 2018. *Electrophoresis* **2019**, *40*, 124–139. [CrossRef]
48. Chvojka, T.; Jelínek, I.; Opekar, F.; Štulík, K. Dual photometric-contactless conductometric detector for capillary electrophoresis. *Anal. Chim. Acta* **2001**, *433*, 13–21. [CrossRef]
49. Vochyánová, B.; Opekar, F.; Tůma, P. Simultaneous and rapid determination of caffeine and taurine in energy drinks by MEKC in a short capillary with dual contactless conductivity/photometry detection. *Electrophoresis* **2014**, *35*, 1660–1665. [CrossRef]
50. Tan, F.; Yang, B.; Guan, Y. Simultaneous light emitting diode-induced fluorescence and contactless conductivity detection for capillary electrophoresis. *Anal. Sci.* **2005**, *21*, 583–585. [CrossRef]
51. Shen, F.; Yu, Y.; Yang, M.; Kang, Q. Dual confocal laser-induced fluorescence/moveable contactless conductivity detector for capillary electrophoresis microchip. *Microsyst. Technol.* **2009**, *15*, 881–885. [CrossRef]
52. Liu, C.; Mo, Y.Y.; Chen, Z.G.; Li, X.; Li, O.L.; Zhou, X. Dual fluorescence/contactless conductivity detection for microfluidic chip. *Anal. Chim. Acta* **2008**, *621*, 171–177. [CrossRef] [PubMed]
53. Ryvolová, M.; Preisler, J.; Foret, F.; Hauser, P.C.; Krásenský, P.; Paull, B.; Macka, M. Combined contactless conductometric, photometric, and fluorimetric single point detector for capillary separation methods. *Anal. Chem.* **2010**, *82*, 129–135. [CrossRef] [PubMed]
54. Zhang, D.L.; Li, W.L.; Zhang, J.B.; Tang, W.R.; Chen, X.F.; Cao, K.W.; Chu, Q.C.; Ye, J.N. Determination of unconjugated aromatic acids in urine by capillary electrophoresis with dual electrochemical detection—Potential application in fast diagnosis of phenylketonuria. *Electrophoresis* **2010**, *31*, 2989–2996. [CrossRef]
55. Oyaert, M.; Delanghe, J. Progress in Automated Urinalysis. *Ann. Lab. Med.* **2019**, *39*, 15–22. [CrossRef] [PubMed]
56. Shirreffs, S.M.; Maughan, R.J. Urine osmolality and conductivity as indices of hydration status in athletes in the heat. *Med. Sci. Sports Exerc.* **1998**, *30*, 1598–1602. [CrossRef]

57. Hernández-Luisa, F.; Abdala, S.; Dévora, S.; Benjumea, D.; Martín-Herrera, D. Electrical conductivity measurements of urine as a new simplified method to evaluate the diuretic activity of medicinal plants. *J. Ethnopharmacol.* **2014**, *151*, 1019–1022. [CrossRef]
58. Mantim, T.; Saetear, P.; Teerasong, S.; Chan-Eam, S.; Sereenonchai, K.; Amornthammarong, N.; Ratanawimarnwong, N.; Wilairat, P.; Meesiri, W.; Uraisin, K.; et al. Reagent-free analytical flow methods for the soft drink industry: Efforts for environmentally friendly chemical analysis. *Pure Appl. Chem.* **2012**, *84*, 2015–2025. [CrossRef]
59. Dias, A.C.B.; Borges, E.P.; Zagatto, E.A.G.; Worsfold, P.J. A critical examination of the components of the Schlieren effect in flow analysis. *Talanta* **2006**, *68*, 1076–1082. [CrossRef] [PubMed]
60. Ruzicka, J.; Hansen, E.H. *Flow Injection Analysis*, 2nd ed.; John Wiley & Sons, Inc.: New York, NY, USA, 1988; p. 27.
61. Bonsnes, R.W.; Taussky, H.H. On the colorimetric determination of creatinine by the Jaffe reaction. *J. Biol. Chem.* **1945**, *158*, 581–591.
62. Sakai, T.; Ohta, H.; Ohno, N.; Imai, J. Routine assay of creatinine in newborn baby urine by spectrophotometric flow-injection analysis. *Anal. Chim. Acta* **1995**, *308*, 446–450. [CrossRef]
63. Ohira, S.-I.; Kirk, A.B.; Dasgupta, P.K. Automated measurement of urinary creatinine by multichannel kinetic spectrophotometry. *Anal. Biochem.* **2009**, *384*, 238–244. [CrossRef] [PubMed]
64. Siangproh, W.; Teshima, N.; Sakai, T.; Katoh, S.; Chailapakul, O. Alternative method for measurement of albumin/creatinine ratio using spectrophotometric sequential injection analysis. *Talanta* **2009**, *79*, 1111–1117. [CrossRef] [PubMed]
65. Kouri, T.T.; Kähkönen, U.; Malminiemi, K.; Vuento, R.; Rowan, R.M. Evaluation of sysmex UF-100 urine flow cytometer vs chamber counting of supravitally stained specimens and conventional bacterial cultures. *Am. J. Clin. Pathol.* **1999**, *112*, 25–35. [CrossRef] [PubMed]
66. Delanghe, J.R.; Kouri, T.T.; Huber, A.R.; Hannemann-Pohl, K.; Guder, W.G.; Lun, A.; Sinha, P.; Stamminger, G.; Beier, L. The role of automated urine particle flow cytometry in clinical practice. *Clin. Chim. Acta* **2000**, *301*, 1–18. [CrossRef]
67. Tanyanyiwa, J.; Galliker, B.; Schwarz, M.A.; Hauser, P. Improved capacitively coupled conductivity detector for capillary electrophoresis. *Analyst* **2002**, *127*, 214–218. [CrossRef]

Sample Availability: Samples of the compounds are not available from the authors.

© 2020 by the authors. Licensee MDPI, Basel, Switzerland. This article is an open access article distributed under the terms and conditions of the Creative Commons Attribution (CC BY) license (http://creativecommons.org/licenses/by/4.0/).

MDPI
St. Alban-Anlage 66
4052 Basel
Switzerland
Tel. +41 61 683 77 34
Fax +41 61 302 89 18
www.mdpi.com

Molecules Editorial Office
E-mail: molecules@mdpi.com
www.mdpi.com/journal/molecules